Current Topics in Microbiology and Immunology

Ergebnisse der Mikrobiologie und Immunitätsforschung

55

Edited by

Springer-Verlag Berlin · Heidelberg · New York 1971

ISBN-13:978-3-642-65226-4 e-ISBN-13:978-3-642-65224-0
DOI: 10.1007/978-3-642-65224-0

 Universitätsdruckerei H. Stürtz AG, Würzburg

Arthropod Cell Cultures and Their Application to the Study of Viruses

Emilio Weiss

Chairman, Editorial Committee

With 151 Figures

Springer-Verlag Berlin · Heidelberg · New York 1971

Current Topics in Microbiology and Immunology

Ergebnisse der Mikrobiologie und Immunitätsforschung

55

Edited by

W. Arber, Berkeley · W. Braun, New Brunswick · R. Haas, Freiburg · W. Henle, Philadelphia · P. H. Hofschneider, München · N. K. Jerne, Basel · P. Koldovský, Prague · H. Koprowski, Philadelphia · O. Maaløe, Copenhagen · R. Rott, Gießen · H. G. Schweiger, Wilhelmshaven · M. Sela, Rehovoth · L. Syruček, Prague · P. K. Vogt, Seattle · E. Wecker, Würzburg

Springer-Verlag New York · Heidelberg · Berlin 1971

ISBN-13:978-3-642-65226-4 e-ISBN-13:978-3-642-65224-0
DOI: 10.1007/978-3-642-65224-0

Softcover reprint of the hardcover 1st edition 1971

 Universitätsdruckerei H. Stürtz AG, Würzburg

Arthropod Cell Cultures and Their Application to the Study of Viruses

Emilio Weiss

Chairman, Editorial Committee

With 151 Figures

Springer-Verlag New York · Heidelberg · Berlin 1971

EARL C. SUITOR, JR.
1930—1969

Preface

To lose is human,
to win is fortune,
but to try is our destiny.

EARL C. SUITOR, JR.

The idea of a Symposium on "Arthropod Cell Cultures" started in July of 1969 shortly after the untimely death of our colleague, EARL C. SUITOR, JR., at the age of 38. At first we thought an afternoon or evening session would be sufficient, but we were soon convinced that the scope of the Symposium should be greatly enlarged. Interest in this topic was increasing at an astonishing rate. Since EARL SUITOR had made a distinct imprint in this new field, many scientists who knew him well or just casually wished to honor him in this manner.

EARL SUITOR was born and raised in New England and received a B.S. degree in bacteriology from the University of Massachusetts in 1952. As a spirited young man, he enlisted in the U.S. Navy to see the world. Instead, he was assigned for most of his four-year "hitch" to the Naval Medical Research Institute. I met him there in 1954, an enthusiastic and imaginative young fellow with many interests, an avid reader of the classics, an occasional writer of poetry, who blended his interest in scientific literature with that of Science Fiction. In 1956, EARL left the Navy to attend George Washington University, where he earned an M.S. degree in 1958 and a Ph.D. degree in 1963. He returned to my laboratory in 1957 as a civilian and part-time student. In 1963, the Navy finally acceded to his wishes to be sent overseas. He spent several months in the laboratory of Dr. THOMAS GRACE in Canberra, Australia, to learn the concepts and techniques of arthropod cell culture. He then joined the staff of the Naval Medical Research Unit No. 2 in Taiwan. In 1966 he returned to the Naval Medical Research Institute to head the Division of Arthropod Microbiology.

His first research interest, as my junior associate and graduate student, was in the field of rickettsiae and rickettsia-like organisms. He isolated an agent from ticks which he named *Wolbachia persica* or more affectionately "the monster". In time, this rickettsia-like microorganism became the most thoroughly studied agent of this group and the best evidence that the so-called "rickettsia-like" organisms are unrelated to rickettsiae. EARL SUITOR's contributions in the field of arthropod cell culture reflect his pioneering attitude and are frequently mentioned in this volume. They include several important "first" achievements, such as the successful cloning of an arthropod cell line and use of arthropod cell cultures for virus growth and plaque production, isolation of virus mutants, and cultivation of an animal parasite. He collaborated and corresponded very extensively with other scientists and gave freely of his time and advice. He was intensely loyal to his friends and colleagues. This volume is a fitting tribute to his memory.

Bethesda, April 1971 EMILIO WEISS

Introduction

This volume is based on papers presented at a Symposium held at the Naval Medical Research Institute on March 17 and 18, 1970. It consisted of four sessions: the first two were devoted to cultivation and physiology of arthropod cells, the second two to the growth of viruses and other pathogens in arthropod cell cultures. The sessions were chaired, respectively, by JAMES L. VAUGHN, THOMAS D. C. GRACE, WILLIAM F. SCHERER, and KARL MARAMOROSCH. The EARL C. SUITOR, JR., Memorial Lecture was given by WILLIAM TRAGER.

The papers varied greatly in contents and scope. Some contained original experiments, others reviews, and many contained a mixture of both. It was not the intention of the Editorial Committee to publish the proceedings of this Symposium but rather to use this material to present current knowledge in a systematic fashion. This required considerable revision and organization of the papers into chapters. We are particularly indebted to the four session chairmen for their valuable help in this task. We are also very grateful to Capt LLOYD F. MILLER for the administrative arrangements that made this Symposium possible, to the other members of the organizing committee who helped us in many ways, and to the members of our Department who contributed to the smooth operation of the meeting.

Bethesda, April 1971

FRANCIS B. GORDON
EMILIO WEISS
EUGENE ZEBOVITZ

Table of Contents

Chapter 1

The Culture of Cells from Insects and Ticks

Chapter 2

Analysis of Cells from Established Insect Cell Lines

Chapter 3

Physiology of Cultivated Arthropod Cells

---

1 USPH post-doctoral trainee (HD-00210).
2 Career Scientist of the Health Research Council of New York City (I-513).

Chapter 5

Propagation of Arboviruses in SINGH's *Aedes* Cell Lines

Chapter 6

Growth of Arboviruses in Arthropod Cell Cultures: Comparative Studies

Chapter 7

Growth of Viruses in Arthropod Cell Cultures: Applications

Chapter 8

Homoptera Cell Culture and Its Application to the Study of Plant Pathogens

H. HIRUMI

Chapter 9

Lepidoptera Cell Culture and Its Application to the Study of Plant Viruses and Animal Parasites

Chapter 10

***Drosophila* Cell Culture and Its Application for the Study of Genetics and Virology**

Organizing and Editorial Committee

List of Authors

BALL, G. H., Department of Zoology, University of California, Los Angeles, CA 90024/USA

BARIGOZZI, C., Istituto di Genetica, Università di Milano, Italy

BELLONCIK, S., Department of Microbiology and Immunology, Laboratory of Comparative Virology, Faculty of Medicine, University of Montreal, Montreal, Quebec/Canada

BOWNE, W., Institute for Medical Research, Camden, NJ 08103/USA

BRADT, C., Institute for Medical Research, Camden, NJ 08103/USA

BROOKS, M. A., Department of Entomology, Fisheries, and Wildlife, University of Minnesota, St. Paul, MN 55101/USA

BUCKLEY, S. M., The Rockefeller Foundation and the Yale Arbovirus Unit, Department of Epidemiology and Public Health, Yale Medical School, New Haven, CT 06510/USA

CHAO, J., Department of Zoology, University of California, Los Angeles, CA 90024/USA

CHARNEY, J., Institute for Medical Research, Camden, NJ 08103/USA

CLARKE, R. B., National Institute of Allergy and Infectious Diseases, Bethesda, MD 20014/USA

CONOVER, J. H., Department of Pediatrics, The Mount Sinai School of Medicine, New York, NY 10029/USA

DEIBEL, R., Division of Laboratories and Research, New York State Department of Health, Albany, NY 12201/USA

DOLAN, T., Department of Microbiology, School of Medicine, University of North Dakota, Grand Forks, ND 58201/USA

DOUGHERTY, K., Insect Pathology Laboratory, Agricultural Research Service, U.S. Department of Agriculture, Beltsville, MD 20705/USA

ECHALIER, G., Laboratoire de Biologie Animale, Faculté des Sciences de Paris, Paris/France

ELLIS, B. J., Faculty of Entomology, The Ohio State University, Columbus, OH 43210/USA

FISCHER, R. G., Department of Microbiology, School of Medicine, University of North Dakota, Grand Forks, ND 58201/USA

GARZON, S., Department of Microbiology and Immunology, Laboratory of Comparative Virology, Faculty of Medicine, University of Montreal, Montreal, Quebec/Canada

GRACE, T. D. C., Division of Entomology, C.S.I.R.O., Canberra, A.C.T. Australia 2601

GREENBERG, B., Department of Biological Sciences, University of Illinois at Chicago Circle, Chicago, IL 60680/USA

GREENE, A. E., Institute for Medical Research, Camden, NJ 08103/USA

HANN, W. D., National Institute of Allergy and Infectious Diseases, Bethesda, MD 20014/USA
Present address: Department of Biology, Bowling Green State University, Bowling Green, OH 43402/USA

HANNOUN, C., Laboratoire des Arbovirus, Institut Pasteur, Paris/France

HINK, W. F., Faculty of Entomology, The Ohio State University, Columbus, OH 43210/USA

HIRSCHHORN, K., Department of Pediatrics, The Mount Sinai School of Medicine, New York, NY 10029/USA

HIRUMI, H., Boyce Thompson Institute for Plant Research, Yonkers, NY 10701/USA

HODES, H. L., Department of Pediatrics, The Mount Sinai School of Medicine, New York, NY 10029/USA

HSU, S. H., U.S. Naval Medical Research Unit No. 2, Taipei, Taiwan, Republic of China

IBRAHIM, A. N., Public Health Science Division, Gulf South Research Laboratories, New Orleans, LA 70126/USA

JENKIN, H., The Hormel Institute, University of Minnesota, Austin, MN 55912/USA

JOHNSON, H., Department of Microbiology, School of Medicine, University of North Dakota, Grand Forks, ND 58201/USA

KURSTAK, E., Department of Microbiology and Immunology, Laboratory of Comparative Virology, Faculty of Medicine, University of Montreal, Montreal, Quebec/Canada

KURTTI, T. J., Department of Entomology, Fisheries, and Wildlife, University of Minnesota, St. Paul, MN 55101/USA

LEVENTHAL, E. A., Department of Genetics, University of Wisconsin, Madison, WI 53706/USA

LOULOUDES, S. L., Insect Pathology Laboratory, Agricultural Research Service, U.S. Department of Agriculture, Beltsville, MD 20705/USA

MAKINO, S., The Hormel Institute, University of Minnesota, Austin, MN 55912/USA

MARKS, E. P., Metabolism and Radiation Research Laboratory, Agricultural Research Service, U.S. Department of Agriculture, Fargo, ND 58102/USA

MAZZONE, H. M., Forest Insect and Disease Laboratory, U.S. Department of Agriculture, Forest Service, Hamden, CT 06514/USA

MCHALE, J. S., Public Health Science Division, Gulf South Research Laboratories, New Orleans, LA 70126/USA

MITSUHASHI, J., Division of Entomology, National Institutes of Agricultural Sciences, Nishigara, Kita-ku, Tokyo, Japan.
On leave to: Division of Entomology, C.S.I.R.O., Canberra, A.C.T. Australia 2601

NICHOLS, W. W., Institute for Medical Research, Camden, NJ 08103/USA

OHANESSIAN, A., Laboratoire de Génétique des Virus, 91 GIF-sur-Yvette, France

PELEG, J., Israel Institute for Biological Research, Ness-Ziona, Israel

REHACEK, J., Department of Microbiology, University of North Dakota, Grand Forks, ND 58201/USA
On leave from: Institute of Virology, Slovak Academy of Sciences, Bratislava, Czechoslovakia

REHACEK, Z., Department of Microbiology, School of Medicine, University of North Dakota, Grand Forks, ND 58201/USA

SCHLESINGER, R. W., Rutgers Medical School, Department of Microbiology, New Brunswick, NJ 08903/USA

SCHNEIDER, I., Department of Entomology, Walter Reed Army Institute of Research, Washington, DC 20012/USA

SEECOF, R. L., Department of Biology, City of Hope Medical Center, Duarte, CA 91010/USA

SHENK, T., Rutgers Medical School, Department of Microbiology, New Brunswick, NJ 08903/USA

SHINEDLING, S. T., Department of Biological Sciences, University of Illinois at Chicago Circle, Chicago, IL 60680/USA

SINGH, K. R. P., Virus Research Centre, Indian Council of Medical Research, Poona 1, India
Present address: World Health Organization—Research Unit on Genetic Control of Culicine Mosquitoes, 2 and 3, Ring Road, Kilokri, New Delhi—14, India

STEVENS, T. M., Rutgers Medical School, Department of Microbiology, New Brunswick, NJ 08903/USA

STOLLAR, V., Rutgers Medical School, Department of Microbiology, New Brunswick, NJ 08903/USA

SWEET, B. H., Public Health Science Division, Gulf South Research Institute, Box 26500, New Orleans, LA 70126/USA

TEPLITZ, R. L., Department of Pathology, City of Hope Medical Center, Duarte, CA 91010/USA

THOMPSON, K., Department of Microbiology, School of Medicine, University of North Dakota, Grand Forks, ND 58201/USA

TOWNSEND, D., The Hormel Institute, University of Minnesota, Austin, MN 55912/USA

TRAGER, W., The Rockefeller University, New York, NY 10021/USA

UNTHANK, H. D., Public Health Science Division, Gulf South Research Institute, Box 26500, New Orleans, LA 70126/USA

VAUGHN, J. L., Insect Pathology Laboratory, Entomology Research Division, Agricultural Research Service, U.S. Department of Agriculture, Beltsville, MD 20705/USA

WHITE, J. F., Department of Zoology, Eastern Illinois University, Charleston, IL 61920/USA

WHITNEY, E., Division of Laboratories and Research, New York State Department of Health, Albany, NY 12201/USA

YANG, T.-K., The Hormel Institute, University of Minnesota, Austin, MN 55912/USA

YUNKER, C. E., Rocky Mountain Laboratory, National Institute of Allergy and Infectious Diseases, Hamilton, MT 59840/USA

ZEPP, H. D., Department of Pediatrics, The Mount Sinai School of Medicine, New York, NY 10029/USA

Chapter 1

The Culture of Cells from Insects and Ticks

I. Cultivation of Dipteran Cells *in Vitro*

IMOGENE SCHNEIDER

A. Introduction

The utilization of insect cell cultures for studying various aspects of the pathogenic cycles of viral, protozoan and bacterial agents having an insect vector or reservoir has long been recognized (GLASER, 1917; TRAGER, 1938). Unfortunately, such studies have been seriously impeded by the difficulties of initiating and maintaining insect cell cultures *in vitro*. As recently as 1966 only two insect cell lines were available, namely *Antheraea eucalypti* SCOTT and *Aedes aegypti* (L.), both having been established by the same investigator (GRACE, 1962, 1966). As evidence of the considerable progress made since then, the one dipteran line established by GRACE has now been joined by eleven additional lines.

Encouraging as these results are, the number of lines currently available is still woefully small in relation to the number of known and potential insect vectors of pathogens capable of infecting higher animals as well as insects themselves. In a number of instances, the lack of appropriate cell cultures has left investigators with little alternative but the use of an atypical pathogen-vector culture system (see reviews by VAUGHN, 1968; GRACE, 1969). The tenuous nature of many of the conclusions drawn from such studies is readily apparent, especially in instances of highly specific pathogen-vector relationships. Such drawbacks can only be eliminated by access to a much wider range of insect cell lines, particularly from those dipteran species which are implicated in the dissemination of disease.

The present report has two main purposes: (1) to provide a general review of the dipteran cell lines now available, covering the techniques used to initiate and establish the cultures and the characteristics of the cell lines so obtained and (2) to describe progress in culturing cells from 3 dipteran species, 2 well known vectors, *Culex tritaeniorhynchus* GILES and *Glossina morsitans* WESTWOOD, and one of peripheral interest, *Culex salinarius* COQUILLETT.

B. Established Dipteran Cell Lines

The first dipteran cell line was established by GRACE (1966) from larval tissues of axenically grown *A. aegypti* (L.). Primary cultures consisting of the disrupted bodies and alimentary canals from 4th instar larvae about to pupate were started in small tubes containing GRACE's medium (1962) and incubated at 30° C. Cell migration and division readily took place and successful subcultures were made after 4 weeks.

The majority of cells grow in suspension, are spindle shaped and very large, e.g., 40–50 μ long and 8–10 μ wide (Fig. 1a). Many round cells are also present, varying from 20–60 μ in diameter. The cells are highly polyploid, the most frequent number of chromosomes being 96 (32 n). Although polyploid cells are present in mosquito larvae during the 4th larval instar, the chromosomes complement rarely reaches such a high number (HOLT, 1917; CLEMENTS, 1963). Since chromosome numbers were not investigated until 6 months after the line had been established it is not known whether diploid cells were initially present. It seems likely, however, that the high numbers found in the cultured cells reflect to some extent a selective or adaptive process.

Although the original *A. aegypti* line required insect hemolymph as a supplement, two or more sublines have been established in which the hemolymph has been replaced by fetal bovine serum (NAGLE et al., 1967; CONVERSE and NAGLE, 1967; HSU et al., 1970). All subsequent dipteran lines have been initiated and maintained in media supplemented with vertebrate sera. The advisability of this practice has been questioned (SCHNEIDER, 1969) but the alternative of using homologous hemolymph, at least for the smaller insects is not feasible. The use of heterologous hemolymph also has some drawbacks from the standpoint of altering the cells' biochemistry and physiology.

In 1967, SINGH reported the establishment of cell lines from *A. aegypti* and *A. albopictus* SKUSE (Figs. 1b, 1c). Several hundred freshly hatched larvae were cut into pieces, incubated in 0.25% trypsin (1 : 250, Difco) and after a thorough washing, placed in the culture medium of MITSUHASHI and MARAMOROSCH (1964), hereafter referred to as M-M medium. This medium was originally designed for leafhopper cells but its composition (lactalbumin hydrolysate and TC Yeastolate in lieu of individual amino acids and vitamins, respectively, with only the sugar and inorganic salts specified as such) apparently makes it quite useful as a basic insect tissue culture medium. (However, see comments under "General Discussion").

The *A. albopictus* cells readily attached to the glass surface of the flasks and formed monolayers composed of 3 types of cells. The majority are either round, 6–20 μ in diameter, or spindle shaped, 7–10 μ wide and 15–90 μ long. Round, binucleate cells, 37–53 μ in diameter, are also found. The growth pattern of the *A. aegypti* cells was very dissimilar to that given above. Shortly after the cultures had been set up, small hollow vesicles or spheres developed at the ends of the cut fragments and increased both in size and in number. These spheres were composed of a monolayer of epithelial-like cells, penta- or hexagonal in shape and varying from 16–50 μ in width. After 8 to 10 days the fragments with the spheres were removed, cut into pieces and seeded into new bottles. Two lines were eventually established, one in which the cells grow in a monolayer and the other in which growth takes place almost wholly in the form of these hollow spheres, the spheres being subcultured as such

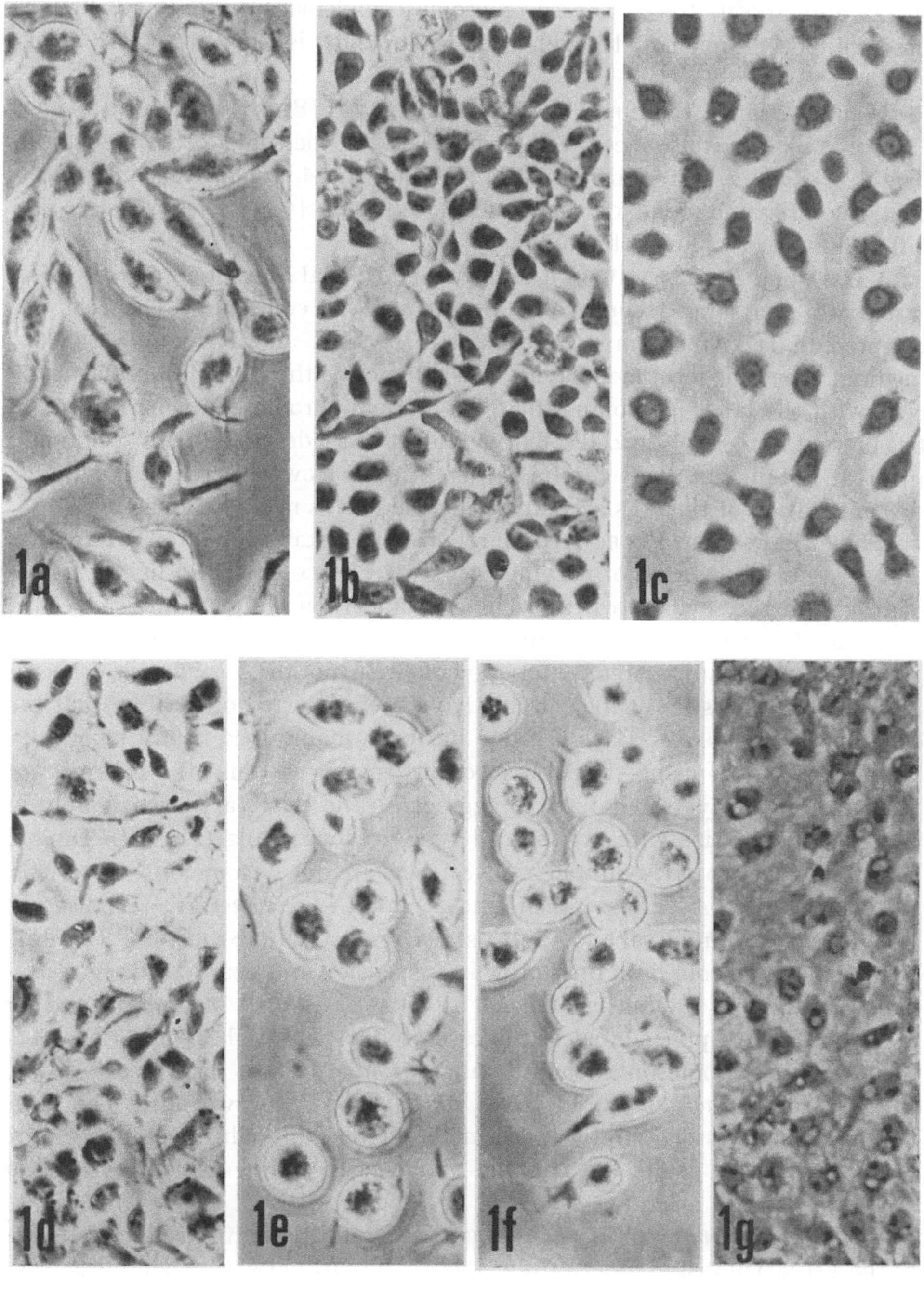

Fig. 1 a-g. Dipteran cells from established lines maintained in author's laboratory: a GRACE's line of *A. aegypti;* b and c SINGH's lines of *A. aegypti* and *A. albopictus*, respectively; d PELEG's line of *A. aegypti;* e and f SWEET's and DUPREE's lines of *A. vexans* and *C. inornata*, respectively; and g SCHNEIDER's line of *An. stephensi*. All 600×

(BHAT and SINGH, 1969). The chromosome complement of these cell lines is predominantly diploid. Although some polyploidy is present, it rarely exceeds 4 n.

Using the same technique, lines of *Aedes vittatus* (BIGOT) and *Aedes w-albus* (THEOBALD) have also been established (SINGH, personal communication). The *A. w-albus* cells are grown in M-M medium; the *A. vittatus* cells in SCHNEIDER's modification of GRACE's medium (SCHNEIDER, 1969). These cells are also either diploid or of low polyploidy.

VARMA and PUDNEY (1969) established additional lines of *A. aegypti* using essentially the same technique as SINGH and his colleagues. The cells are cultured in M-M medium or a composite of M-M and KITAMURA's (1965) medium. The cells form monolayers consisting of epithelial and fibroblast type cells with the latter predominating. The normal chromosome complement of 6 is present in most of the cells. VARMA and PUDNEY described their techniques in great detail, but their results rather briefly. However, the photographic documentation of the cells accompanying the text is most convincing.

Somewhat earlier, PELEG (1968b) reported the establishment of yet another *A. aegypti* cell line. Eggs of varying ages were surface sterilized, lightly homogenized and the shells discarded. The cells and tissue fragments were seeded into a modified KITAMURA medium supplemented with chick embryo extract. Subcultures were initiated after a period of 4 to 6 months. Few details concerning the characteristics of the cell line have been reported other than that the cells form a monolayer. This cell line has been maintained in my laboratory for almost a year using the M-M medium. In this medium the cells do not form true monolayers but tend to aggregate. The cells vary considerably in both shape and size with no one cell type predominating (Fig. 1 d). Diploid cells are in the majority although 4n–8n cells are not uncommon.

The cell lines of *Aedes vexans* (MEIGEN) and *Culiseta inornata* (WILLISTON) were initiated from pupal and adult tissues, respectively (SWEET and DUPREE, 1968). The cells from both lines grow in suspension and bear a striking resemblance to those of the line established by GRACE. GRACE's medium is used for both lines, but supplemented in the case of *A. vexans* with an additional 1% l-glutamine. Chromosome counts range from 152–368 for *A. vexans* and 138–202 for *C. inornata* (SCHNEIDER, unpublished observations). The large number of chromosomes in these cells makes it very difficult to obtain accurate counts. Hence it is likely that the above counts may be off by 5% or more.

Cell lines of *Anopheles stephensi* LISTON were initiated from minced tissues of the 1st larval instar (SCHNEIDER, 1969). The early growth pattern was very similar to that of the SINGH *A. aegypti* cells. GRACE's medium, modified with respect to the concentrations of the sugars and inorganic salts was supplemented with abbreviated NCTC 135 medium and fetal bovine serum. The cells form monolayers with some tendency toward aggregating at central foci, are epithelial in appearance and range from 4–9 μ in diameter and 12–20 μ in length (Fig. 1g). Diploid cells predominate although 4n cells are present.

Table 1. *A summary of established cell lines from dipteran species. In some instances more than one cell line from a single species has been established by the same investigator(s). These have been considered as one with the exception of the Aedes aegypti lines reported by* SINGH *and* BHAT *which differ considerably from one another in their pattern of growth*

Genus species	Stage of primary explant	Medium	Type of growth	Chromosome complement	Reference
1) *Aedes aegypti* L.	4th larval instar	GRACE	Suspension	Polyploid (16n or higher)	GRACE, T. D. C. (1966)
2) *A. aegypti*	1st larval instar	MITSUHASHI and MARAMOROSCH	Monolayer	Diploid	SINGH, K. R. P. (1967)
3) *A. aegypti*	1st larval instar	MITSUHASHI and MARAMOROSCH	Hollow vesicles	Diploid	BHAT, U. K. M. and SINGH, K. R. P. (1969)
4) *A. aegypti*	Embryos	KITAMURA/TRAGER	"Monolayer"	Diploid	PELEG, J. (1968a)
5) *A. aegypti*	1st larval instar	MITSUHASHI and MARAMOROSCH/KITAMURA/VP12	Monolayer/vesicles	Diploid	VARMA M. G. R. and PUDNEY, M. (1969)
6) *Aedes albopictus*	1st larval instar	MITSUHASHI and MARAMOROSCH	Monolayer	Diploid	SINGH, K. R. P. (1967)
7) *Aedes vexans*	Pupae	GRACE (modified)	Suspension	Polyploid (16n to 128n?)	SWEET, B. H. and DUPREE, L. T. (1968)
8) *Aedes vittatus*	1st larval instar	GRACE as modified by SCHNEIDER	1) Monolayer 2) Vesicles	Diploid	BHAT, U. K. M. and SINGH, K. R. P. (1970)[a]
9) *Aedes w-albus*	1st larval instar	MITSUHASHI and MARAMOROSCH	Monolayer	Diploid/polyploid	SINGH, K. R. P. and BHAT, U. K. M. (1970)[a]
10) *Anopheles stephensi*	1st larval instar	GRACE as modified by SCHNEIDER	Monolayer	Diploid	SCHNEIDER, I. (1969)
11) *Culiseta inornata*	Adult	GRACE (modified)	Suspension	Polyploid (16n to 64n)	SWEET, B. H. and DUPREE, L. T. (1968)
12) *Drosophila melanogaster*	Embryos	Based on hemolymph analysis; some specifics not given	1) Suspension 2) Monolayer	1) Polyploid 2) Diploid	ECHALIER, G. and OHANESSIAN, A. (1968)

[a] Personal communication

The first cell line of *Drosophila melanogaster* MEIGEN established by ECHALIER and OHANESSIAN (1968) was initiated from embryos, not less than 12 hrs old, in a medium based on an analysis of larval hemolymph. The cells are highly polyploid, grow in suspension and in appearance and behavior are very similar to GRACE's cell line. Thus, they differ in all of these respects from any of the other dipteran cell lines derived either from embryonic or early post-embryonic material. Subsequent lines of diploid, fibroblast-like cells which grow in a monolayer have apparently been established (ECHALIER and OHANESSIAN, 1969). Further details have not been forthcoming and to my knowledge the exact composition of the culture medium for any of these lines has not been published.

The more pertinent details of the above section have been summarized in Table 1.

C. Primary Cultures from Three Additional Dipteran Species

1. Culex tritaeniorhynchus and C. salinarius

In anticipation of conducting comparative studies involving *in vivo* versus *in vitro* invasion and replication of a number of arboviruses, cells from several species of the genus *Culex* differing markedly in their vector capacity were selected for culturing.

Previous reports by KITAMURA (1964, 1965, 1966) and GUBLER (1968) indicated that primary cultures could readily be obtained from ovarian and midgut cells of adult *Culex pipiens molestus* FORSKÅL. Moreover, the fact that KITAMURA (1965) was able to maintain such cells for over 3 months in an active state of proliferation suggested that the transition from primary to secondary cultures might be achieved without undue difficulty. Since cells from the early larval stages equal, if not exceed, those from the adult stages in their ability to adapt to conditions *in vitro*, the former were chosen as the source of primary explants in the present study.

Materials and Methods. Egg rafts were collected over a 16 hr period and the embryos allowed to develop to the point of red eye spot formation. Prior to this stage in development the embryos will not withstand surface sterilization. The rafts were placed in a 2.5% sodium hypochlorite solution for 2 min. and the container agitated to facilitate separation of the eggs. The eggs were then surface sterilized by immersion for 1 hr in 10% benzalkonium chloride followed by an additional 20 min. in 70% ethanol. After repeated rinses in sterile distilled water, the eggs were transferred to moistened, sterile filter paper in a petri dish and placed in a 22° C incubator.

Newly hatched larvae with unsclerotized head capsules (Fig. 2a) were transferred to a depression slide and minced in RINALDINI's (1954) salt solution (RSS) containing 0.2% trypsin (1 : 250, Difco). Each culture was initiated with a minimum of 200 (*C. salinarius*) or 500 (*C. tritaeniorhynchus*) larvae. The fragments were transferred to fresh RSS-trypsin solution and incubated an additional 15–30 min. at room temperature. After the addition of fetal bovine serum (FBS), the fragments were centrifuged, resuspended in 1.5 ml of culture medium and seeded into glass T-15 flasks. The cultures were maintained at $27 \pm 0.5°$ C with a gaseous phase of ambient air.

The medium was identical to that used for culturing *A. stephensi* cells (SCHNEIDER, 1969) with the following exceptions. The concentration (in mg/100 ml) of sucrose was decreased to 900; that of glucose increased to 300, and of NaCl to 440. Supplements to the medium were 1% 10× abbreviated NCTC 135 medium (see SCHNEIDER, 1969) and 15% inactivated FBS. Phenol red (0.01%) was used to monitor a pH of 6.7–6.8.

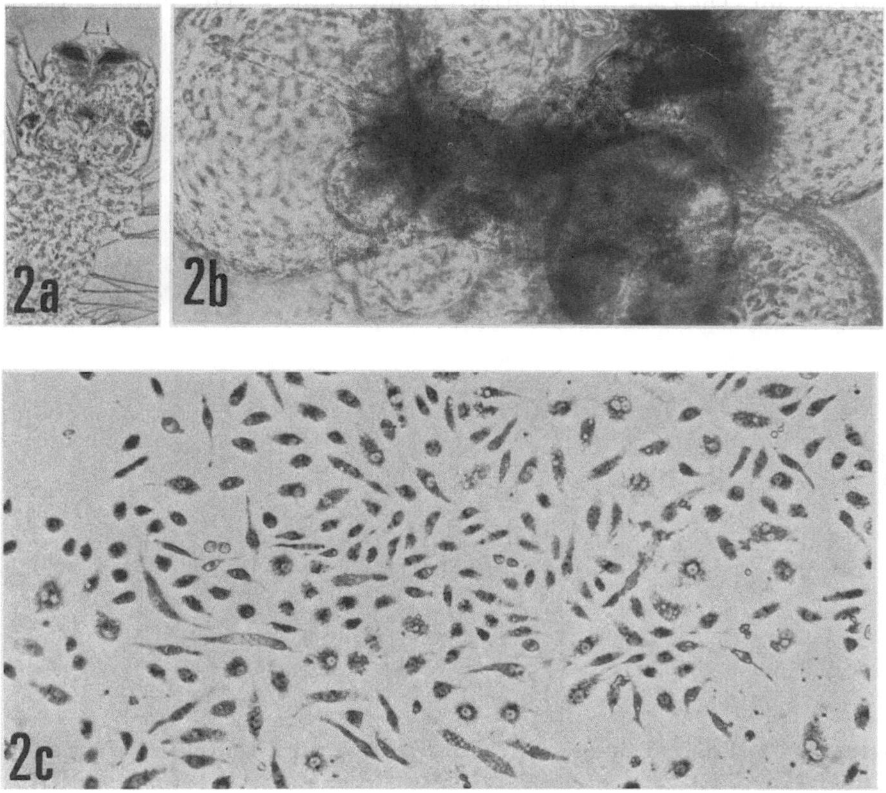

Fig. 2. a. Newly hatched *Culex tritaeniorhynchus* larva with unsclerotized head capsule, 120×; b monolayer spheres of epithelial-like cells issuing from *C. tritaeniorhynchus* larval fragments after 12 days in culture, 300×; c cells of *C. tritaeniorhynchus* in 4th subculture, 300×

Cytological study of the cells was facilitated by dispersing precleaned sterile 9×35 mm coverslips over the bottom of plastic T-30 flasks prior to cell seeding. The cells were fixed in FAA (formalin, ethanol, acetic acid, and water in volume ratios of 18 : 49 : 3 : 30) and stained *in situ* with Ehrlich's hematoxylin. Chromosome counts were made on cells blocked in metaphase with colchicine.

Results. Forty-four primary cultures of *C. tritaeniorhynchus* were initiated from August to November 1969. The pattern of development was identical to that first described by SINGH (1967) for *A. aegypti*. Cell growth, in the form of spherical monolayers issuing from the ends of the larval fragments, was evident within a week after the cultures had been started (Fig. 2b). The medium was partially withdrawn and renewed every 7 to 10 days. After 4 to 6 weeks in primary culture, the fragments were removed from the flasks and the cell spheres excised with tungsten needles. Following mild trypsin treatment of the spheres (0.1 % trypsin in RSS for 3 min. at room temperature) the cells were reseeded back into the same flasks containing $^{2}/_{3}$ old and $^{1}/_{3}$ fresh medium. Subcultures were made 1 to 2 weeks later, the inoculum containing a minimum of 10^4 cells/ml. The cells do not attach firmly to the surface of the glass flask and can be removed by pipetting.

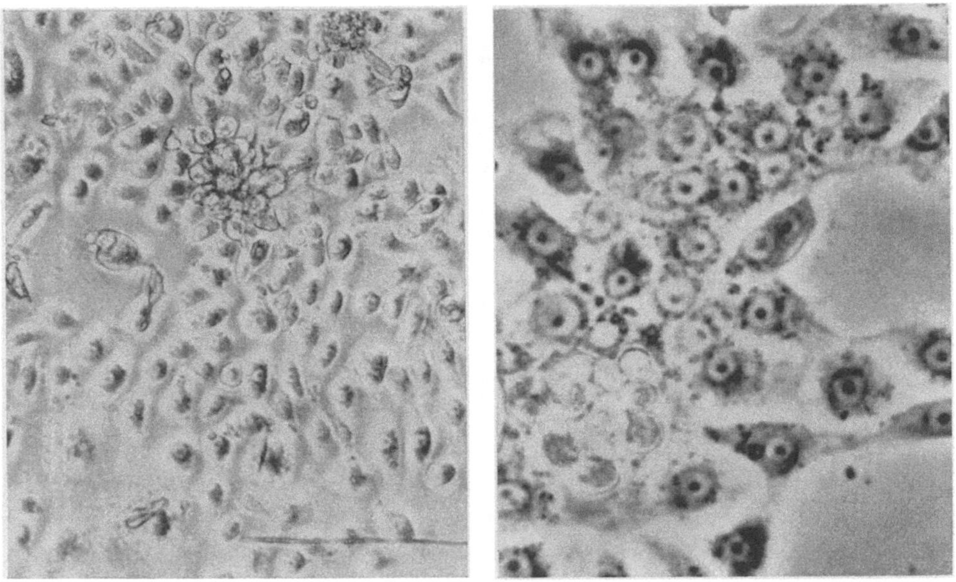

a b

Fig. 3 a and b. Primary cultures of *Culex salinarius* after 5 weeks *in vitro*. a 300 ×; b 600 ×

The cells range from 5–17 μ in diameter and 28–60 μ in length. Epithelial-like, round and fibroblast cells are present with the latter predominating (Fig. 2c). Confluent monolayers are not formed, but rather the cells grow in large, scattered patches. The mean mitotic index, based on a minimum count of 1000 cells is rather low (0.7%) in 48 hr cultures but gradually increases to 1.6% at 120 hr. Preliminary chromosome counts indicate that the majority of cells are diploid ($2n=6$).

Eighteen primary cultures of *C. salinarius* were initiated between December 1969 and February 1970. Development of these cultures differed from those of *C. tritaeniorhynchus* only in that the cell spheres usually reached greater diameters and could be dislodged from the larval fragments by gentle pipetting of the medium. In primary cultures, the cells are predominently epithelial in appearance and quite uniform in size, ranging from 10–24 μ in diameter and 25–45 μ in length (Figs. 3a, b). Subculturing has not yet been attempted.

2. Glossina morsitans

Our knowledge of the developmental physiology of the life cycle of trypanosomes in the vector tsetse fly has many serious gaps. The *Trypanosoma brucei*-type cycle begins with the ingestion of stumpy forms which initially develop into typical trypanosome forms in the midgut of the fly and advance to the crithidial and metacyclic forms which multiply in the salivary glands. The metacyclic trypanosome is the only form capable of infecting a susceptible mammalian host (Wenyon, 1926). Metabolic studies of the midgut forms and of the blood streams forms have been fairly extensive (von Brand, 1966) but very little is known about the metabolism of the crithidial and metacyclic forms (Vickerman, 1966). This situation is likely to remain unchanged until it is possible to culture these forms *in vitro*.

In a very extensive study, Trager (1959) used organs from pupal and adult *Glossina palpalis* Rob.-Desv. as primary explants and succeeded in two instances in culturing *Trypanosoma vivax* Ziemann to the infective stage. In the present study, explants from larval and pupal *G. morsitans* were used with emphasis on extensive cell growth rather than on long term maintenance of isolated organs.

Materials and Methods. Females of the genus *Glossina* are ovoviviparous, the larva maturing within the uterus and pupation usually taking place within an hour after the larva has been expelled from the female. To obtain early larval stages the females were surface sterilized for 20 to 40 min. in 70% ethanol and then the larva was removed from the uterus with care to avoid rupturing the gut. Pupae were similarly sterilized. Healthy pupae immediately sank to the bottom of the dish; defective pupae remained floating in the ethanol.

Tissues and organs from both stages were lightly teased apart and placed in T-15 flasks with 3 ml of Schneider's medium (1964) supplemented with 15% FBS. Two to 4 larvae or a single pupa provided sufficient material for a culture. The cultures were maintained at $27 \pm 0.5°$ C with a gaseous phase of ambient air.

Results. A minimum of 12 cultures were initiated from each of the three larval instars. The most successful cultures were obtained from larvae in which the polypneustic lobes were present but not fully developed. Cells from this stage readily attached to the bottom of the flask and monolayers, not fully confluent, were attained within a few days. Although the cells retained a healthy appearance for extensive periods, all attempts to subculture them failed. Explants from the later larval stages or from pupae less than a week old proved refractory for culturing.

Extensive growth, however, was obtained with explants from pupae 10 to 16 days old. In particular, growth in the form of hollow cellular spheres was most abundant from fragments of pupal gut (Fig. 4). Within 2 weeks, these spheres frequently attained diameters of 3 to 4 mm. The spheres began to pulsate by the 4th or 5th day of culture, a property extending to small groups of cells which attached to the bottom of the flask after the spheres had been teased apart. The cells are epithelial in appearance, contain many dense granules and vary from 30 to 40 μ in diameter and 55 to 70 μ in length.

Somewhat disconcerting was the realization that these cells may be very sensitive to radiation emitted by the microscope lamp. If examined under phase contrast for more than a few minutes, vacuolization of the cells begins, followed by pronounced degenerative changes. Recovery of such cells is very poor. If the deterioration is occurring as the result of radiation from the lamp, it is unlikely that wavelengths in the infrared portion of the spectrum are responsible. The lamp is cooled by forced air with 2 heat filters interspersed between the lamp and the condenser. After 10 min. exposure, the air temperature at the plane of the cells on the microscope stage rose 0.2° C. That such a

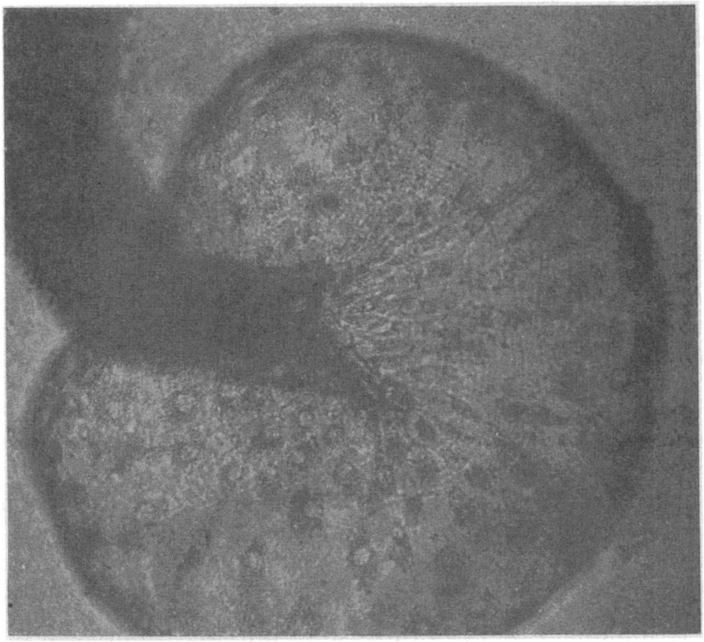

Fig. 4. Cellular sphere developing from fragment of pupal gut of *Glossina morsitans* after 3 weeks in culture. 150 ×

minute change in temperature could have so pronounced an effect on the cells does not seem very plausible. This problem has now been largely mitigated by placing all *Glossina* cultures in T-15 flasks which have been coated with amber stain in 521 brushing medium and fired for 10 min. at 553° C.

D. General Discussion

The discussion will focus on two aspects of insect culture which have not, as yet, received the attention they deserve. Other topics were discussed by previous reviewers (DAY and GRACE, 1959; MARTIGNONI, 1960; JONES, 1962, 1966; SCHNEIDER, 1967; GRACE, 1968; VAUGHN, 1968).

1. Origin of Cells in Culture

One of the major obstacles to the culture of insect tissue *in vitro* is the very small size of the insects themselves. Although a few insect cell lines have been established from isolated organs, primarily ovaries, or from hemocytes, the majority of dipteran lines currently available originated from primary explants consisting of hundreds of individuals. To ascertain the identity of those cells which proliferate in such cultures presents a formidable problem. A few approaches, however, may prove feasible, especially for lines established from the late embryonic or early post-embryonic stages. At this stage of development *in vivo*, multiplication has largely ceased except for 3 cell types, namely nerve cells, hemocytes, and the cells of the imaginal discs. Distinguishing between these 3 types might not be too difficult assuming, of course, that dedifferentiation of the cells has not been too extensive. Histochemical and cytochemical studies as well as micro-cinematographic recordings may be of value. A more direct method, however, could be applied to those primary explants which give rise to spherical cell monolayers. Excised cell spheres could be transplanted into suitable hosts and the identity of the cells determined by their subsequent differentiation *in vivo*. One might also explore the possibility of obtaining "tissue specific" antigens from donor insects, preparing antisera against them and using fluorescent antibody staining to detect cells carrying such antigens in primary cultures. OKADA (1965) has used this technique with considerable success in identifying cells derived from a single source in heterotypic reaggregates of mammalian cells. But such studies may not be feasible with insect cultures for a number of reasons, the most prominent again being the small size of the insects.

2. Culture Media and Cell Nutrition

There seems to be a disturbing tendency for culture media to be selected more with emphasis on availability than suitability. This applies not only to culture media for dipteran species but for other insect species as well. In all likelihood, the very great plasticity of insect cells to adapt to conditions *in vitro* has tended to obscure the dubious value of this practice.

The composition of insect hemolymph varies considerably, in inorganic cation-anion ratios, concentrations of carbohydrates, amino acids, organic acids, proteins and the like, not only among orders but also among families, genera, species and even between different stages of an individual insect (FLORKIN and JEUNIAUX, 1964). Hemolymph analyses for all of the insect species used to initiate cell cultures are not, of course, available, nor are all of those reported as adequate and/or accurate as they might be. Yet there is little evidence (with a few notable exceptions) to suggest that such information, when available, is considered in the design of culture media.

WAYMOUTH (1957) in reference to designing vertebrate culture media has stressed the potential pitfalls of basing such solutions on analyses of blood composition. Her contention that "the blood stream is the sink as well as the source for a wide array of cell products and metabolites" is also valid for the hemolymph of insects. Nonetheless, such analyses would seem to provide a more logical starting point for the design of culture media than the rather haphazard approach now so much in evidence.

Equally valuable, if not more so, would be an evaluation of the nutritional requirements of insect cells once placed in culture. GRACE and BRZOSTOWSKI (1966) analyzed the amino acids and sugars utilized by *A. eucalypti* cells grown *in vitro* over a 7 day period. They found that 14 of 21 amino acids in the culture medium decreased in amount, the other 7 either remained at the same level or increased in concentration. The sugars, glucose, fructose and sucrose were utilized to the extent of 90, 60 and 23%, respectively. And in the present volume, separate reports by VAUGHN (p. 42, 92) and by JENKIN et al. (p. 97), describe certain aspects of lipid metabolism in cell lines of *A. eucalypti* and *A. aegypti*, respectively. Additional studies of this nature should be conducted, not only on the established lines but also on cells in the very early subcultures; e.g., before, not after, the metabolism of the cells has been altered, perhaps profoundly, by the culture conditions.

To continue the current practice of relying on the eight or ten insect culture media so far devised for establishing cell lines from very diverse species may seriously compromise the value of these cells for future studies, including those with the arthropod-borne viruses.

II. Culture of Cells of the Flesh Fly, *Sarcophaga bullata*

SIDNEY T. SHINEDLING and BERNARD GREENBERG

A. Introduction

Fly cells maintained *in vitro* would be useful for the study of fly-virus-man interactions. Ideally, these cells should come from an established tissue culture line. We have thus far had no success in establishing such a line, but one of us (GREENBERG, 1969) had used tissues of *Musca domestica* and *Musca sorbens* in a preliminary study. Now we have been able to consistently obtain primary cultures from pupae of the fly, *Sarcophaga bullata*. These pupae are large—about 4 mm by 12 mm, and each weighs about 0.17 g, when the maggots are well nourished—so that a single animal can provide a sufficient number of cells to start a single culture. Microscopic techniques are not required.

B. Materials and Methods

The culture medium used was a mixture of equal parts of GRACE's (1962) insect tissue culture medium (Grand Island Biological Company, GIBCO), SCHNEIDER's (1964) *Drosophila* medium (Revised) (GIBCO), and Medium 199 (GIBCO), supplemented with 1.0 mg/ml Yeastolate (Difco Laboratories), 10% fetal calf serum (GIBCO or Reheis Chemical Company), and 100 μg/ml streptomycin-100 units/ml penicillin mixture (GIBCO). The pH was adjusted to 6.9–7.1 with 5% sodium bicarbonate. Also tested in this study were SCHNEIDER's *Drosophila* medium alone, supplemented as the above combination, or either medium with 20% fetal calf serum, 5 mg/ml Yeastolate, 5 mg/ml lactalbumin hydrolysate (General Biochemicals), or other combinations of these supplements.

Cultures were prepared in 4 ounce glass prescription bottles, on cover slips in Leighton tubes, or occasionally, for photographic purposes, in sealable glass tissue culture vessels.

Maggots were reared in sacrificed laboratory rats, kept frozen till needed. One-half to two-day old pupae were washed free of gross contamination with laboratory detergent, and rinsed thoroughly. A 15 min. immersion in 2% Clorox (sodium hypochlorite 0.1%), followed by three rinses in double-distilled water sterilized the outside surfaces of the pupae.

For dissection, a pupa was held by forceps in a small, disposable Petri dish containing culture medium and opened longitudinally with a No. 11 scalpel. The expressed lump of pupal material was agitated in the medium to remove unattached cells and tissue, and the remaining pupa was then discarded. The cells of several pupae were transferred to a culture vessel by either syringe or Pasteur pipette. Approximately 200,000 cells/ml, or one pupa/ml were found to give good surface coverage. The cultures were incubated at room temperature, 22–24° C. Medium was changed at intervals of a few days to a month. Partial changes, or changes more frequent than once every two weeks did not seem to benefit the cultures. Observations were made with a Wild M40 inverted phase contrast microscope. Transfers were attempted using aspiration, scraping with a rubber policeman, or 0.25% trypsin.

C. Results

Typically, cells attached in the first half hour and began to aggregate within the first 6 hours (Fig. 5). By the third to fifth day a partial monolayer had formed (Fig. 6), which then slowly began to degenerate. In the denser cultures, mitotic figures could be demonstrated at 8 weeks, though cell numbers were diminished. Some cells showed somatic pairing.

The cultured cells were highly variable morphologically (Figs. 7 and 8). Distinct types were observed, but variations of these were common. The regularly obtainable cells, ranging from about 15 to 60 μ in width, fell roughly into five categories:

(1) Smoothly bounded, round, single cells.

(2) Clusters of small, broad, fusiform cells.

(3) Large clumps of small round cells with occasional larger cells. Cells on the borders of these clumps often extended rays onto the vessel surface and the clumps were firmly attached. Mitoses were seen almost exclusively among these cells.

(4) Bipolar or multipolar, fibroblast-like cells with smooth, round or ovoid cell bodies. These cells and other cell types were often joined, singly or in groups, by networks of long, straight, syncytoid rays. In some cases these ray networks included most of the cells in the culture (Fig. 9).

(5) Medium to large, round or multipolar, single cells with granular cytoplasm, sometimes binucleate. An anomalous, possibly degenerating, variation of this type, possibly a complex of more than one cell was observed. This type of cell was very large (up to 160 μ in width) and was often highly vacuolated,

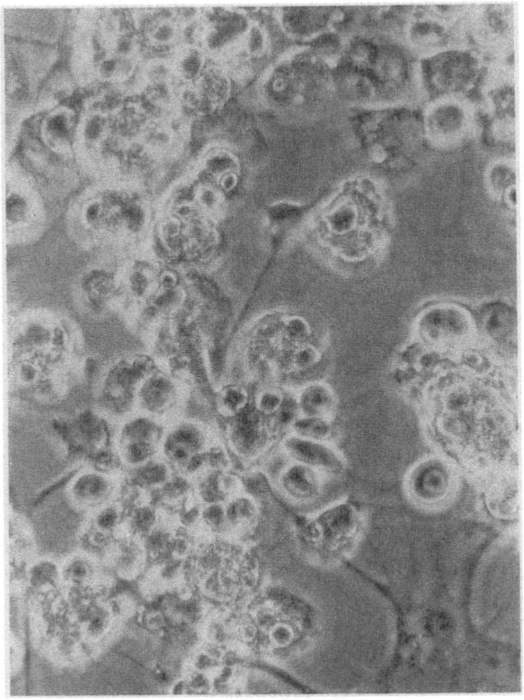

Fig. 5. Aggregated cells at six hours. ×450

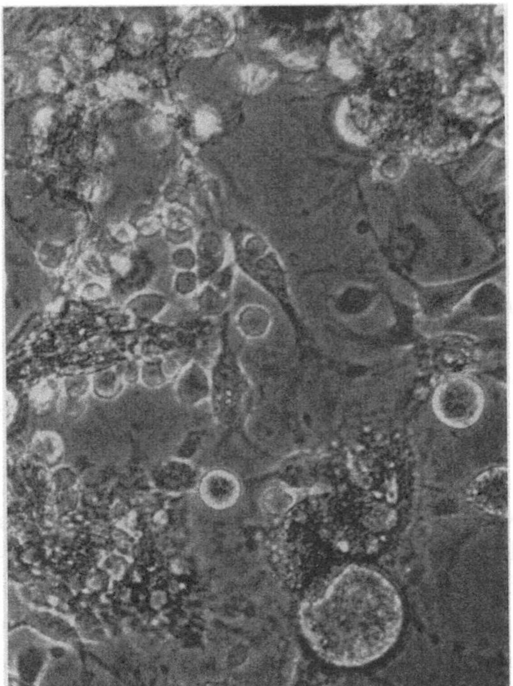

Fig. 6. Partial monolayer, at five days. ×450

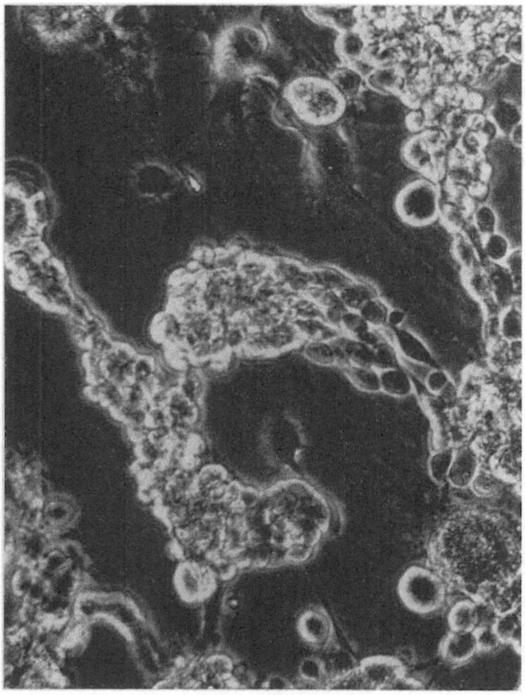

Fig. 7. Spindle-shaped cells between clumped round cells. Other types may also be seen. ×415

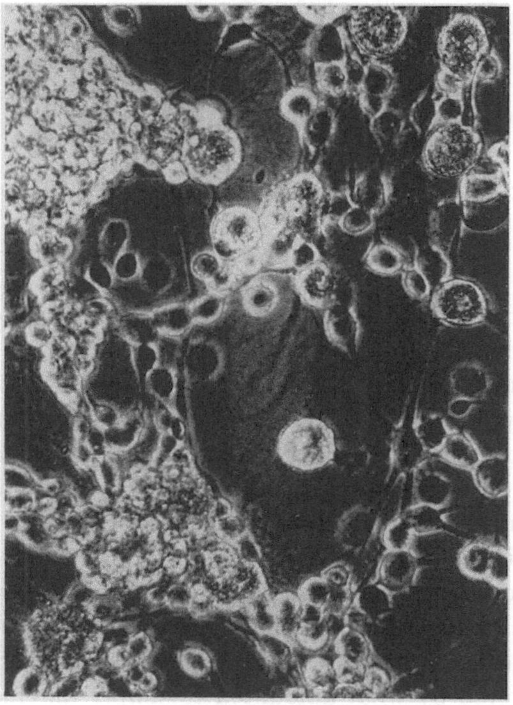

Fig. 8. Clumps, fibroblast-like cells, and large round cells with granular cytoplasm. ×360

or multinucleated (3 to 50 nuclei), or both (Figs. 10, 11). Frequently, large inclusions with a layered appearance were seen in stained preparations. These cells also sometimes showed large globules of lipid (demonstrated by staining with Oil Red 0).

A sixth cell type was seen only in one group of cultures and has not yet been cultured again. These cells were small and spindle shaped and tended to form a completely confluent monolayer (Fig. 12). This unusual, profuse cell

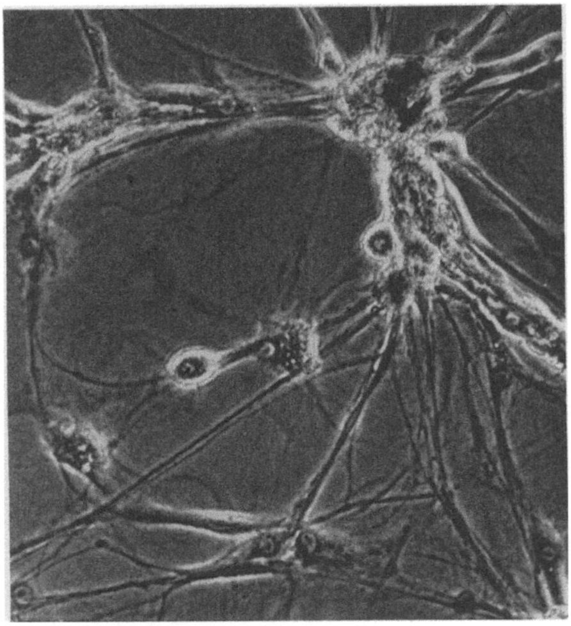

Fig. 9. Cells growing in a syncytoid ray network. ×340

sheet, formed in the first three days after the culture was initiated and began to degenerate within the first week.

Media supplemented with 5 mg/ml Yeastolate proved to be toxic. The other heavily supplemented media (20% calf serum, 5 mg/ml lactalbumin hydrolysate, or both) and the normally and heavily supplemented SCHNEIDER's *Drosophila* medium produced much denser growth. The same cell morphology was observed, but the culture consisted mainly of dense, three dimensional clumps of cells. These cell masses, however, followed much the same course as the sparser monolayers. Our most effective transfers were made by aspiration from these cultures, but neither these nor other transfers ever grew to the density of the parent cultures and it would seem that the course of the culture was unaltered by what amounted to simply moving some of the cells to a new vessel with a partial change of medium.

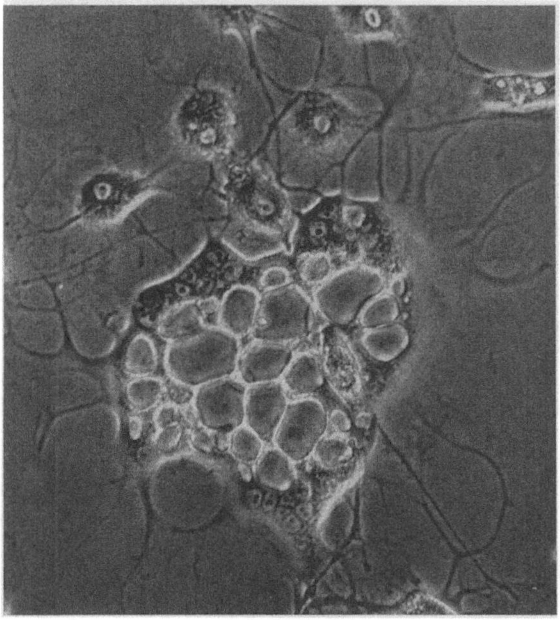

Fig. 10. Giant, highly vacuolated, multinucleate cell (or cells), with other granular type 5 cells. ×350

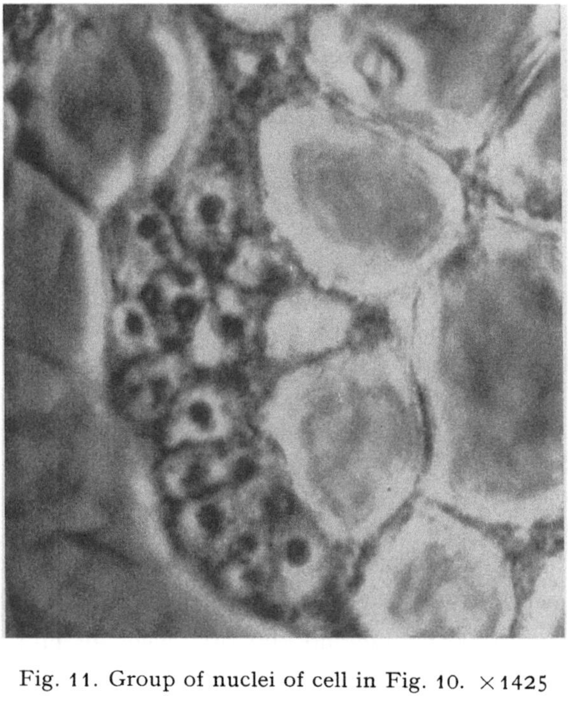

Fig. 11. Group of nuclei of cell in Fig. 10. ×1425

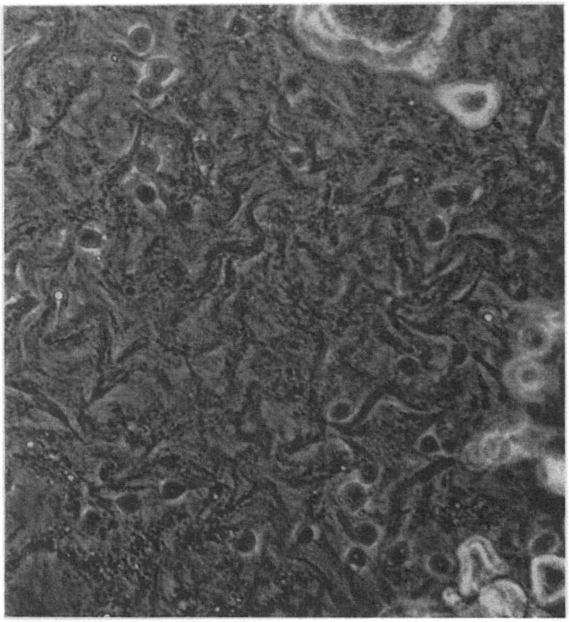

Fig. 12. Confluent monolayer of spindle-shaped cells. ×340

D. Discussion

It is not possible to know, from simple morphological description, the tissue of origin of the cells in these cultures. The origin of two of the types is suggested, however, by the detailed studies of *S. bullata* hemocytes by JONES (1956) and WHITTEN (1964). Several of the plasmatocyte forms described by JONES are very similar to the type 4 fibroblast-like cells. Also, the filamentous rays, networks, and fragments of our study resemble the attached and detached filaments of plasmatocytes described by JONES and the type 5 granular cell is similar to his description of the granular hemocyte. This cell also resembles the amoeboid phagocytes described by HIRUMI and MARAMOROSCH (1964) in their culture of *Anosia plexippus*.

The giant form of the type 5 cell has interesting similarities to the multinucleate granular hemocytes found by WHITTEN. The inclusions of phagocytized material and the capacity to either retract or extend filaments suggest that these may be related to the granular, round or multipolar cells in our cultures and that the lamellar inclusions in the cells may be phagocytic vacuoles.

It should be noted that the smooth, round or ovoid cells present in the early cultures (Fig. 5) resemble WHITTEN's photographs of larval plasmatocytes and that these cells, according to the plasmatocyte transformation scheme of GUPTA and SUTHERLAND (1966), often become granular hemocytes.

The lack of proliferation beyond the first week of cultural life may be considered from three points of view: dedifferentiation, alteration and nutrition.

GRACE (1962) reported a four month-period of quiescence and a culture age of ten months before his *Antheraea* cells began to proliferate indefinitely. This period may be necessary in some instances to allow the cells to dedifferentiate. Applied to our cultures this would suggest the necessity for a heavy inoculum and a rich medium to try to ensure the survival of the culture for many months.

A more active approach is to attempt to alter the cells in culture, namely, to induce malignancy. Experimental use of insect viruses to achieve this would be the most logical approach to alteration of insect cells. HAYFLICK (1967) suggests that even those *in vitro* oncogenic events which appear to be spontaneous may be mediated by virus and gives as examples evidence of alteration of African green monkey kidney cells by low concentrations of adenovirus and association of virus with Burkitt's lymphoma, a malignancy not historically regarded as viral. Other genomic alterations might also be fruitful—X-rays, ultraviolet light and chemical mutagens such as nitrogen mustard and 5-bromouracil.

Finally there is a possibility that the medium we have been using is simply inadequate to support continued mitosis of *Sarcophaga* cells and that some nutritional or hormonal factor which the cells need is being removed by diffusion, rather than supplied, by the medium. A possible hint in this direction is that most mitoses we have seen were in clumps of cells, where a cell would be somewhat "protected" from the medium by its neighbors. (The alternate explanation, that clumps will only appear in those areas where mitoses are occurring, should not be overlooked.) We have attempted to empirically provide the cells with their nutritional needs, but they are taken from a specialized phase in the life of the fly and may have specific requirements that we have not met.

III. Establishment and Characterization of Two New Cell Lines (CP — 1268 and CP — 169) from the Codling Moth, *Carpocapsa pomonella* (with a Review of Culture of Cells and Tissues from Lepidoptera)

W. F. HINK and B. J. ELLIS

A. Introduction

Cell lines have been established from 8 species of Lepidoptera representing 5 different families (Table 2). The line from *Antheraea eucalypti* (GRACE, 1962) was the first insect cell line to be established and is still in culture after $9^1/_2$ years. The primary explants from which the eight lines originated were larval, pupal, and adult ovaries; larval and pupal hemocytes; and embryonic tissue. Larval, pupal and adult explants seem to follow the same general pattern when cultivated. For a period of several weeks, cell migration or proliferation

Table 2. *Cell lines from Lepidoptera*

Insect	Preparation of culture	No. of sub-cultures	Time in culture	Growth medium[a]	Reference
Bombyx mori (silkworm)	Larval ovaries	12[b]		Salts, glucose, CH, YE, glutamine, choline, 10% HH	Vago and Chastang (1958)
Bombyx mori	Larval gonads (minced)	22[b]		Trager solution A (salts, maltose), 10% HH	Gaw, Liu, and Zia (1959)
Bombyx mori	Larval ovaries (trypsinized)		3 yr[b] 5 yr[c]	GMA, 5% *A. eucalypti* hemolymph	Grace (1967)
Antheraea eucalypti	Pupal ovaries (trypsinized)		2 yr[b] $9^1/_2$ yr[c]	GMA, 5% HH	Grace (1962)
Chilo suppressalis (rice stem borer)	Larval hemocytes		2 yr and 8 mo[b]	CSM-2F (salts, glucose, fructose, peptone, LH, YL, choline, FBS, TC-199)	Mitsuhashi (1967a)
Heliothis zea (corn earworm)	Adult ovaries (trypsinized)	115[c]	3 yr[c]	Yunker medium (GMA, FBS, EU, BA) crystallized BA	Hink and Ignoffo (1970)
Trichoplusia ni (cabbage looper)	Adult ovaries (minced)	183[c]	2 yr[c]	TNM-FH (GMA, FBS, EU, YL, LH, BA)	Hink (1970)
Carpocapsa pomonella (codling moth)	Embryo (teased apart)	57[c]	17 mo[c]	TNM-FH	Hink and Ellis (this section)
Spodoptera frugiperda (fall armyworm)	Pupal ovaries (treated with trypsin and hyaluronidase)	30[c]	16 mo[c]	GMA, FBS, EU, 4.4% *B. mori* hemolymph	Vaughn (personal communication
Samia cynthia	Pupal hemocytes	35[b]	15 mo[b]	GMA, FBS, 0.5% HH	Chao and Ball (p. 28)

[a] The following abbreviations are used: *GMA* Grace's (1962) insect tissue culture medium; *HH* homologous hemolymph; *YE* yeast extract; *CH* casein hydrolysate; *FBS* fetal bovine serum; *EU* chicken egg ultrafiltrate; *BA* bovine albumin; *YL* Yeastolate; *LH* lactalbumin hydrolysate.

[b] At time of original publication.

[c] As of 3/15/70.

occurs. Following this initial growth, cell multiplication ceases or decreases drastically. After a period of 6–9 months, cell growth resumes and subculturing may be initiated.

It is significant that 7 of 8 cell lines are cultured in Grace's (1962) medium plus various supplements. Grace's medium contains concentrations of amino acids, organic acids, fructose, and sucrose similar to those present in the hemolymph of the silkworm, *Bombyx mori* (WYATT, 1956). The growth of cells from diverse species in this medium suggests that the nutritional requirements of cultured insect cells may not be as critical as early investigators had assumed. Further evidence for the flexibility of growth requirements is the adaptation of the *A. eucalypti* cell line to growth in several media (NAGLE et al., 1967; YUNKER et al., 1967; MITSUHASHI and GRACE, 1969).

Of the eight established lines, four were either established or are currently being cultured in hemolymph-supplemented (HS) media and four were established and are maintained in hemolymph-free (HF) media. The most important advancement, in terms of technical and practical considerations, is the removal of insect hemolymph from cell culture media.

The use of heterologous hemolymph in culture systems is desirable when an insect furnishing the explant is not a good source of homologous hemolymph. In studies of the specificity of hemolymph, SEN GUPTA (1964) cultured larval intestinal tissue of *Galleria mellonella*, *Antheraea pernyi*, and *Melolontha melolontha* in media supplemented with homologous hemolymph and with heterologous hemolymph from the other two species. No significant differences in growth were observed indicating lack of specificity in the requirements for hemolymph. Heterologous hemolymph will also support growth of cell lines from *B. mori* (GRACE, 1967) and *Spodoptera frugiperda* (VAUGHN, personal communication). RAHMAN et al. (1966), however, assayed replacements for the 3% *A. pernyi* hemolymph supplement to the medium supporting growth of the *A. eucalypti* cell line. Hemolymph from five species of Lepidoptera was toxic. Only the hemolymph from *Antheraea polyphemus* was an adequate substitute. The conclusion must be drawn that heterologous hemolymph will support growth, but each culture system must be tested individually.

The realization that hemolymph may not be a necessary requirement for growth of lepidopterous tissue *in vitro* was suggested by SEN GUPTA (1961). The growth of intestinal, ovarian, and testicular tissues of *G. mellonella* larvae was compared in HS medium and medium supplemented with calf serum, yeast extract, trehalose and organic acids. Growth in HS medium was more pronounced but serum-supplemented medium did support growth and survival for periods comparable to those obtained with HS medium. Variation in the percentage of hemolymph supplement between 5 and 20% did not affect growth. Cell migration from *G. mellonella* and *B. mori* ovaries was similar in a medium consisting of WYATT's (1956) salts plus lactalbumin hydrolysate and yeast extract and in a medium composed of WYATT's salts and sugars plus a dialysate of *B. mori* hemolymph (AIZAWA et al., 1961; AIZAWA and SATO, 1963). Apparently lactalbumin hydrolysate and yeast extract were effective replacements for hemolymph.

Calf serum can, to some extent, be used as a replacement for insect hemolymph (VAGO and CHASTANG, 1962a). Cultured *B. mori* ovaries were utilized to assay various proportions of *B. mori* hemolymph and calf serum media supplements. Five combinations were used: (1) 15% hemolymph, (2) 10% hemolymph and 5% calf serum, (3) 5% hemolymph and 10% calf serum, (4) 2% hemolymph and 13% calf serum, and (5) 15% calf serum. The first 4 media supported cellular outgrowth, multiplication and culture longevity equally well. In medium containing only calf serum, initiation of cellular outgrowth was retarded, cell multiplication was slower and cultures did not survive as long as when cultured in the presence of hemolymph. In further tests of hemolymph substitutes, gonads of *Lymantria dispar*, *B. mori*, and *G. mellonella* were cultured in (1) medium supplemented with 15% *B. mori* hemolymph, (2) medium with 2% hemolymph and 13% calf serum and (3) medium with lyophilized chick embryo extract (VAGO and CHASTANG, 1962b). Cell migration from all explants began about 6 hours after initial culturing and all cultures in serum plus hemolymph or HS media were in better condition than those in media containing chick embryo extract.

In media for established cell lines, various substances are used to partially or completely replace hemolymph. Fetal bovine serum (FBS) is used at concentrations ranging from 4.4% for the *S. frugiperda* cell line (VAUGHN, personal communication) to 20% for the *Chilo suppressalis* cell line (MITSUHASHI, 1967a). Different concentrations of FBS affect the growth of cells (SOHI and SMITH, 1970). When FBS is omitted from the media, GRACE's *A. eucalypti* and *B. mori* lines will not grow. Maximum growth is obtained in media supplemented with 5–10% FBS. YUNKER et al. (1967) adapted the *A. eucalypti* cell line, normally cultured in the presence of hemolymph, to a HF medium. Whole chicken egg ultrafiltrate, FBS and bovine plasma albumin were employed to replace hemolymph. NAGLE et al. (1967) cultured this same line as a suspension culture in a FBS-supplemented medium that is otherwise chemically defined. MITSUHASHI and GRACE (1969) also adapted these cells to a HF serum-supplemented medium (CSM-2A) that is very different from the HS medium in which the line had been grown for many years.

Four of the cell lines (*C. suppressalis*, *Heliothis zea*, *T. ni* and *C. pomonella*) originated from explants placed directly into HF media containing FBS and other substances such as egg ultrafiltrate, lactalbumin hydrolysate, Yeastolate and bovine plasma albumin. The use of these diverse and chemically undefined ingredients has, in most cases, eliminated the requirement for hemolymph.

Cell lines grown in HF media multiply as rapidly as those grown in HS media (Table 3). This suggests that the growth of established cell lines is independent of hemolymph.

There is a wide range in the growth rates of different cell lines and even of a single line (*A. eucalypti*) in the laboratories of various investigators. The most rapidly growing line is *T. ni* (TN-368) with a population doubling time of 16 hrs and a subculture interval of 2–3 days (HINK, 1970). Most other lines have population doubling times of 24–108 hrs and are subcultured at approximately weekly intervals. The maximum populations are generally $1–3 \times 10^6$ cells/ml of medium. Exceptions are $2–4 \times 10^5$ cells/ml for *C. suppressalis*

Table 3. *Growth of cell lines from Lepidoptera*

Insect	Insect hemolymph supplemented	Sub-culture interval	Population doubling time (hr)	Maximum population increase per ml medium ($\times 10^5$) from		to	days	Reference
A. eucalypti	+	8-10 days	60-72			—		Grace (1962)
A. eucalypti	—	weekly	—	2	→	22	23	Nagle et al. (1967)
A. eucalypti	—	weekly	60	0.4-0.6	→	9.5	13-14	Yunker et al. (1967)
A. eucalypti	+	weekly	100	0.4-0.6	→	2.5	11-12	Yunker et al. (1967)
A. eucalypti	—	weekly	42-48	2	→	10	6-7	Hink, unpublished data
A. eucalypti	+	7-10 days	—	1-2	→	24	20-22	Sohi and Smith (1970)
A. eucalypti	+	—	36	1-2	→	20-30	6	Mitsuhashi and Grace (1969)
A. eucalypti	—	—	60	1-2	→	10	6	Mitsuhashi and Grace (1969)
B. mori	+	6 days	48			—		Grace (1967)
B. mori	+	7-10 days	—	2	→	16	16	Sohi and Smith (1970)
C. suppressalis	—	1-2 weeks	108	0.7	→	4	15	Mitsuhashi (1967a)
T. ni (TN-368)	—	2-3 days	16	1-2	→	20-30	3-4	Hink (1970)
C. pomponella (CP-1268)	—	weekly	24	2-3	→	100	7-8	Hink and Ellis (this section)
C. pomponella (CP-169)	—	weekly	24	3-4	→	50	8-9	Hink and Ellis (this section)
H. zea (IMC-HZ-1)	—	weekly	—	1-2	→	10-20	7-8	Hink and Ignoffo (1970)
S. cynthia	+	—	—			—		Chao and Ball (p. 28)
S. frugiperda	+	—	24			—		Vaughn, personal communication

(MITSUHASHI, 1967a) and, in one study, for *A. eucalypti* (YUNKER et al., 1967) and 5×10^6 cells/ml for CP-169 and 10^7 cells/ml for CP-1268. The latter two cell lines will be described in this paper.

B. Materials and Methods

The culture medium for *C. pomonella* is the one employed for the *T. ni* (TN-368) cell line (HINK, 1970). It is designated TNM-FH and contains 90.0 ml GRACE's insect tissue culture medium, 8.0 ml fetal bovine serum, 8.0 ml whole chicken egg ultrafiltrate, 0.5 gm crystallized bovine plasma albumin, 0.3 gm lactalbumin hydrolysate, and 0.3 gm TC Yeastolate. The pH is 6.65; the osmotic pressure is 370 milliosmols. All cultures are grown in 30 ml polystyrene disposable tissue culture flasks (Falcon Plastics) at a temperature of 27° C.

For each primary culture ten *C. pomonella* eggs containing fully formed embryos within 24–48 hrs of emergence were used. The eggs were surface sterilized in 0.5% sodium hypochlorite for 15 min., rinsed in a sterile salt solution (HINK, 1966) and transferred to depression plates containing TNM-FH. The embryos were torn apart with teasing needles and transferred to tissue culture flasks containing 5.0 ml of medium.

The culture from which both cell lines were obtained was initiated in December of 1968. The first subculture, which developed into the CP-1268 line, was made two months later by removing 3.0 ml of cell suspension and adding it to 2.0 ml of fresh medium. The original volume of the primary culture was re-established. One and a half months after the first subculture, a second line (CP-169) was initiated from the same primary culture by a similar procedure. Initially, the interval between subcultures was 10 days for both lines, but was reduced to 7 days after 13 or 14 subcultures.

Both cell lines are maintained under a normal atmosphere in stationary flasks. For subculturing the attached cells are loosened by placing the flasks on a platform rotary shaker for 1 hr at 100 rpm and 0.2–0.5 ml of the resulting cell suspension is transferred to 4.5–4.8 ml of fresh medium. The volume of the inoculum varies, depending on the final cell count of the parent culture. Each subculture is started with $2\text{–}4 \times 10^5$ cells/ml and a total volume of 5.0 ml. Currently (3/17/70), the CP-1268 line has been subcultured 58 times and the CP-169 line 57 times.

Chromosome studies were done with cells in the logarithmic phase of growth, using a modified Moorhead method (MOORHEAD et al., 1960). The cells were exposed to colchicine (1×10^{-6} gm/ml of culture medium) for 5–6 hrs, centrifuged at $100 \times g$ for 10 min. and the tissue culture medium replaced with 5.0 ml of 1% sodium citrate. After incubation at 30° C for 15 min., 2–3 drops of fresh Carnoy's fixative (1 part glacial acetic acid and 3 parts absolute methanol) were added. The cell suspension was centrifuged at $100 \times g$ for 10 min., the pellet resuspended in 5.0 ml of fresh fixative, the suspension incubated at 4° C for 20 min. and the cells washed twice and resuspended in 2–3 ml of fixative. The cell suspension was applied to cooled slides and the slides were passed through a flame to ignite the fixative and to spread the chromosomes. The slides were then air dried and stained for 30–45 min. with 20% Giemsa in phosphate buffer, pH 6.86.

The haploid number of chromosomes for *C. pomonella* was determined from squash preparations of testicular spermatocytes from 5th instar larvae. Testes were removed from larvae, submerged in 1% sodium citrate and incubated for 30 min. at 30° C. Then the sodium citrate was removed and replaced with Carnoy's fixative. After 5 min., the fixative was withdrawn. The tissues were stained with orcein stain (BRELAND, 1961) without lactic acid, squashed, and examined.

C. Results and Discussion

1. Growth of the Primary Culture

The growth pattern of the *C. pomonella* primary culture differed from other primary cultures of lepidopterous tissues which have developed into cell lines. The tissues of *C. pomonella* embryos did not attach to the culture vessel. Instead, after one month in culture, cells began to migrate from the interior of the suspended tissues and many layers of cells were formed on the exterior surface of the fragments. These cells descended to the bottom of the flask and formed colonies of attached cells. Within two months, the primary culture contained many suspended cells and a confluent monolayer which could be subcultured. The two-month period from initiation of the primary culture until cells began rapid proliferation was relatively short compared to $10^1/_2$ months for *A. eucalypti*, $9^1/_2$ months for *B. mori*, 7 months for *T. ni* and *H. zea*, and 6 months for *C. suppressalis* (GRACE, 1962, 1967; HINK, 1970; HINK and IGNOFFO, 1970; MITSUHASHI, 1967a).

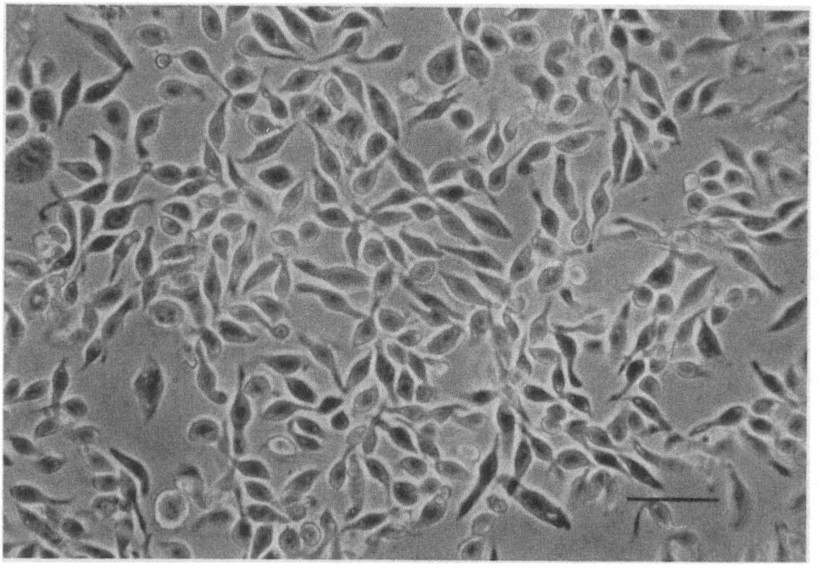

Fig. 13. Phase contrast appearance of the codling moth cell line, CP-1268. A 2-day old culture in the 42nd passage. Scale, 50μ

2. Morphology

The cell lines consist of morphologically diverse suspended and attached cells (Figs. 13 and 14). Both lines contain spindle-shaped, round or subspherical cells, cells with a single protoplasmic extension and polymorphic cells with irregular shapes or many protoplasmic extensions. Approximately 1 % of the total population of both lines is composed of the polymorphic cells.

The majority (61 %) of the cells in the CP-1268 line are spindle-shaped with widths of 9.4–15.9 μ and lengths of 25.5–47.6 μ. Approximately 23 % are round or subspherical with diameters of 12.9–19.3 μ and 15 % of the cells have a single extension and measure 8.6–10.1 × 24.2–34.6 μ. The population of the CP-169 line is composed of 70 % round or subspherical cells measuring 11–20.2 μ in diameter, 22 % cells with single extensions and sizes of 11–15.9 μ × 18.9–40 μ, and 7% spindle-shaped cells with widths of 10.1–13.1 μ and lengths of 26.5–31.3 μ. There are two morphological characteristics that distinguish the lines from

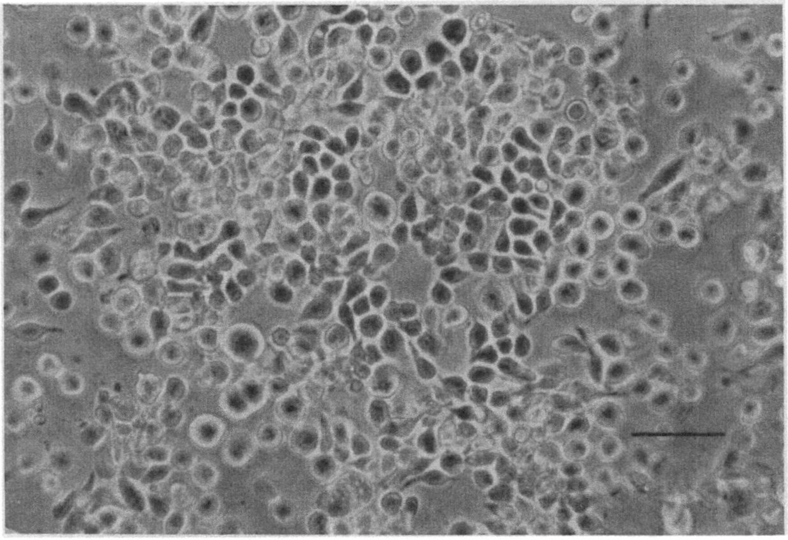

Fig. 14. Phase contrast appearance of the codling moth cell line, CP-169. A 5-day old culture in the 41st passage. Scale, 50μ

each other: First, the majority of the cells in the CP-1268 line are spindle-shaped and the majority of the cells in line CP-169 are round subspherical. Second, the spindle-shaped cells from CP-1268 are longer than those from CP-169. Differences in sizes of spindle-shaped cells are also observed in cell lines of *A. eucalypti*, *B. mori*, *C. suppressalis* and *T. ni*. Those from *T. ni* are the largest and vary in length from 86.3–123.7 μ (HINK, 1970), in *A. eucalypti* they are 47.4–98.9 μ (HINK, unpublished data), in *B. mori* their length is 50–70 μ (GRACE, 1967), and in *C. suppressalis* they are about 34 μ long (MITSUHASHI, 1967a). While there is no significant size difference in the spherical-shaped cells of these lines, the size of the spindle-shaped cells may offer a criterion for distinguishing the lines.

3. Growth Characteristics

Prior to logarithmic growth, the lag phase is about 24 hrs for CP-1268 and 96 hrs for CP-169 (Fig. 15). The population doubling times during logarithmic growth are approximately 24 hrs for both lines. CP-1268 reaches a maximum population of 10^7 cells/ml, a 40-fold increase over the initial population, in 7–8 days. The maximum population of CP-169 is 5×10^6 cells/ml in 8–9 days, which is a 14-fold increase over the initial population. These established cell cultures of *C. pomonella* can be stored at 5° C for $1^1/_2$ months without loss of viability.

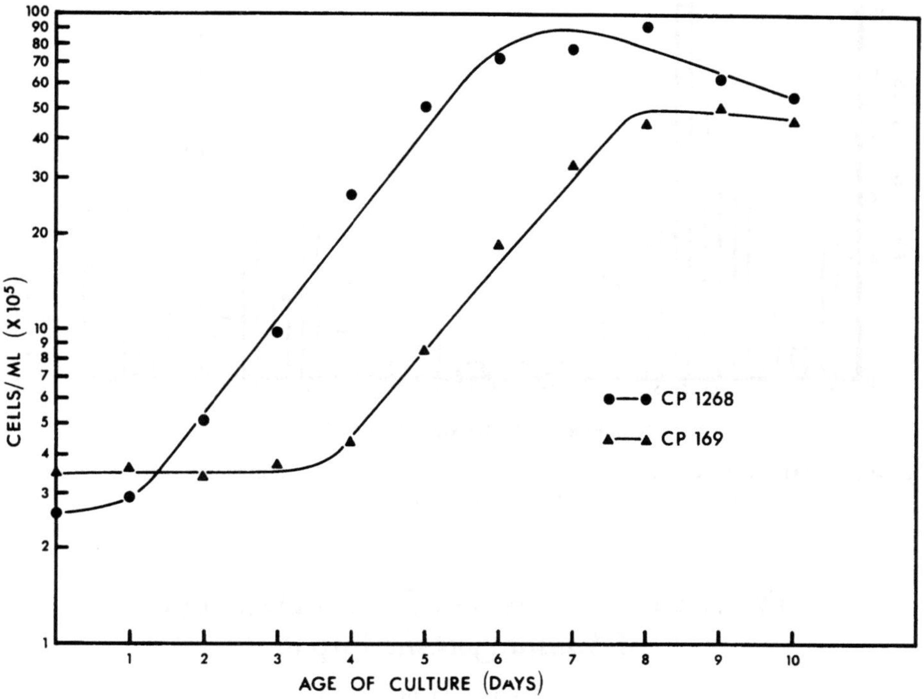

Fig. 15. Growth curves of codling moth cells, lines CP-1268 and CP-169

4. Chromosome Analysis

The haploid number of chromosomes for *C. pomonella*, as determined from spermatocytes, is 27.

The distribution of chromosome numbers of the CP-1268 line is bimodal (Fig. 16). One mode is at the diploid number of 54 and the other is near the tetraploid number of 108. Fifty-one percent of the cells examined had 51–57 chromosomes and 35% had 102–110 chromosomes. Cells appearing to have more than 250 chromosomes were not included in the tabulation of the frequency distribution because they could not be accurately counted and there is no way of disproving massive fragmentation in the absence of visible centromeres. Karyotypic analysis was not attempted because of the small size of the chromosomes and frequent lack of distinguishable centromeres.

The chromosome complement of the CP-169 line was much more difficult to quantitate. Exact counts were impossible to obtain because of greater chromosome numbers and difficulties in obtaining adequate spreads. The modal number of chromosomes is between 101–110 with 29% of the plates within this range. Approximately 9% of the cells are diploid and 72% contain more than 100 chromosomes. Thus line CP-169 exhibits a higher degree of heteroploidy than the CP-1268 line.

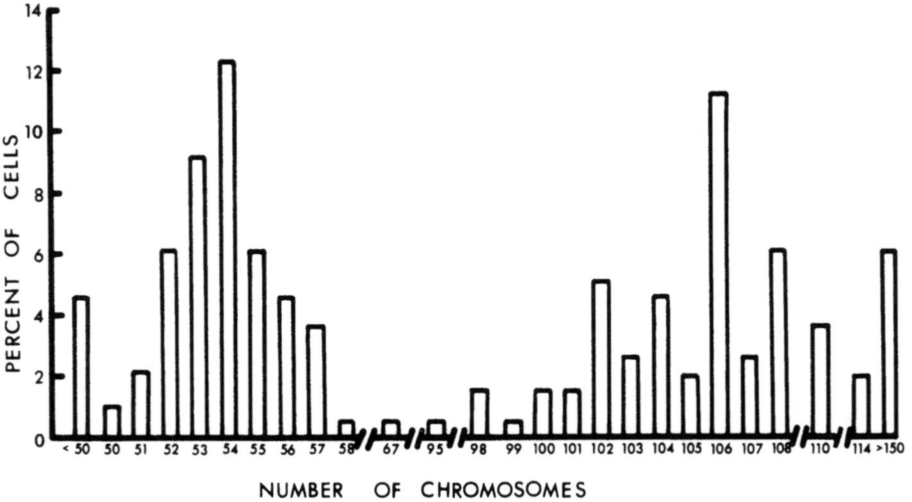

Fig. 16. Distribution of chromosome numbers in line CP-1268 after 1 year of *in vitro* culture. A total of 195 chromosome spreads were counted

IV. A Cell Line Isolated from Hemocytes of *Samia cynthia* Pupae

JOWETT CHAO and GORDON H. BALL

A. Introduction

Although the *in vitro* culture of insect hemocytes for varying periods of time has been reported for about 50 years, the establishment of hemocyte cell lines has been accomplished only recently. MITSUHASHI (1966, 1967a) succeeded in obtaining such a line from the diapausing larvae of the rice borer, *Chilo suppressalis*. In this report the establishment of another hemocyte cell line is described. It is isolated from the diapausing pupae of a moth, *Samia cynthia*.

B. Materials and Methods

A diapausing cynthia pupa was surface sterilized with 75% ethanol and a puncture made at the wingpad with a sterile needle. Hemolymph was dripped directly into a 25 cm² Falcon flask holding 3 ml of GRACE's medium (1962) plus 10% fetal bovine serum (FBS). The flask was placed horizontally, allowing the hemocytes to attach to the bottom surface. The supernatant was then withdrawn and the attached hemocytes washed 3 times with complete GRACE's medium containing 10% FBS plus 0.5% cynthia pupal hemolymph (both heat inactivated), to remove any polyphenol oxidase remaining from the hemolymph of the pupa. Later, this step was eliminated without ill effects in primary isolations or subcultures.

Generally, suspended cells were subcultured by transferring 0.5 ml of the culture to 2.5 ml of fresh medium once in 2 weeks. Often the donor flask was repeatedly washed and the medium replaced to permit culturing the attached cells. The pH of the medium was near 6.6 and remained so during the normal period of culture. Most of the cultures were kept at a temperature of about 25° C.

Other media alone or in combinations, supplemented with varying amounts of FBS and homologous hemolymph were also treated for their suitability for isolation and growth of the pupal hemocytes. They were SINGH's medium (1967), media 199 and 109, and SINGH's medium with the inorganic salts replaced by those of the GRACE's medium (1962). Smears and chromosome spreads of cultured cells were stained with Giemsa stain. The growth of the cells was observed with an inverted microscope with phase contrast optics.

C. Results

Most of the hemocytes became attached to the bottom surface of the flask immediately after the cynthia hemolymph was added to the flask (Figs. 17–20). Depending upon the type of pupa used two patterns of cell growth in the primary culture were observed. In one pattern, straight strands of cells and cells of epithelial type grew from hemocytes of pupae collected within 1 to 4 months prior to cell isolation (Table 4, Fig. 21). These cells apparently remained in good condition for many days but all died upon trypsinization and transfer. However, a different pattern of cell growth was observed in the cultures of hemocytes isolated from pupae which had gone through a long period of diapause, e.g. 10 to 14 months (Table 4). The attached cells multiplied more readily and became a cell sheet in a week, completely covering the flask bottom (Fig. 22). At the same time, small round cells (13 to 24 μ) gradually appeared in free suspension (Figs. 23, 27 and 28). These cells have been transferred 40 times in the past 17 months and are considered to be an established line. In old cultures of this line, some cells assumed elliptical, angular or spindle shape and others became attached again and grew like fibroblasts (Figs. 24 and 25).

Attempts to isolate the attached cells as a separate line failed because free cells eventually reappeared regardless of thorough washing of the attached cells. Repeated washing followed by trypsinization and transfer gave the same result.

The free cells could easily be adapted to grow in hemolymph-free GRACE's medium either by weekly replacement of used medium with small amounts of medium containing only FBS or by adding a large inoculum directly into hemolymph-free medium. One of the adapted lines is now in its 20th subculture; the cell morphology and growth pattern are unaltered compared to those of the main line. The cultured cells have about 52 chromosomes (Fig. 26). The diploid chromosome number of the somatic cells of *S. cynthia* is 26 (HARVEY, 1916). Cells cultured in medium 199 or 109 supplemented with FBS and hemolymph degenerated in several days. Those cultured in supplemented SINGH's medium or in various mixtures of SINGH's and GRACE's media survived one or two transfers.

Table 4. *The isolation and establishment of Samia cynthia hemocyte cell lines*

Pupae collected	Hemocytes isolated	Treatment of hemocytes after attachment	Primary culture	Subculture	Cell line
1967 October	1968 September	Medium exhaustion and replacement	Attached and free cells	Free cells	40 transfers in 17 months
1967 October	1968 September	Trypsinization and transfer	Attached and free cells	Free cells	Not followed
1967 October	1968 September	Trypsinization and transfer	Attached and free cells	Free cells	17 months
1968 October	1968 November	Medium exhaustion and replacement	Attached cells only	No growth	—
1968 October	1969 January	Medium exhaustion and replacement	Attached cells only	No growth	—
1969 October	1969 December	Medium exhaustion and replacement	Attached cells only	No growth	—
1968 October	1969 December	Medium exhaustion and replacement	Attached and free cells	Free cells	3 transfers in 2 months
1968 October	1970 February	—	Attached and free cells	Free cells	2 transfers

Figs. 17–20. Newly isolated hemocytes of cynthia pupae. Fig. 17 and 18, phase contrast. 235 ×. Figs. 19 and 20, oil immersion. 1100 ×

Fig. 21. Primary cultures showing attached hemocytes isolated from pupae in early state of diapause. 235 ×

Fig. 22. Primary culture showing attached and free hemocytes isolated from pupa in late state of diapause. 235 ×

Figs. 23 and 24. Hemocytes growing in suspension. 235 ×

Fig. 25. Hemocytes in 3 month old culture after several replacements of culture medium. 235 ×

Fig. 26. Chromosome spread of cultured hemocytes. 1100 ×

Figs. 27 and 28. Cultured hemocytes, one almost completely divided. 1100 ×

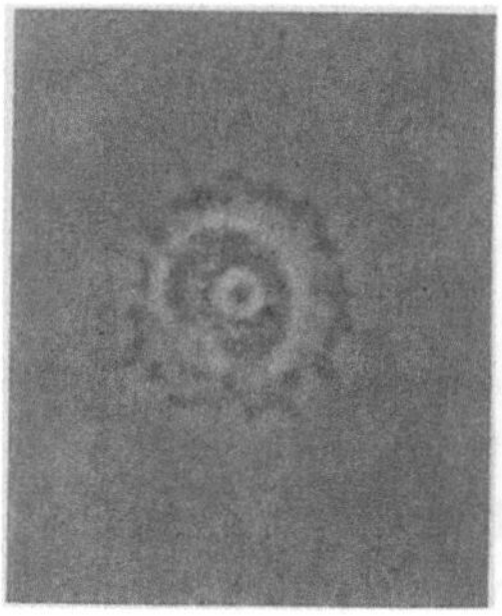

17

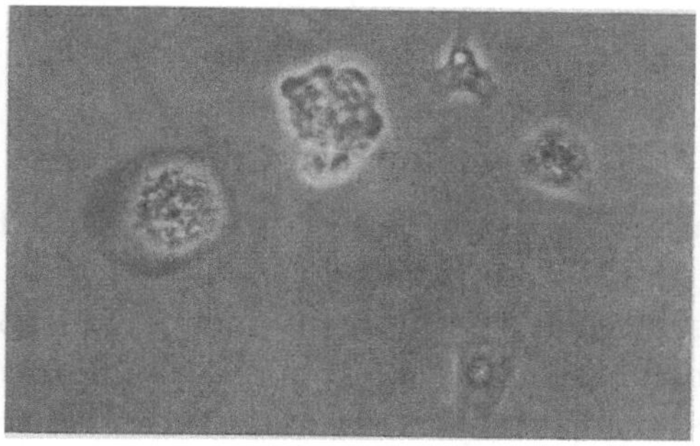

18

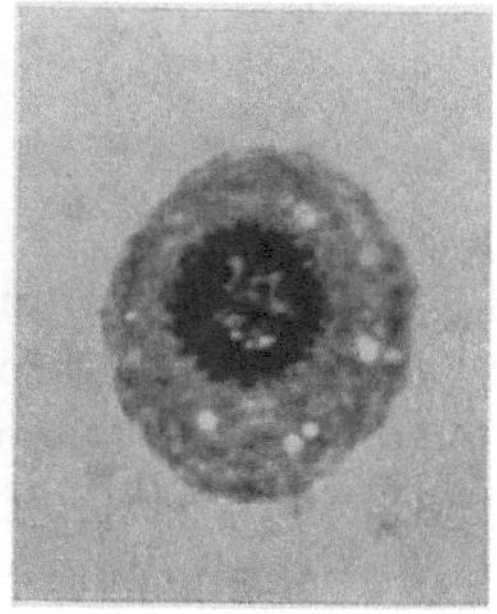

19

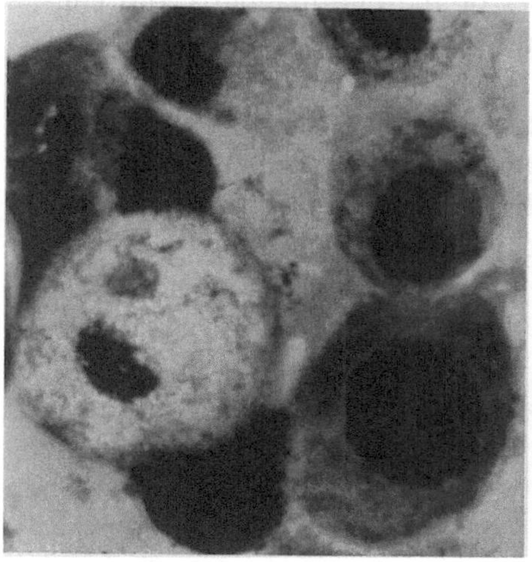

20

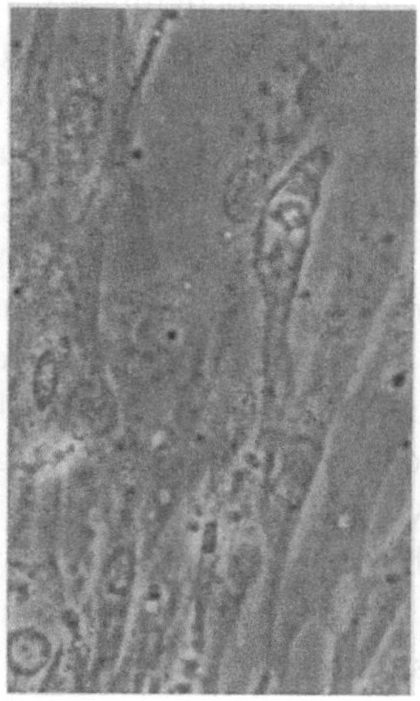

21

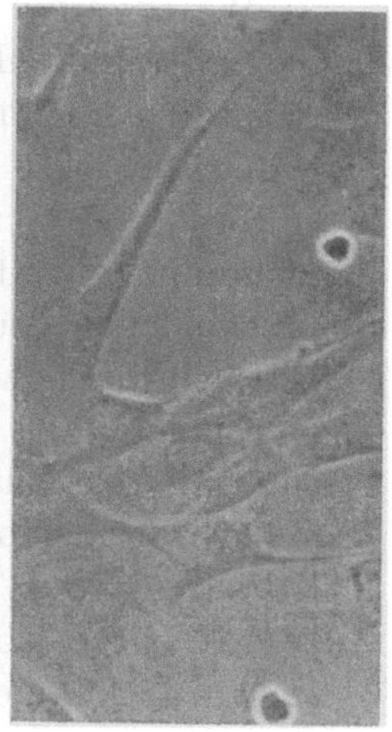

22

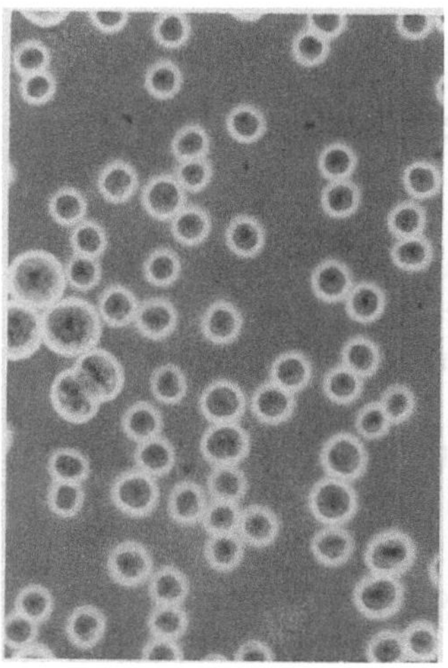

23

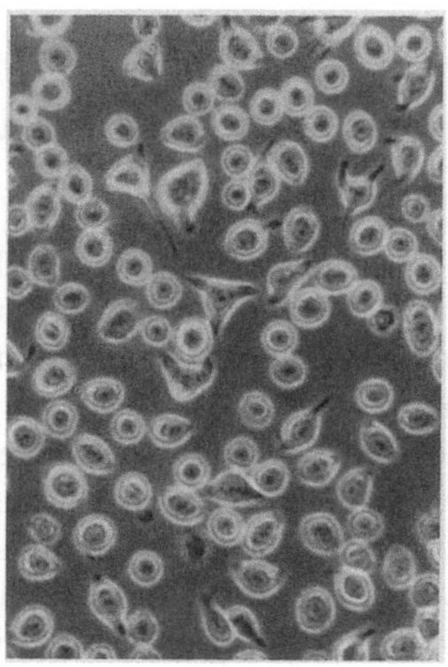

24

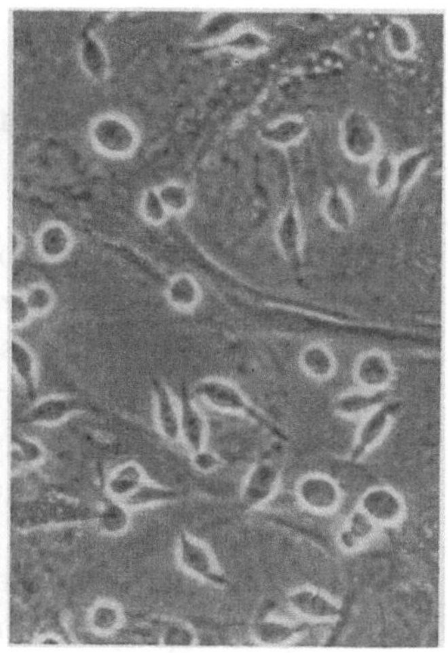

25

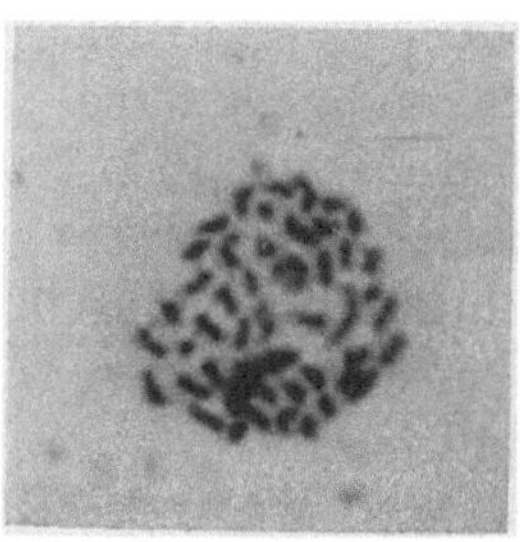

26

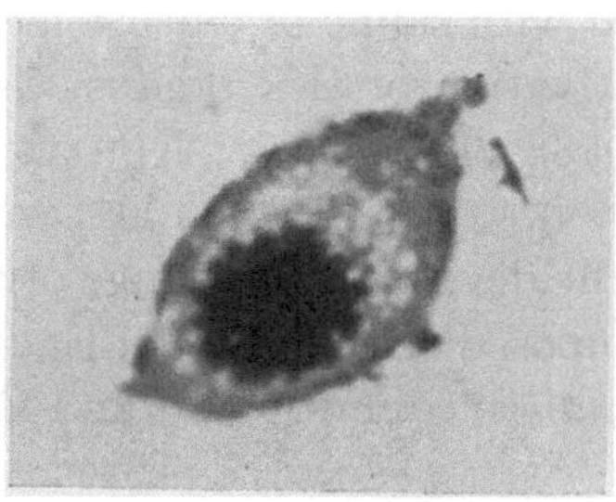

27

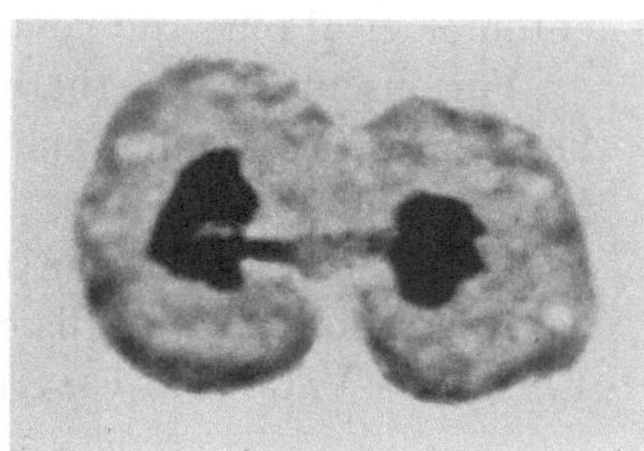

28

D. Discussion and Conclusions

The many failures in growing insect hemocytes in the past suggest the difficult nature of the task. Subcultures grew well from hemocytes isolated from pupae undergoing diapause for many months but not from those still in the early state of diapause (Table 4). This result indicates that the age of the diapausing pupae is an important factor for the success of growing lepidopterous pupal hemocyte cell lines.

We have not studied the hemocyte types of the cynthia pupa and consequently can only present a tentative conclusion on the origin of the cell line. According to MITSUHASHI (1966, 1967a), of the seven types of *C. suppressalis* hemocytes, only prohemocytes and plasmatocytes multiplied in the primary culture. With increase in the number of cell generations, prohemocytes became predominant and plasmatocytes gradually disappeared. He also thought that prohemocytes could transform into plasmatocytes because intermediate forms between these two kinds of hemocytes were present. Our observations on the morphology and growth pattern of cynthia hemocytes in culture lead us to conclude that our cell line contains a single type of hemocyte (Fig. 19), probably the prohemocyte, which nevertheless changes shape and growth characteristics in culture. This conclusion is based on the morphologic similarity of the cells in suspension derived after a long series of transfers of a few cells at a time, a technique similar to cloning.

Acknowledgement. This work was supported by Grant AI-00087 from the National Institutes of Health.

V. Present Status of Tick Tissue Culture

JOSEF REHACEK

Although all developmental stages of ticks of both sexes supply cells which can be cultivated *in vitro*, only adult organs and tissues from metamorphosing nymphs or from adults following molting, either starved or engorged, have been employed in experiments.

Decontamination of ticks from crude environmental debris such as dust, hair of host animals and tick feces can be accomplished with tap water. The ticks are disinfected by immersion for a few minutes in 70% ethanol followed by washing with sterile water or saline and the tick cells are prepared for seeding under sterile conditions. Dissection of ticks is accomplished in a Petri dish on a layer of mixed paraffin and beeswax. The tick is fixed in a dish with entomological pins and covered with saline or culture medium.

Various media for the cultivation of tick organs, tissues, and cells have been employed (REHACEK, 1958, 1962, 1965a; REHACEK and PESEK, 1960; MARTIN and VIDLER, 1962; YUNKER and CORY, 1967) without any preference based on the type of tissue or the developmental stage of the tick. Most studies have been done with media composed of a mixture of EAGLE's medium (EAGLE, 1955) plus VAGO and CHASTANG's medium (VAGO and CHASTANG, 1960) in a ratio of 1 : 1 with the addition of 0.1 % dried calf fraction (REHACEK, 1962, 1965a). VARMA and PUDNEY (1967) found that the dried calf serum could be successfully substituted with 10 % fetal bovine. Promising results were achieved using HANKS' balanced salt solution to which were added the amino acids and vitamins of EAGLE's medium with 20 % ox serum (MARTIN and VIDLER, 1962). Lactalbumin hydrolysate in HANKS' balanced salt solution plus 10 % rabbit serum, 10 % whole chicken egg ultrafiltrate and 10 mg/ml of bovine plasma albumin was also found to be a useful medium (YUNKER and CORY, 1967). A recently established medium was devised by the chemical analysis of amino acids, sugars and salts from hemolymph of engorged *Boophilus microplus* females (REHACEK and BRZOSTOWSKI, 1969a) and used successfully with cells of *Rhipicephalus sanguineus* ticks. The same authors (REHACEK and BRZOSTOWSKI, 1969b) observed the uptake from this medium of several amino acids such as leucine, methionine, threonine, phenylalanine, proline, glutamic acid, and aspartic acid to a significant extent by tick cells while other amino acids did not change in concentration. Glucose and inositol were utilized by cells in cultures and this finding was confirmed later for glucose by VARMA and PUDNEY (1967).

All the media contain penicillin and streptomycin, and in some cases, neomycin. The pH of these media varies between 6.8 and 7.2. The cultures are incubated at 27–31° C.

Most of the experiments have been performed with developing adult tissues and cells from nymphs undergoing metamorphosis several days after engorgement. The best source for a high number of growing cells are the nymphs in which the developing adults are clearly visible; e.g. the developing frontal part of the imaginal body and the legs. The preimaginal tissues in early development as well as in later stages, after a cuticle starts to develop, produce poor outgrowths. Tick materials in this state consist of cells, tissues, and hemocytes from the whole preimaginal tick organism, with the exception of tissues and cells of malpighian tubules and digestive tract which are discarded during dissection. This material is either used as an explant or certain cells are separated mechanically by gentle pipetting or by the use of trypsin, or by a combination of both methods.

By careful agitation of tissue fragments and trypsinization, it is possible to obtain a suitable cell population from various organs plus hemocytes. Such a suspension, when seeded, results in the establishment of a cell monolayer in which the cells divide for about 2 months and are maintained in good condition for a long period (VARMA and PUDNEY, 1967). The yield of trypsin dissociated cells is increased significantly by using a magnetic stirrer during trypsinization (VARMA and WALLERS, 1965). The number of cells depends primarily on the tick species and rate of its development. These types of cultures are often used for the cultivation of viruses and rickettsiae (REHACEK, 1965b; REHACEK et al., 1968). It is not necessary to use trypsin for the dissociation of tissues as careful cutting by fine scissors, followed by gentle pipetting, results in a suitable suspension of small fragments and single cells (Fig. 29).

Martin and Vidler (1962) investigated the behavior of various cell types appearing in cultures of *Rhipicephalus appendiculatus* ticks. They found that explants in a medium of Hanks' balanced salt solution with Eagle's amino acids and vitamins plus 20% ox serum survived *in vitro* for a long period of time. Contraction with cell outgrowths were observed in cultures for as long

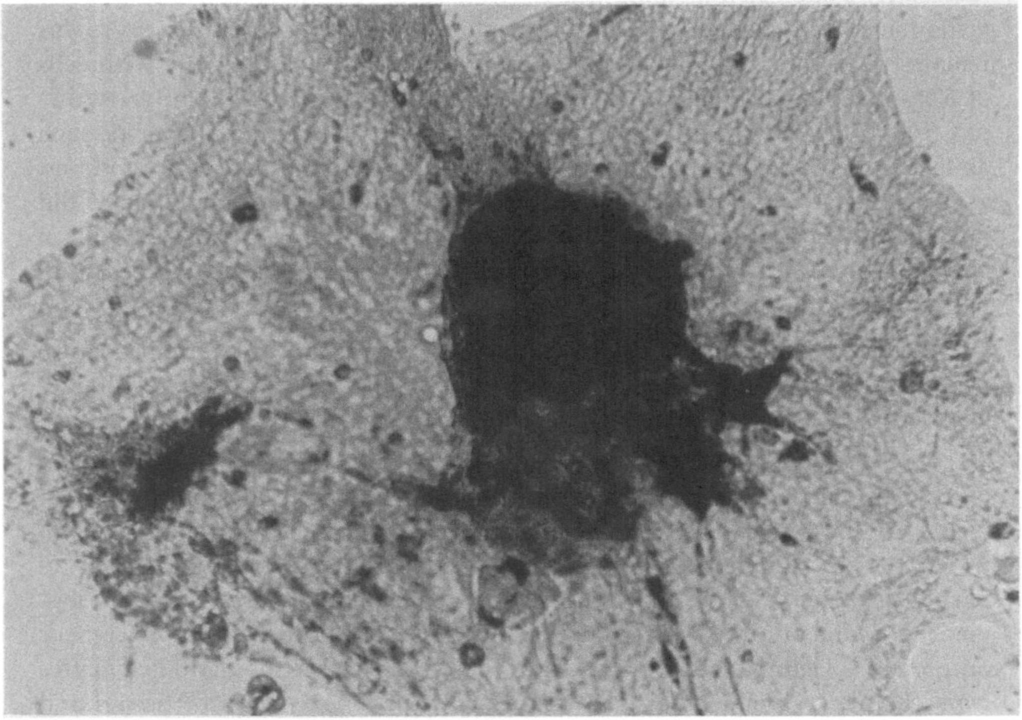

Fig. 29. *Dermacentor andersoni*, culture of preimaginal tissue, 5 days old

as 175 days. By using lactalbumin hydrolysate in Hanks' solution with 10% rabbit serum and 10% of whole chicken egg ultrafiltrate plus 10 mg/ml of bovine plasma albumin as the medium (Yunker and Cory, 1967), one explant of *Dermacentor andersoni* produced cellular outgrowths for 246 days with survival as evidenced by contraction for as long as 263 days. The nine-month survival period of a cell monolayer in good condition has been accomplished with *R. sanguineus* ticks in a medium based on chemical analysis of tick hemolymph (Rehacek and Brzostowski, 1969b) (Fig. 30).

Using only mechanical preparation of the tissue, it is possible to separate hemocytes, which are included in the small fragments of tissues, from the other cells. This method consists of careful and gentle cutting by fine dissecting scissors of the dissected organs of preimaginal ticks into tiny fragments in a small volume of media. Additional medium is added to the suspension, agitated very gently, and shortly thereafter, the supernatant is collected and seeded into suitable containers. The hemocytes adhere immediately to the surface of the vessel and are maintained in good condition for more than two months (Fig. 31). It is assumed that these cells will be useful for the cultivation

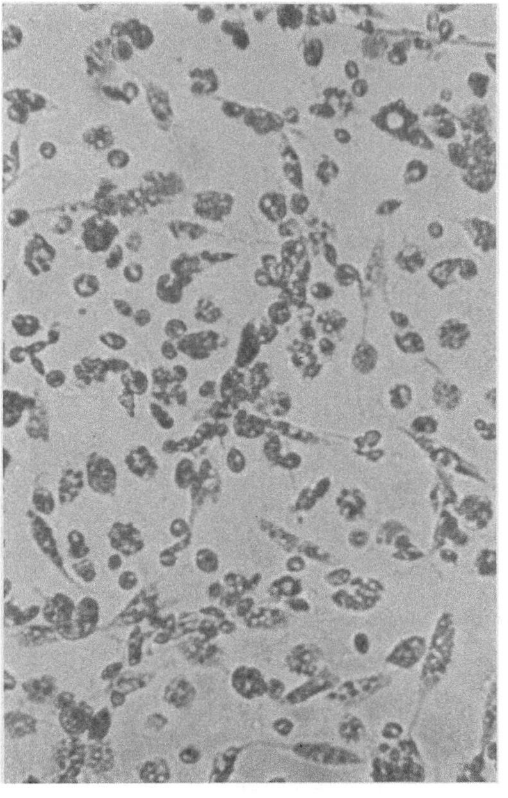

Fig. 30. *Rhipicephalus sanguineus*, culture of preimaginal tissue and cells, 3 months old

of microorganisms as, for example, for the growth of *Rickettsia prowazeki* (Fig. 32) (Rehacek et al., 1968).

The sediment taken from such a suspension and consisting of tiny fragments of cut organs, when seeded into vessels, also adheres very quickly to the surface and gives rise to cell migration and multiplication. These cells are epithelial as well as fibroblast-like and establish a monolayer. Such types of cultures are also quite suitable for cultivating various microorganisms, as shown with Colorado tick fever virus (Yunker and Cory, 1967) and various rickettsiae (Rehacek et al., 1968).

Under certain as yet unclear conditions the explants from such a suspension may establish hollow vesicles in cultures. This type of cell multiplication is well known in insect tissue cultures (Larsen, 1967; Bhat and Singh, 1969).

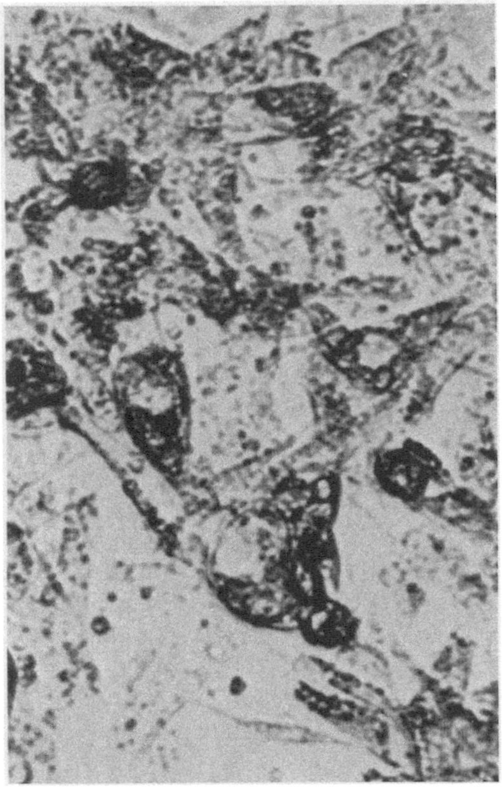

Fig. 31. *Dermacentor andersoni*, hemocytes of preimaginal tick stage, 13 days old

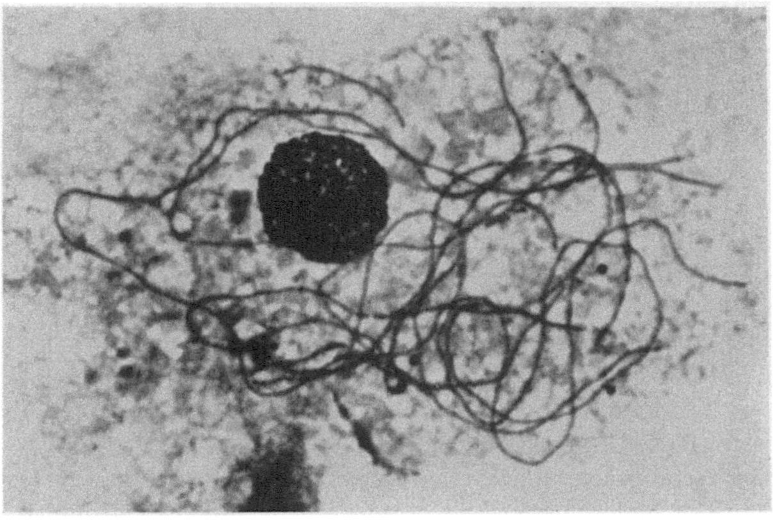

Fig. 32. *Hyalomma dromedarii*, hemocyte with *Rickettsia prowazeki*, 8 days old

In cultures from developing adults the vesicles are seen from the first day after seeding the explants. They are of a different type, rounded, oval and dumb-bell form, and are either attached to the mother explant or floating freely in the medium (Figs. 33 and 34). Often these forms are attached to the bottom of the vessel. The vesicles from primary cultures are large, about

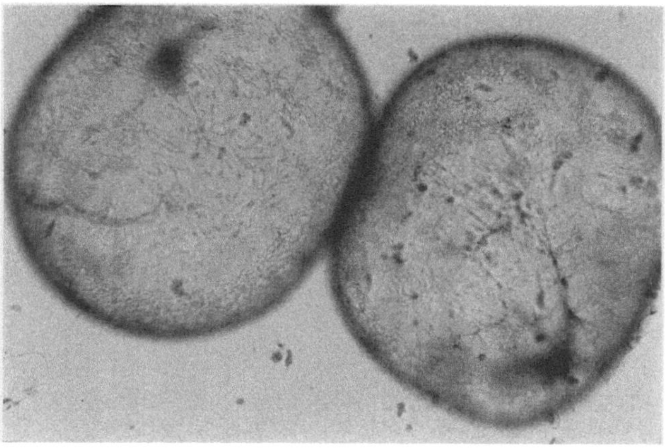

33

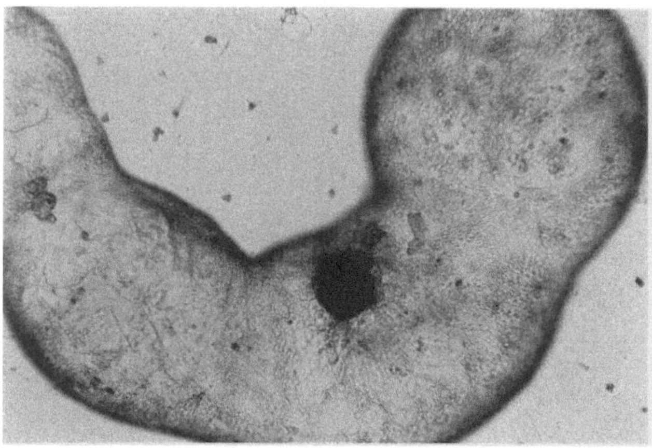

34

Figs. 33 and 34. *Dermacentor andersoni*, vesicles from fragments of preimaginal tissue, 12 days old

2 mm in size, often exceeding by several times the size of the mother explants and are visible with the naked eye as small bladders. The large vesicles, when entirely developed, are composed of single layers of epithelial-like cells. A thin cuticle-like membrane covers the outer surface. The culture medium and sometimes also a conglomerate of various cells can be seen inside the vesicles. The vesicle cells multiply, and on section through the vesicle sheet we repeatedly observed cells in division. Very often the cells inside the vesicles are connected by cellular bridges. The hemocytes, sometimes present here are either floating in the medium within the vesicles or are attached to the vesicle sheet. Vesicles

gradually increased in size and when they reach a maximum they become darker in color and begin to degenerate. Attempts were made to pass the vesicles, together with the other tissues and cells, by gently scraping them from the flask walls with rubber policemen and gently disrupting them with

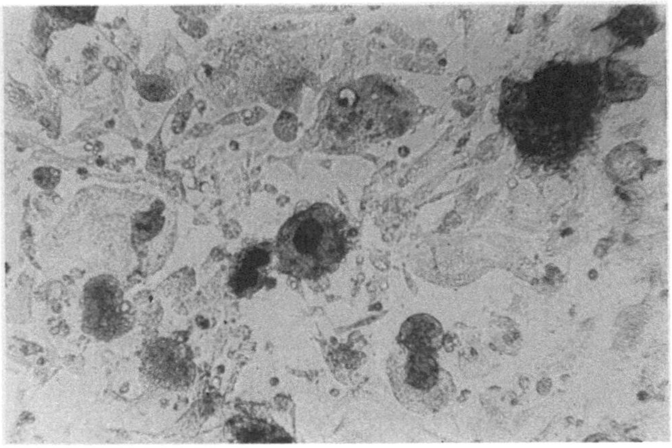

Fig. 35. *Dermacentor andersoni*, second passage of culture from preimaginal tissues consisting of growing tissue explants, vesicles and hemocytes, 7 days after passage

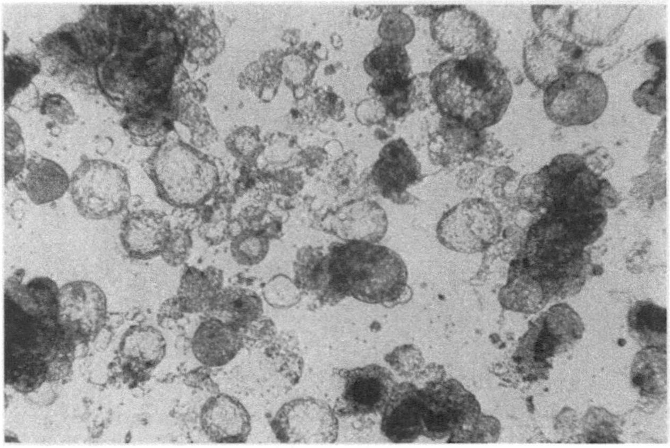

Fig. 36. *Dermacentor andersoni*, fourth passage of culture of preimaginal tissues and cells consisting of growing explants and mostly vesicles, 9 days after passage

pipettes. The first three passages attempted at weekly intervals were successful and resulted in good cell growth exceeding those of primary cultures (Fig. 35). However, with each passage the vesicles established were smaller and in higher number, with the fifth passage in our most recent experiments consisting almost only of small vesicles with almost no cellular explants (Fig. 36). The nature of these vesicles, as a type of cell multiplication observed in tick cultures, is not understood and is possibly a form of cytopathology.

Another source of material for tick tissue cultures is the adult tick body. It is possible to use starved or engorged ticks, but in our experience best results have been achieved by using partly engorged ticks, as they are easier to dissect and the cell yield is greatest. These ticks are metabolizing rapidly and may be developing eggs, producing higher hormonal activities which actively influence the cells' behavior *in vitro*.

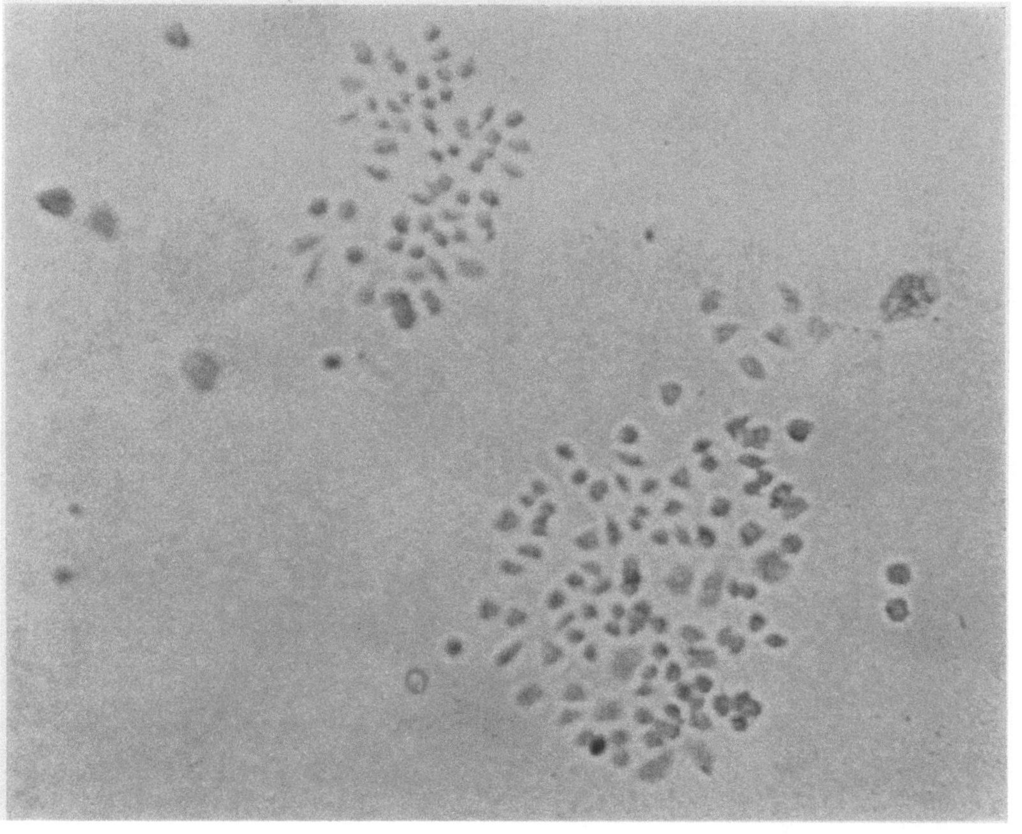

Fig. 37. *Dermacentor andersoni* females, epithelial cells of ovarian tissue origin, 30th day

Cultivation of the complete body contents of *R. appendiculatus* females in a medium composed of HANKS' solution, amino acids and vitamins of EAGLE's medium plus 20% ox serum resulted in contractile movements of explants for 163 days, and multiplication of epithelial cells, of probable midgut origin, for a period of 115 days. The fibroblast-like cells developed in moderate numbers for a period of 44 days and these cells appeared intermittently up to the 66th day. Sections prepared of cultures on the 170th day revealed healthy salivary gland cells and midgut epithelial cells (MARTIN and VIDLER, 1962). The organs usually demonstrate their viability by contractions, migrations and multiplication of epithelial and fibroblast-like cells with the formation of vesicles.

Vesicles were observed from almost all the organs during various periods and on occasion they were observed in the first days after seeding. They have also been seen to develop after a few months following seeding of organ explants.

The explants of male tick organs have survived for as long as two months following migration and outgrowth of cells, particularly from sexual organs.

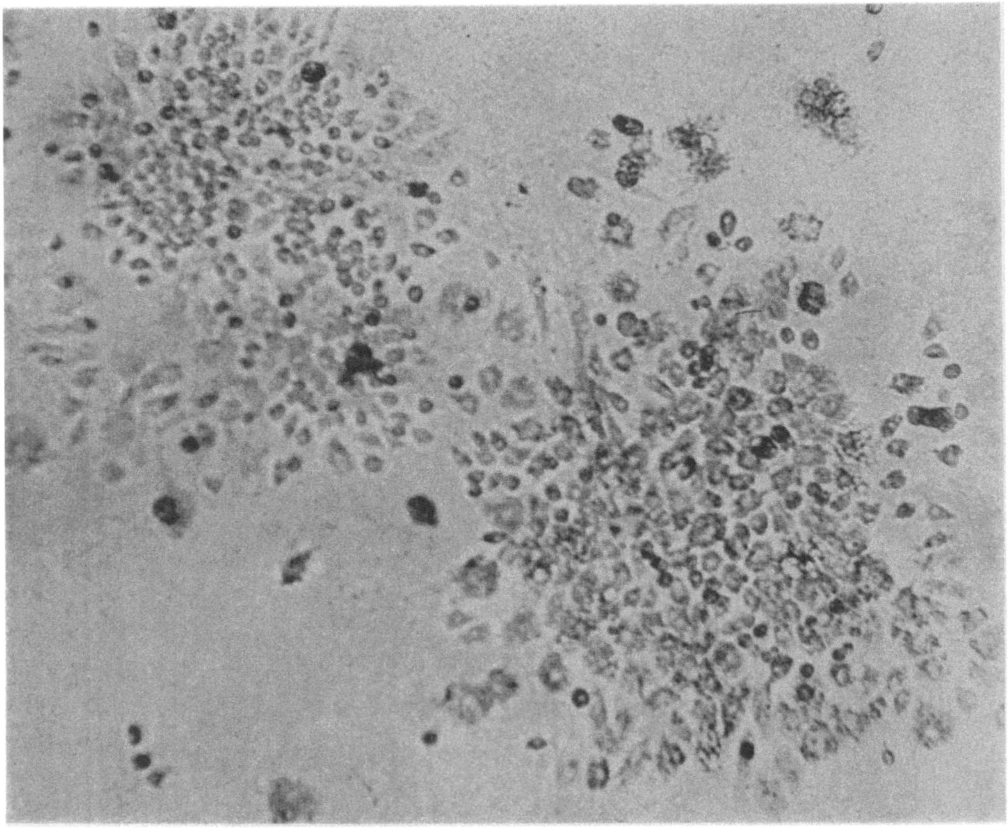

Fig. 38. The same culture as in Fig. 37, 12 days older

The spermatocytes dedifferentiated to fibroblast-like cells. We intend to continue our investigations with male genital glands, because other types of cells also have been found to migrate and multiply from them.

The second most useful material for tick tissue cultures is represented by the adult females of various species of ixodid ticks. All the organs including connective tissues and hemocytes were employed in these experiments. Most of the experiments resulted in long persistence of the explanted organs. The ovaries of various tick species persisted up to 82 days in the experiments by Hoffman and Kohler (1968). We have observed that ovarian tissues of partly fed females produced well-formed epithelial cells which migrated and multiplied. These investigations were achieved with all three types of media described in this review, and with various genera and species of ixodid ticks.

We recently repeated these investigations with *D. andersoni* ticks received from Dr. BRAM of the U. S. Department of Agriculture, Beltsville, Maryland. The medium used was lactalbumin hydrolysate in HANKS' balanced salt solution with 0.1 % bovine plasma albumin and 10% fetal calf serum heated at 56° C for 60 min. The ovaries of partially engorged females, which had been held for one week at room temperature, were cut into small pieces and cultivated at 28° C in T—flasks of different sizes. The viability of the ovaries was demonstrated by contractile movements, the establishment of vesicles from the first day after seeding, and by migration of a few epithelial-like cells in the first days of cultivation. About ten days after initiating the cultures, discrete small colonies of these cells appeared and started slowly to increase in size and in number (Figs. 37 and 38). Two weeks later these colonies became very dense and covered almost all of the surface of the culture vessel. The older cells remained attached to the glass surface as epithelial cells and new rounded cells produced an overlayer. Forty-six days after seeding the cultures, the cells were scraped very carefully with a rubber policeman and transferred to a new T-flask. The cells adhered to the glass surface and started to grow slowly. The following two passages were done after 3 weeks. Even though these cells multiply very slowly, it appears that they will be one of the most convenient cell types for our attempts to establish a cell line.

Chapter 2

Analysis of Cells from Established Insect Cell Lines

I. Introduction

J. L. VAUGHN

Within the last five years the number and variety of insect cell lines has increased rapidly; approximately 20 different cell lines were discussed in the preceeding chapter. As the availability of insect cell lines increases, and as their use becomes more widespread, the accidental contamination of these lines with microorganisms or cells from other lines becomes more likely. Such contamination of cell lines from higher animals has already been thoroughly documented, and evidence exists that it has already occurred in insect cell lines (see the Discussion by GREENE and CHARNEY in this chapter). Methods have been developed for detecting such contamination in cell lines from higher animals. It is now essential that these methods be tested and modified, where necessary, to permit the identification of insect cell lines at least to the species of origin.

Information concerning the physiological characteristics, the chromosome complements, and the serological relationships of the various cell lines is needed to provide markers for identifying insect cell lines. The section by MCHALE and SWEET illustrates the difficulties in using only cultural characteristics to distinguish two very similar cell lines.

The other sections describe methods currently available and the information thus far obtained with them for some established insect cell lines. Even a casual reading of the chapter calls attention to the problems involved and the amount of research still to be done to solve them. Thus, the reasons for this chapter are three-fold: (1) to stimulate further research in this very important area of insect cell culture; (2) to call to the attention of insect cell biologists the gravity of the problems of contamination; and (3) to provide them with some techniques for checking the purity of their cultures.

II. Morphological and Cultural Characteristics of *Culiseta inornata* and *Aedes vexans* Mosquito Cell Lines

JANIECE S. MCHALE and B. H. SWEET

The two mosquito cell lines discussed in this paper were initiated on hemolymph-free medium, one from pupal tissue of *Aedes vexans* and another from newly emerged adult tissue of *Culiseta inornata* (SWEET and DUPREE, 1968; SWEET and MCHALE, 1970). GRACE's medium (Grand Island Biological Co.) supplemented with 10% heat-inactivated fetal bovine serum and anti-

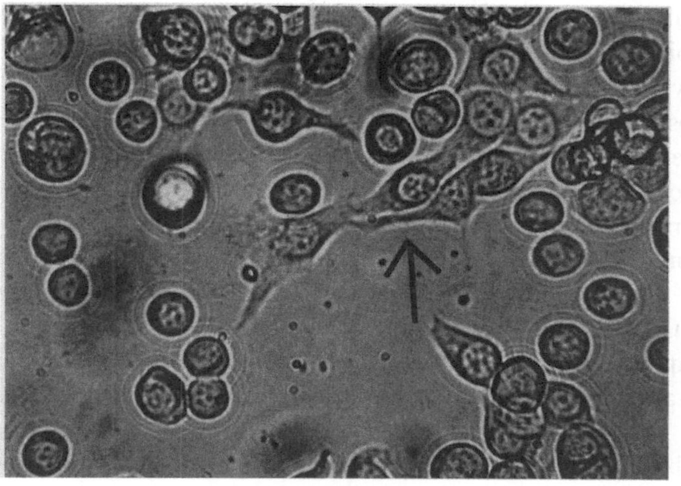

a

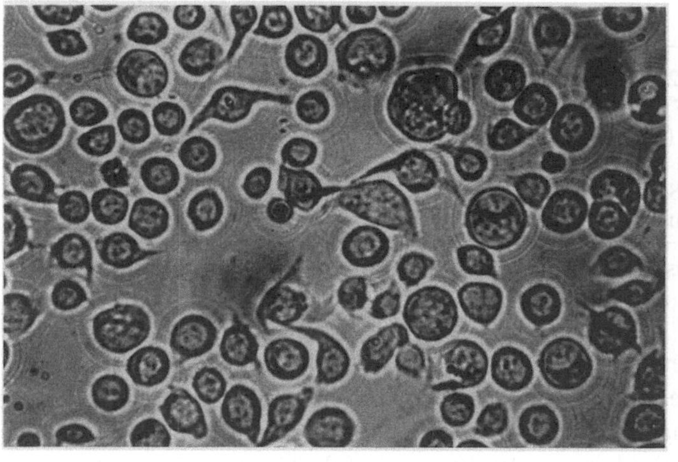

b

Fig. 39a and b. *In situ* phase contrast photomicrographs of living cells of (a) *Aedes vexans* and (b) *Culiseta inornata*. 720×. Note the variety of cell types (round, spindle) which grow in suspension in these cultures as well as the occasional fibroblast-like attached cells (arrow in a). These lines were started and maintained on hemolymph-free medium

biotics was used routinely. Growth from explants was not observed for a period of 1–3 months, after which small round cells appeared free-floating in the medium. The number of cells increased rapidly, and a culture-split ratio of 1 : 10 or 1 : 15 has been carried out weekly for two years for both cell lines. Optimum growth in both cell lines occurs at 28° C, followed by limited growth at 32° C and 20° C and cell death occurs at 37° C. Cells can be stored in the frozen state at −70° C and −120° C with 10% glycerol added to the medium.

The effect of centrifugation on the subsequent proliferation of cells of *C. inornata* indicated that these cells are relatively fragile. Centrifugation at 300× g and above resulted in noticeable cell loss. In stationary flask cultures a maximum density of 400,000 cells/ml of medium has been achieved, and we are developing techniques for mass culture of these cells in roller bottles which should provide at least a two-fold increase in the cell density.

The gross morphology of the cells is illustrated in Fig. 39. Most of the cells are in suspension, although some attach lightly to the flask. Note the different cell types found in these cultures; round, spindle shaped, and a few fibroblast-like cells (arrow, Fig. 39a). Although we had no other insect lines in-house when these cultures were initiated, we recognized, through the literature, certain similarities to cultures of Grace's *Aedes aegypti* (Grace, 1966) and *Antheraea eucalypti* (Grace, 1962). Grace's *A. aegypti* cultures, however, have giant spindle-shaped cells not seen in the other lines, and there are certain antigenic differences among these cell lines which will be discussed in the next section.

Table 5. *Multiplication*[a] *of two mosquito cell lines in different media*

Medium	*Aedes aegypti* (RecA-HF) Passage 85	*Aedes vexans* Passage 62
Grace (1962)	+	+
Singh (1967)	+	—
Schneider (1969)	+	—
Mitsuhashi and Maramorosch (1964)	—	—
Peleg (1968b)	Dead	—
Lactalbumin hydrolysate	Dead	Dead

[a] Increase (+) or decrease (—) in cell count over initial inoculum after 10 days.

We attempted to induce these suspended cell cultures to form monolayers by: (1) selecting the few lightly attached cells when cultures were split and (2) changing the medium to one used routinely for monolayer type cultures. Neither method was entirely successful. The attached cell populations could not be subcultivated and Singh's medium (1967), routinely used for insect cultures which grow in monolayers, inhibited multiplication of both *C. inornata* and *A. vexans* cell lines, although some viability was present after a week.

A further analysis of the effects of various media was carried out with *A. vexans* cells and cells derived from Suitor's clone of Grace's *Aedes aegypti* (RecA-HF) (Suitor et al., 1966a). These cultures are routinely carried on Grace's medium without hemolymph. The media used are listed in Table 5. All were supplemented with 10% serum and antibiotics. None of the test media supported the healthy cell morphology and rapid multiplication found with Grace's medium. *A. vexans* cells could not multiply in any of the test media. Some increases in cell count over the inoculum were seen with *A. aegypti* cells in Singh's and in Schneider's *Drosophila* medium, but the cells were necrotic and did not have the characteristic appearance. Singh's (1967), Mitsuhashi and Maramorosch (1964), and Peleg's (1968b) media all support the growth of insect monolayer cultures derived from *A. aegypti* and *A. albopictus*.

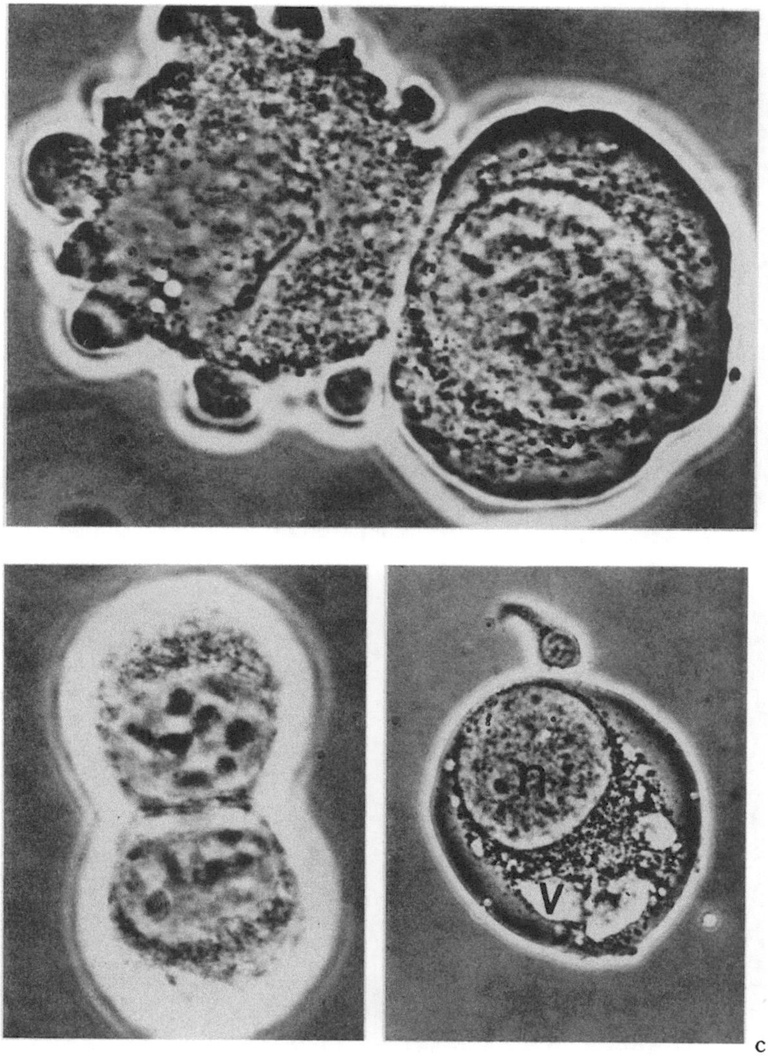

a b c

Fig. 40a–c. Phase contrast photomicrographs of *C. inornata* cells in coverslip wet mount preparations. See text. a Two cells in different planes of focus, 3000×. b Telophase 3000×. c A very small cell adjacent to an average size cell. Nucleus (*n*), vacuole-like bodies (*v*) 1400×

A possible critical component, common to the media inhibiting growth of either cell line, is lactalbumin hydrolysate. Tests with *A. vexans*, *C. inornata*, GRACE's *A. aegypti*, and *An. eucalypti* cells revealed that lactalbumin hydrolysate was not toxic at a concentration of 0.5 %, but at 1 % it was selectively toxic for *A. vexans* cultures. This may be a possible marker to distinquish between the morphologically similar lines of *A. vexans* and *C. inornata*.

It is our hypothesis that the difference in monolayer and suspended cultures is one of initial selection of cell type and not an environmentally induced variation of cellular morphology. Suspended type cultures are more sensitive

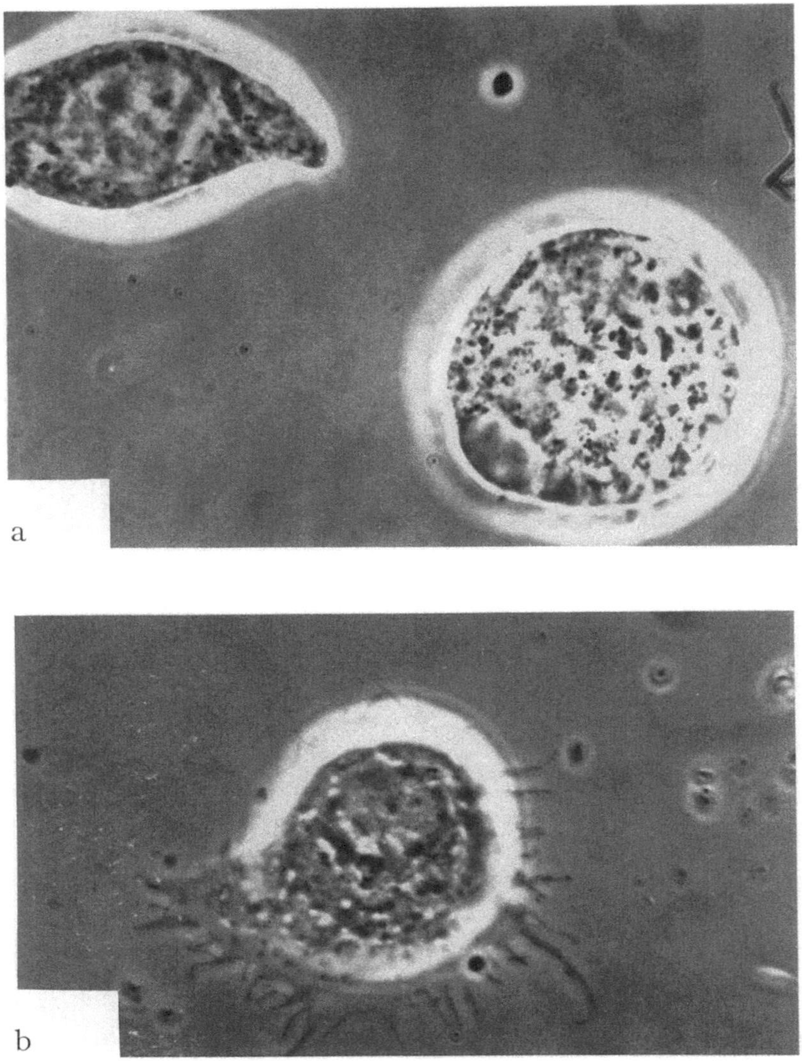

Fig. 41a and b. Phase contrast photomicrographs of *C. inornata* cells in coverslip wet mount preparations. a Typical spindle-shaped and round cells. b Micro-extensions often observed at high magnification. 3000 ×

than monolayer cultures to changes in media and temperature. Media used in initiating cultures may select against suspended cell types if they are insufficient in some critical factor such as amino acid concentration or if they contain some factor selectively toxic to suspended cultures such as lactalbumin hydrolysate. Monolayer cultures would then be favored. If Grace's medium is used to initiate culture, either monolayer or suspended type cultures could result unless an optimal medium for suspended cells allows them to predominate at the expense of attached cell types. Preliminary experiments involving the initiation of cultures on various media tend to support our hypothesis.

We are grateful to Dr. IMOGENE SCHNEIDER of Walter Reed Army Institute of Research for making chromosome preparations and counts of *A. vexans* and *C. inornata* cultures. The cultures are highly polyploid. *A. vexans* cells have chromosome numbers ranging from 152 to 216, with a mean of approximately 190. The range in *C. inornata* cells is even greater. Individual chromosomes were too small and difficult to separate for karyotypic analysis. A careful examination of *A. vexans* cells revealed that a chromosome complement of 2 n or 4 n could be found in perhaps one cell in 1000.

A detailed cytological study of *C. inornata* is presented in Figs. 40 and 41. Note typical cells at higher magnification (40a, 41a), a mitotic figure (40b), single nuclei (40a, 40c), vacuole-like bodies (40c), micro-extensions (41b) and blebs (40a) on the membrane of one cell and the nucleus of the other. These cells are similar in morphology to cultures described by other workers as "hemocyte-like". If the cells are hemocytes, however, they are hemocytes of the species of origin, since hemolymph from another insect species was never added. We have not been able to demonstrate phagocytosis conclusively in these cells using standard particle uptake techniques.

The similar morphology among these suspended cell lines makes it even more necessary to establish characteristic markers. Primary emphasis should be placed on cloning, followed by characterization of the clones.

Acknowledgments. Supported in part by USPHS Grant No. A 108208 and USPHS Grant No. 1-501-FRO-5672-01. The technical assistance of Messrs: HOWARD D. UNTHANK, L. D. DUPREE, MILTON L. MOUTON, KENNETH J. HARDY, RONALD LEGER, and Mrs. DORIS GARNIER is gratefully acknowledged.

III. Antigenic Relationships of Mosquito Cell Lines as Determined by Immunodiffusion Techniques

A. N. IBRAHIM and B. H. SWEET

Studies were undertaken in our laboratories to identify mosquito cell lines and/or to determine the antigenic relationships among them by gel precipitin techniques using both immunodiffusion and immunoelectrophoresis. Table 6 shows the list of monolayer and suspended cell cultures used in this study, the primary source from which they were derived, and the tissue culture passage level.

Cell packs, derived from cultures thoroughly washed free of media, were homogenized and used as antigens before and after sonication. Sera were prepared against all the listed cell lines, except that of PELEG (1968b) from *Aedes aegypti*, and our *A. vexans* line, by immunizing rabbits with multiple injections at 10 day intervals. In the case of the two SINGH cell lines (1967), *A. aegypti* and *A. albopictus*, excellent antibody responses were obtained after 4 intravenous inoculations of aqueous suspensions. The suspended cell lines required a 5th inoculation of antigen with Freund's complete adjuvant.

None of the preimmunization serum samples reacted with any of the test antigens, nor did antisera prepared against the media in which the cells were propagated. None of the antisera prepared against the cells reacted with the growth medium or a continuous porcine cell line (Y-15).

Table 7 summarizes the results obtained with rabbit antisera prepared against the monolayer mosquito cells reacting with the monolayer and the suspended cell antigens, both sonicated and unsonicated. Antisera prepared to the two Singh lines *A. aegypti* and *A. albopictus* cross-reacted only with each other, either before or after sonication. These same sera also reacted with the sonicated antigen derived from Peleg's cells but did not react with any other of the antigens tested.

Table 6. *Insect cell lines used in immunodiffusion studies*

Type of culture	Designation of cell line species	Reference	Primary source	Cell culture passage
Monolayer	*A. aegypti*	Singh (1967)	Larvae	82
Monolayer	*A. albopictus*	Singh (1967)	Larvae	162
Monolayer	*A. aegypti*	Peleg (1968b)	Embryo	110
Monolayer	*An. stephensi*	Schneider (1969)	Larvae	25
Suspended	*A. aegypti*	Grace (1966)	Larvae	185+
Suspended	*A. aegypti* (clone)	Suitor et al. (1966a)	Larvae	86
Suspended	*C. inornata*	Sweet et al. (1968, 1970)	Adult	52
Suspended	*A. vexans*	Sweet et al. (1968, 1970)	Pupae	30
Suspended	*Anth. eucalypti*	Grace (1962)	Pupal ovaries	177

Rabbit anti-*Anopheles stephensi* serum reacted only with its homologous unsonicated antigen. However, it reacted with all monolayer test antigens which were sonicated and with that of *Antheraea*. It would appear that this antiserum is type specific and can identify its homologous antigen if the test is performed with unsonicated antigens. However, sonication of the monolayer cells may liberate or expose intracellular antigens responsible for cross-reactions.

When anti-*A. albopictus* serum was tested against cellular antigens from Singh's *A. albopictus* and *A. aegypti* lines, a pattern of complete identity between the antigens of the two cell lines appeared. No reaction appeared with any other of the cellular antigens or control media. In a similar manner, anti-*A. aegypti* (Singh) serum reacted only with the two Singh lines. When cross-absorption tests were carried out, no reaction appeared; thus, the two antigen preparations appeared identical under the conditions of the test.

Figure 42 is a schematic representation of immunoelectrophoretic patterns. Singh's *A. albopictus* and *A. aegypti* antigen preparations in the upper and

Table 7. *Immunodiffusion reactions of antisera prepared against monolayer mosquito cell lines*

Antisera	Sonicated or unsonicated antigens derived from:				
	monolayer cultures				suspended cultures
	A. aegypti (SINGH)	*A. albopictus* (SINGH)	*A. aegypti* (PELEG)	*An. stephensi* (SCHNEIDER)	*A. aegypti*, *C. inornata*, *A. vexans*, *Anth. eucalypti*
A. aegypti (SINGH) or	+	+	not tested	—	all negative
A. albopictus (SINGH)	+[a]	+[a]	+[a]	—[a]	all negative[a]
An. stephensi (SCHNEIDER)	—	—	not tested	+	all negative
An. stephensi (SCHNEIDER)	+[a]	+[a]	+[a]	+[a]	+ (?) vs. *Anth. eucalypti* only[a]

[a] Results with sonicated antigens.

Table 8. *Immunodiffusion reactions of antisera prepared against suspended cell lines*

Antisera	Sonicated or unsonicated antigens derived from:						
	monolayer cultures			suspended cultures			
	A. aegypti or *A. albopictus* (SINGH)	*A. aegypti* (PELEG)	*An. stephensi* (SCHNEIDER)	*A. aegypti* (GRACE; or SUITOR clone)	*C. inornata*	*A. vexans*	*Anth. eucalypti*
A. aegypti (GRACE; or SUITOR clone)	0	not tested	0	+	+	+	+
	0[a]	0[a]	0[a]	+[a]	+[a]	+[a]	+[a]
C. inornata	0	0	0	+	+	0	+
C. inornata	+[a]	+[a]	±[a]	+[a]	+[a]	0[a]	+[a]
Anth. eucalypti	0	0	0	+	+	+	+
Anth. eucalypti	+[a]	+[a]	0[a]	+[a]	+[a]	+[a]	+[a]

[a] Results with sonicated antigens.

lower wells were exposed to electrophoresis for $2^1/_2$ hrs and then made to react with hyperimmune rabbit sera against SINGH's *A. albopictus* and *A. aegypti*. The latter were placed in the troughs of the upper and lower slides, respectively. Examination of each of the patterns of the upper or the lower figure alone indicates that the two antigen preparations are identical, since precipitin arcs on both sides of the trough of each slide are symmetrical. There are some differences in the two sera, however. Although A antigen appears in both tests, B appears only with the *A. albopictus* serum and C only with the *A. aegypti* serum. Examination of both figures leads to the assumption that each of SINGH's *A. albopictus* and *A. aegypti* possesses at least three antigenic components. The differences in immunoelectrophoretic patterns obtained with

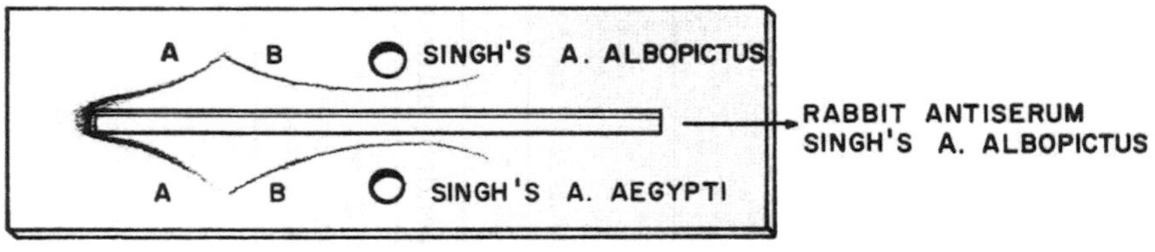

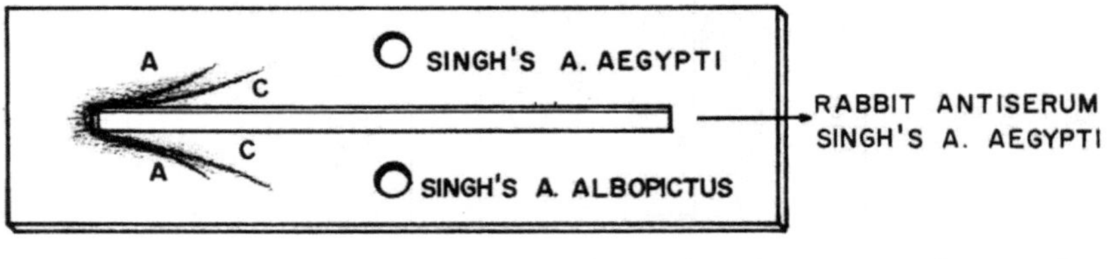

Fig. 42. Immunoelectrophoretic patterns of reactions between antisera and antigens of SINGH's *A. albopictus* and *A. aegypti* cell lines. *A*, *B*, *C*, hypothetical common antigens; see text

the two sera may be attributed to differences in amount or immunogenicity among antigenic components, or variation in response among individual rabbits.

Table 8 summarizes the results of the reactions of hyperimmune rabbit sera prepared against the suspended cell lines with the monolayer and the suspended test antigens before and after sonication. GRACE's (1966) and/or SUITOR's (1966a) clone of *A. aegypti* gave similar reactions. These antisera reacted with all the suspended cell antigens but did not react with any of the monolayer cell antigens either before or after sonication.

Anti-*Culiseta inornata* serum reacted with all of the suspended cell antigens except that of *A. vexans*. However, it reacted with all sonicated antigens, including those derived from monolayer cultures. Results were again negative with *A. vexans*. Reaction with *Antheraea eucalypti* antigen was weak but present.

Rabbit anti-*Anth. eucalypti* antisera behaved in a similar manner as did *C. inornata* antisera, except it did react with *An. vexans* and the monolayer antigens after sonication. It did not react with *An. stephensi* cells.

On the basis of these initial studies, we can summarize as follows:

(1) SINGH's *A. aegypti* and *A. albopictus* cell lines appear to be qualitatively identical but may be quantitatively different, in one antigenic component.

(2) Unexpectedly, no antigenic relationships were shown between the monolayer antigen of the 3 *Aedes* lines and the suspended *A. aegypti* lines or *A. vexans*.

(3) The unsonicated antigens derived from the monolayer *A. aegypti* lines did not react with any of the heterologous antisera. However, after sonication they reacted with antisera of *C. inornata*, *Antheraea* and *An. stephensi*.

(4) One could, by step-wise analysis, differentiate *A. vexans* from *C. inornata*, *A. aegypti*, and *Antheraea*.

(5) Antisera against the suspended *A. aegypti* lines cross-reacted with all the suspended test cell antigens; so did the *Antheraea* antiserum. However, they can be differentiated since the former antisera did not react with the monolayer *Aedes* antigens, whereas the *Antheraea* antiserum did.

These results indicate the complexity of the problem with which one is faced in distinguishing insect lines from each other. Further studies are urgently needed before immunological "markers" for the insect cell lines can be established.

Acknowledgment. Supported in part by USPHS Grant No. A 108208 and USPHS Grant No. 1-501-FRO-5672-02.

IV. Characterization and Identification of Insect Cell Cultures

ARTHUR E. GREENE and JESSE CHARNEY

A. Introduction

During the past two decades we have witnessed an increasing accumulation of biological knowledge in cytogenetics, biochemistry, immunology and biology related to experimental techniques developed in mammalian cell cultures. Comparable achievements are being reported from research studies in poikilothermic cell cultures. It is now obvious that insect cell cultures pioneered by many of the authors in this volume and by the late Dr. SUITOR will also allow us to gain further insight into many biological phenomena associated with insects.

Two of the major problems with which cell biologists have contended over the last two decades are microbial contamination due to bacteria and mycoplasma and erroneous mixtures of cells of different species. Some of our experiences in characterizing and preventing contamination of insect cell cultures are here presented.

B. Materials and Methods

Insect tissue cultures:

1. GRACE's *Antheraea eucalypti* cells were obtained on May 27, 1965, from Dr. GRACE and cultured in GRACE's insect medium with 5% *Antheraea pernyi* hemolymph (GRACE, 1962). "Unadapted moth cells", Freeze 268, designates *Antheraea* cells grown in hemolymph and stored in liquid nitrogen in our laboratory. The cells were frozen in GRACE's insect medium, 1.5% fetal bovine serum and 1.5% hemolymph and 5% glycerol after 14 transfers at the Institute for Medical Research.

2. GRACE's *A. eucalypti* cells adapted to grow in a medium without hemolymph by YUNKER, VAUGHN, and CORY were obtained from Dr. VAUGHN on February 1, 1967. The culture medium was GRACE's insect medium as modified by YUNKER et al. (1967). "Adapted moth cells", Freeze 308, designated the adapted *Antheraea* cells stored in liquid nitrogen for the Cell Bank of the American Type Culture Collection. The cells were frozen in 10% glycerol after 2 passages in our laboratory and a total of 37 passages in hemolymph-free medium.

3. GRACE's *Aedes aegypti* mosquito cell line (GRAGE, 1966) was received on February 23, 1967 from Dr. VAUGHN. The cells were cultured in GRACE's insect medium with 5% fetal bovine serum (inactivated) and 5% *A. pernyi* hemolymph. On February 28, 1967, the cells were frozen in 10% glycerol without serial passage in our laboratory and stored in liquid nitrogen as Freeze 309.

4. GRACE's *A. aegypti* cells, Freeze 335, were initiated from an ampule of Freeze 309. The cells were grown in our laboratory, passaged approximately 30 times, and frozen on November 3, 1967, in GRACE's insect medium, 1% *A. pernyi* hemolymph and 10% glycerol.

5. SINGH's *Aedes albopictus* cells (SINGH, 1967) were obtained on November18, 1968, from Dr. SONJA M. BUCKLEY at the Yale Arbovirus unit, New Haven, Conn., in the 37th passage. The cells were cultured in MITSUHASHI and MARAMOROSCH insect medium with 20% non-inactivated fetal bovine serum (MITSUHASHI and MARAMOROSCH, 1964). The cells in 10% dimethyl sulfoxide (DMSO) were stored in liquid nitrogen and designated as Freeze 402. The cell line was passaged 10 times in our laboratory before freezing.

6. SINGH's *A. aegypti* cells were obtained on November 18, 1968 from Dr. BUCKLEY. The cells were cultured in the same medium as *A. albopictus*, stored in liquid nitrogen and designated as Freeze 415. Cells recovered from frozen ampules were used in both immunological and isoenzyme studies.

Frozen storage of cells. The frozen ampules of insect cells were prepared according to the procedures established by the Cell Culture Collection Committee. Ampules were slow-frozen in a Linde BF-1 apparatus at the rate of 1–2° C/min. to $-50°$ C and then transferred to a Linde 300 Tank containing liquid nitrogen. Efficiency of preservation was determined by the vital stains, neutral red for the *A. eucalypti* cells, and trypan blue for the *A. albopictus* cells, and by testing for multiplication of thawed cells. Neutral red was judged the better stain for determining the viability of the unadapted *A. eucalypti* cells.

Preparation of antisera. In the absence of fresh insects as a source of antigen, cell cultures were used to immunize animals. GRACE's *A. aegypti* cells were harvested from 75 cm^2 plastic flasks. The cells were centrifuged at 1500 RPM for 10 min. to remove the culture medium, hemolymph and fetal calf serum. They were then washed 5 times in GRACE's insect medium without serum or hemolymph. Rabbits received 6 intramuscular injections of 1 ml of cell suspension each at 3 day intervals. A final dose was administered 7 days later and the animals were bled 3 days afterwards. The sera were inactivated at 56° C for 30 min. and stored at $-20°$ C.

Antiserum to GRACE's *A. eucalypti* cells adapted to YUNKER's medium was prepared with a pool of 68×10^6 cells by methods similar to those detailed in the proceeding paragraph.

Guinea pig anti-*A. aegypti* serum prepared against larval protein, and rabbit anti-*A. pernyi* serum were obtained from Dr. VAUGHN. Since these antisera gave weak precipitin tests, they were concentrated five-fold by ammonium sulfate precipitation (CHARNEY and CORIELL, 1964). All antisera were absorbed by addition of $^1/_{10}$ volume of newborn calf serum.

Agar microimmunodiffusion test method was that of CROWLE (1958) with modifications by CHARNEY and CORIELL (1964). The cell extracts were prepared by washing the cells 3 times in saline followed by 3 freeze-thaw cycles of the cell pellet. All antisera used for immunodiffusion tests were absorbed with calf serum and *A. pernyi* hemolymph to remove non-specific reactions to media constituents.

Isoenzyme techniques. Isoenzyme patterns of cells were studied by electrophoresis on polyacrylamide gel in a vertical gel apparatus as previously described (RAYMOND, 1964). Malate dehydrogenase (MDH) patterns were developed with NADP as well as NAD. A number of specimens were analyzed in starch gel electrophoresis by G. LEVAN (personal communication) using the methods described by SHAW and KOEN (1968).

The extracts were prepared by treating the cell sheets in plastic flasks with octyl alcohol overnight at 4° C to disrupt the cells and release the enzymes from the cells (GREENE et al., 1969) or by removing cells from plastic flasks or blake bottles with 0.25 % trypsin, washing the cells 3 times with saline and releasing the isoenzymes by 3 freeze-thaw cycles in a dry ice-ethyl alcohol bath. The crude extracts were centrifuged at 1500 RPM for 15 min. and clarified at 12,000 RPM for 2 hrs at 4° C.

Sterility tests. Insect cell cultures were tested for microbial contamination by culturing the cells and supernate in HAYFLICK broth and agar plates (HAYFLICK, 1965) for mycoplasma. After 5 or 7 days, an aliquot for the mycoplasma broth culture was transferred to plates and observed for 2 weeks. Blood agar plates, trypticase soy broth, brain-heart infusion broth, mold and yeast broth, and tryptose phosphate broth were inoculated with insect cell and medium to detect bacterial, yeast and mold contamination.

C. Results

Microbial sterility studies. The 6 insect cell lines in our laboratory were found free from contamination with bacteria, yeast, molds and mycoplasma. Two of the cell lines, the adapted *A. eucalypti* (CCL 95) and *A. albopictus* (CCL 126) which are candidates for storage in the Cell Bank at the American Type Culture Collection, were also free of microbial contamination in another reference laboratory (LAM, personal communication).

Frozen storage of cells. Our studies indicated that both the *A. eucalypti* and *A. albopictus* cells could be preserved in liquid nitrogen without loss of viability. Both dimethyl sulfoxide and glycerol were equally effective as additives in preventing freezing damage. *A. eucalypti* cells grown in hemolymph and frozen in 5 % glycerol were 78 % viable after recovery from the frozen state shortly after storage. These cells showed no loss of viability when removed after 4 years storage in liquid nitrogen. Two hundred ampules of the adapted *A. eucalypti* were frozen in liquid nitrogen in 1967. The cells were 90 % viable by the neutral red test upon recovery in 1967 and no loss of viability was evident upon recovery in 1970. One hundred and eighty-five ampules of *A. albopictus* were stored in liquid nitrogen in 1969. Upon recovery these cells were approximately 85 % viable and there was no loss of viability after liquid nitrogen storage for one year.

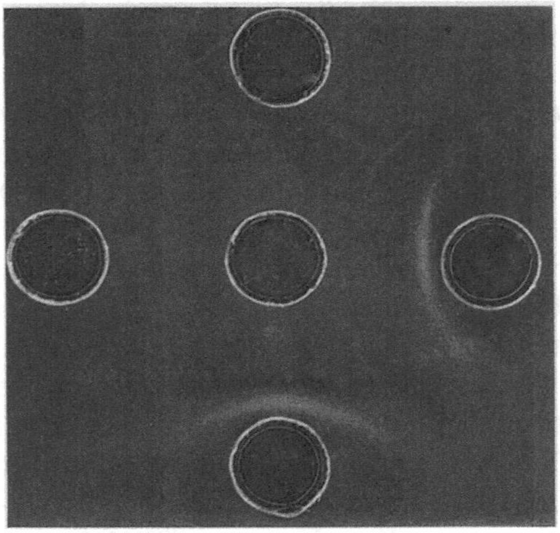

Fig. 43. Center well: *Antheraea eucalypti* cell antigen; bottom well, rabbit anti-unadapted *Antheraea eucalypti* globulin; left well, guinea pig anti-*Aedes aegypti* larval protein globulin; upper well, rabbit anti-adapted *Antheraea eucalypti* globulin; right well, rabbit anti-*Antheraea pernyi* globulin

Agar gel immunodiffusion. Agar gel immunodiffusion was used successfully for serological identification. In Fig. 43 GRACE's *A. eucalypti* cell antigen in the center well formed precipitin lines with concentrated rabbit anti-unadapted *A. eucalypti* globulin and rabbit anti-moth (*A. pernyi*) globulin. A very faint reaction occurred with concentrated rabbit anti-moth (adapted *A. eucalypti*) globulin which does not show on the photograph. No reaction occurred with guinea pig anti-*A. aegypti* larval protein globulin. Similar reactions were observed between the adaped *A. eucalypti* cell antigen placed in the center and the three rabbit anti-moth globulins. No precipitin lines could be observed with guinea pig anti-*A. aegypti* larval protein globulin. This information together with the chromosome analysis described in the accompanying report of NICHOLS, BRADT, and BOWNE (p. 61) indicated that GRACE's *A. aegypti* cells received in our laboratory are moth and not mosquito cells and that contamination of the original cell line had occurred at some stage in its passage through a number of laboratories.

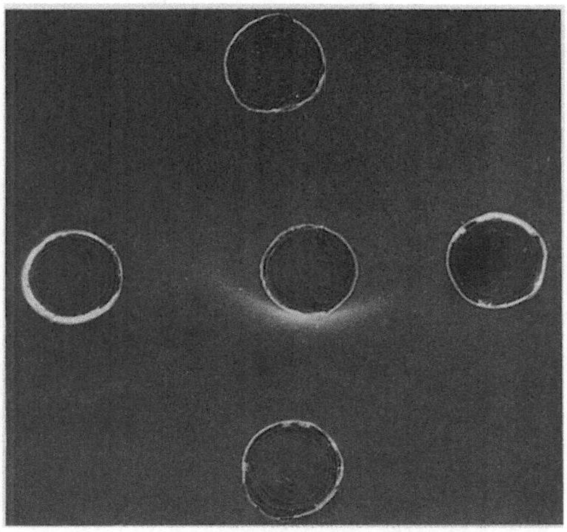

Fig. 44. Center well: Rabbit anti-*Antheraea eucalypti* globulin; bottom well, moth (*Antheraea eucalypti*) cells; left well, human (HeLa) cells; upper well, mouse (L 929) cells; right well, bovine calf kidney cells

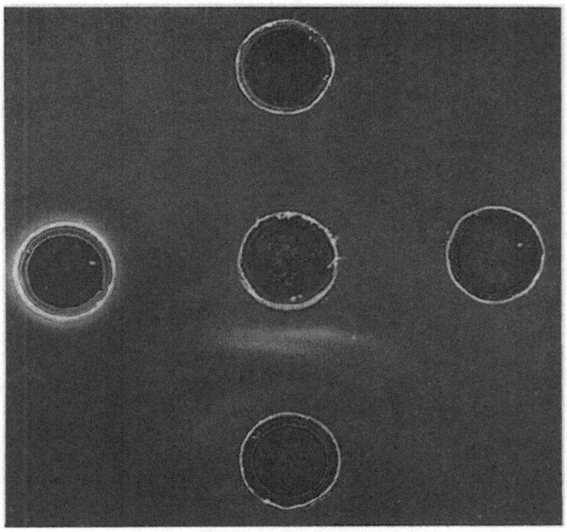

Fig. 45. Center well: Rabbit anti-*Antheraea eucalypti* globulin; bottom well, moth (*Antheraea eucalypti*) cells; left well, rat cells; upper well, fathead minnow cells; right well, gekko lung cells

Rabbit anti-adapted *A. eucalypti* globulin formed a precipitin line with moth cell antigen but not human, mouse or bovine kidney antigen (Fig. 44). Antiglobulin to moth cells reacted with moth cell antigen but not with rat, fathead minnow, or gekko lung cell antigen (Fig. 45). A precipitin reaction was obtained between guinea pig antiglobulin to *A. aegypti* larval protein in the center well and *A. albopictus* cell antigens but not with moth, human and mouse antigens (Fig. 46). This indicates that both *A. eucalypti* and *A. albopictus* cell cultures can be distinguished from each other and from homiothermic and poikilothermic species by immunodiffusion.

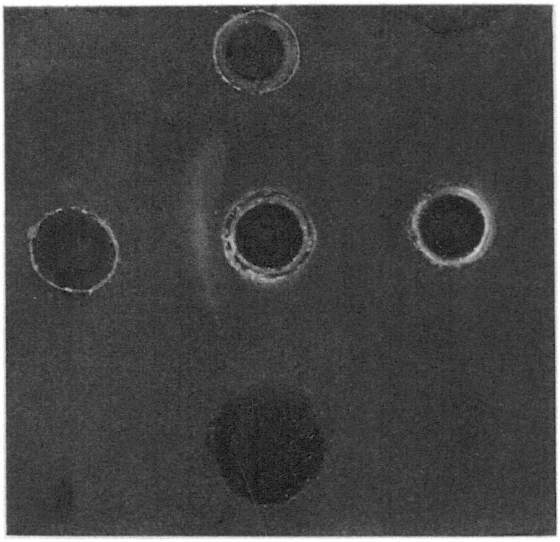

Fig. 46. Center well: Guinea pig anti-*Aedes aegypti* larval protein globulin; bottom well, moth (*Antheraea eucalypti*) cells; left well, mosquito (*Aedes albopictus*) cells; upper well, human (HeLa) cells; right well, mouse (L 929) cells

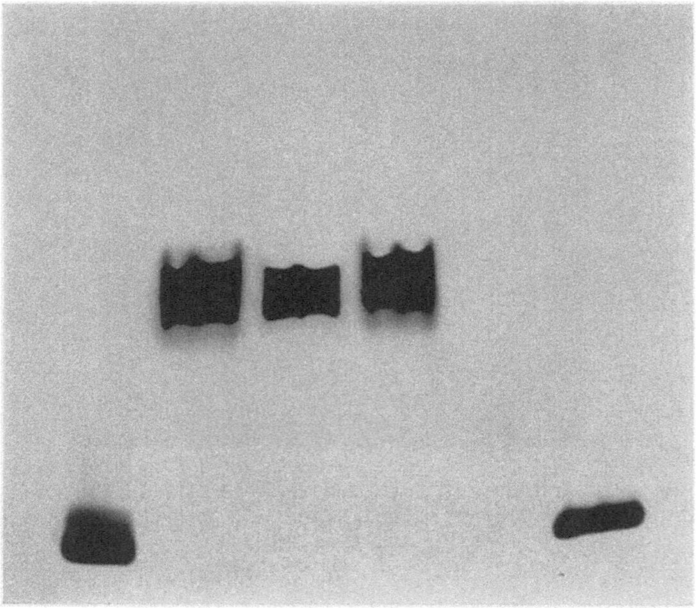

Fig. 47. Glucose-6-phosphate dehydrogenase on polyacrylamide gel. Cell lines from left to right: Human (HeLa); mosquito (*Aedes albopictus*); unadapted moth (*Antheraea eucalypti*); adapted moth (*Antheraea eucalypti*; unadapted moth (*Antheraea eucalypti*); mouse (L 929)

Isoenzyme analysis. Isoenzyme analysis proved to be an excellent method for identifying the species of cell cultures. Fig. 47 demonstrates that it is possible to distinguish between human, mouse and moth cells by glucose-6-phosphate dehydrogenase (G6PD) on polyacrylamide gel. The cell lines of

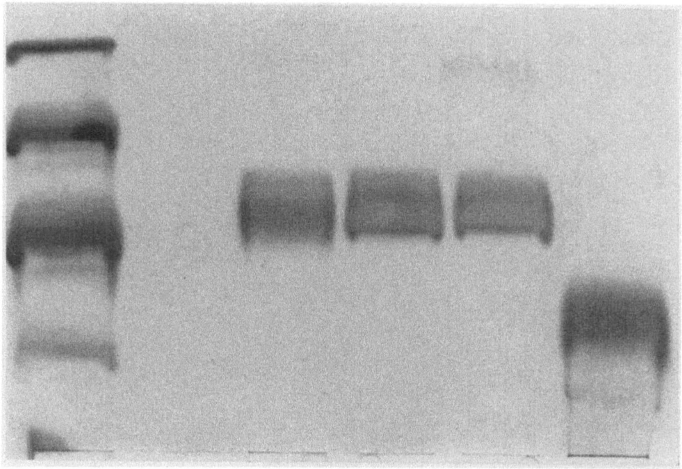

Fig. 48. Lactate dehydrogenase on polyacrylamide gel. Cell lines from left to right: Human (RPMI 2650); mosquito (*Aedes albopictus*) no pattern; unadapted moth (*Antheraea eucalypti*); adapted moth (*Antheraea eucalypti*); unadapted moth (*Antheraea eucalypti*); mouse (L 929)

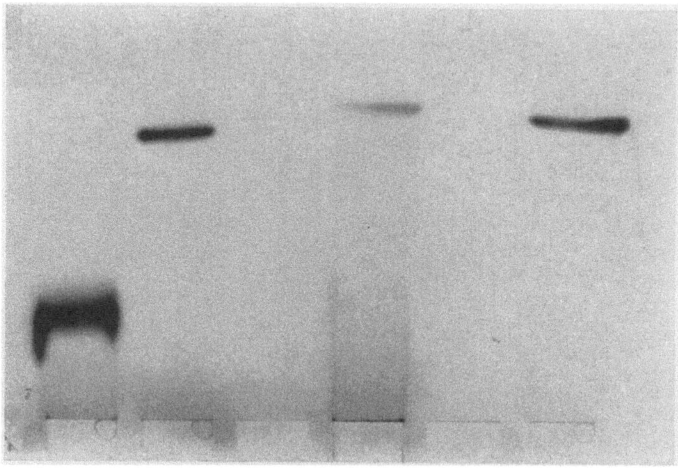

Fig. 49. Malate dehydrogenase on polyacrylamide gel. Cell lines from left to right: Mouse (L 929); human (L-132); moth (*Antheraea eucalypti*); mosquito (*Aedes albopictus*)

the three different species each form one G6PD band but the bands have different mobilities. No pattern was obtained from the *A. albopictus* cells. Similar results were observed on starch gel electrophoresis.

Both unadapted and adapted moth cells gave identical lactate dehydrogenase (LDH) patterns which are easily distinguished from mammalian species (Fig. 48).

Moth cells can be distinguished from mosquito cells by MDH isoenzyme analysis; the moth cells yield a very slightly more mobile anodal band than the mosquito cells. They are followed by the slower human and mouse bands (Fig. 49).

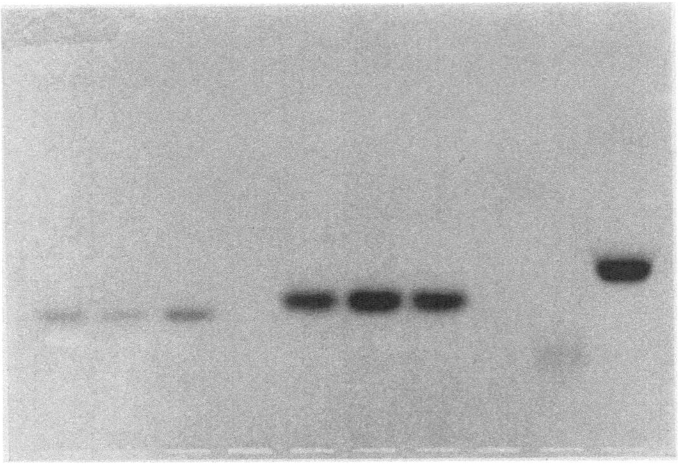

Fig. 50. 6-Phosphogluconate dehydrogenase on starch gel. Cell lines from left to right: *1* human (WI-38); *2* human (WI-38 2RA); *3* human (WI-26 VA4); *5* rat (RR1022, clone 1034); *6* rat (RR1022, clone 1032-P); *7* rat (RR1022, clone 1033); *9* mosquito (*Aedes albopictus*); *10* moth (*Antheraea eucalypti*)

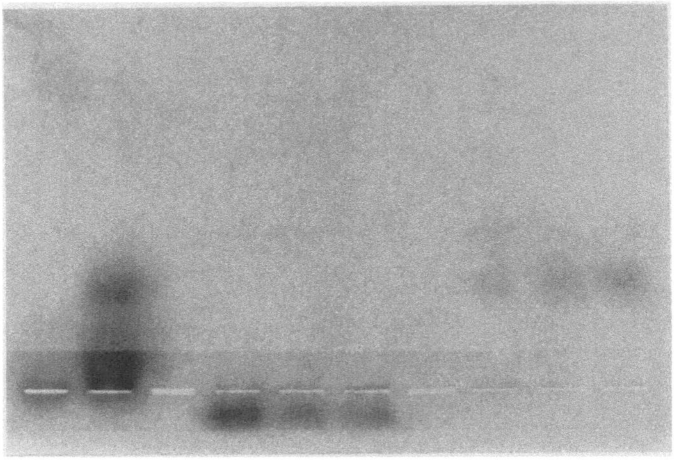

Fig. 51. Acid phosphatase on starch gel. Cell lines from left to right: *1* moth (*Antheraea eucalypti*); *2* mosquito (*Aedes albopictus*); *4* rat (RR1022, clone 1034); *5* rat (RR1022, clone 1034-P); *6* rat (RR1022, clone 1033); *8* human (WI-38); *9* human (WI-38 2RA); *10* human (WI-26-VA4)

6-Phosphogluconate dehydrogenase (6PGD) and acid phosphatase isoenzyme patterns on starch gel electrophoresis are shown in Fig. 50 and Fig. 51. The moth cells have a prominent single 6PGD band that is anodal to both rat and human cell extracts. The mosquito *A. albopictus* band trails the rat and human bands. The acid phosphate pattern indicates two bands for the mosquito cells which move toward the anode and thus allows differentiation between moth, mosquito and mammalian cell cultures.

The identification of Suitor's clone of *Aedes aegypti* as a moth cell was confirmed by G6PD isoenzyme analysis (Fig. 52). Singh's *Aedes aegypti* and *Aedes albopictus* yielded no detectable G6PD (patterns 1 and 2). In contrast Suitor's clone (patterns 3 and 6) contained a G6PD isoenzyme identical in mobility with that of *Antheraea eucalypti* (patterns 4 and 5). Therefore, by immunodiffusion, isoenzyme analysis and chromosome analysis, this clone is shown to be moth rather than mosquito. It is difficult to determine when the *Aedes aegypti* cells were contaminated with the *Antheraea eucalypti* cells since the cells were cultured in four different laboratories before arriving at our laboratory.

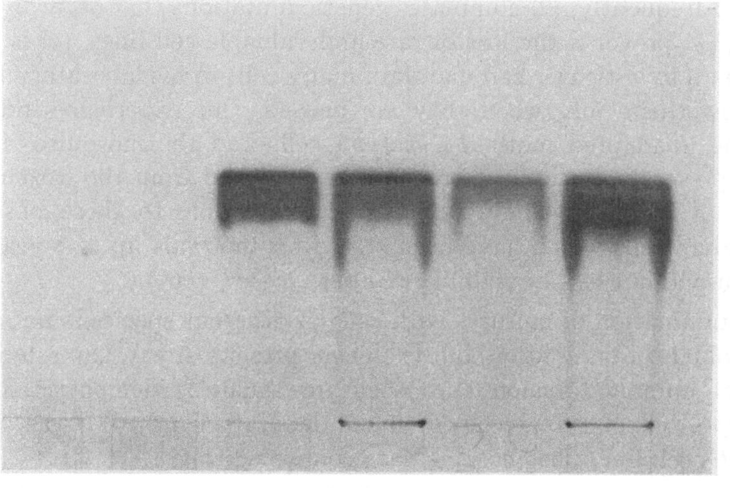

Fig. 52. Glucose-6-phosphate dehydrogenase on polyacrylamide gel. Cell lines from left to right: *Aedes aegypti*, *Aedes albopictus*, Suitor's clone of Grace's *Aedes aegypti* (Freeze 309 IMR); adapted moth (*Antheraea eucalypti*); unadapted moth (*Antheraea eucalypti*); Suitor's clone of Grace's *Aedes aegypti* (Freeze 335, passaged 30 times at IMR)

D. Discussion

The recent development of a number of continuously propagated insect cell cultures offers an opportunity for many studies previously limited by lack of access to insect tissues. Experience with mammalian cell cultures has demonstrated the ease which cell cultures are contaminated with microorganisms or cells of other animal species. Hayflick (1969) observed that 60% of all cultures received in his laboratory between 1955–1965 were contaminated with mycoplasma. In Barile's laboratory (1968) 45% of all continuously propagated cultures tested were positive for mycoplasma and last year in our laboratory 70% of cultures received from outside sources were found to contain mycoplasma, bacteria, yeast or molds.

These observations indicate the need for rigorous efforts to control contamination. Some means that we have found to be helpful include quarantine of newly established or recently acquired cell cultures in an isolation laboratory until they are proven to be free of extraneous microorganisms. Sterility tests should be carried out on all lots of sera and culture media before they are used. Aliquots of culture media should be saved for culture when each experiment is set up. If at all possible, tissue cultures should be made in a sterile room equipped with laminar airflow and HEPA filters to remove dust and bacterial contamination (CORIELL, 1968). The removal of antibiotics from the tissue cultures is one of the most effective methods of eliminating microbial contamination because it encourages good aseptic technique and faulty technique quickly becomes apparent.

Storage of mammalian cells in liquid nitrogen has proved invaluable because it: (1) prevents microbial contamination which is so common when cells are passaged frequently, (2) eliminates genetic mutations that occur in continuous culture, (3) prevents the loss of rare and valuable cell lines, (4) is economical because an investigator can maintain many cells in his laboratory but recover and grow them only when they are needed. Our experience indicates that both the unadapted moth *A. eucalypti* cells and the mosquito *A. albopictus* can be preserved in liquid nitrogen and recovered from the frozen state with no loss of viability. These results are comparable to those observed with mammalian cells which have been stored for intervals up to 8 years in liquid nitrogen without loss of viability (GREENE et al., 1969).

Contamination of cultures with cells of different species is not as common as microbial contamination, but is an ever present threat. Our studies utilizing agar gel immunodiffusion and isoenzyme analysis demonstrated that one *A. aegypti* cell culture received in our laboratory in 1967 was a moth (*A. eucalypti*) cell line. Chromosome data confirmed this.

Rabbit antisera prepared against both unadapted and adapted *A. eucalypti* moth cells reacted with homologous moth cells but not with *A. albopictus* cells. No reaction was observed against a number of mammalian, fish or amphibian cell cultures, thus demonstrating that the agar immunodiffusion technique can distinguish *A. eucalypti* cell cultures from these species.

Guinea pig anti-*A. aegypti* larval protein serum gave precipitin lines with *A. albopictus* cells but not with moth cells. The reaction could be observed visually but was often difficult to photograph. This antiserum did not react against a number of mammalian fish and amphibian cell cultures.

In 1962 VESELL et al. examined the LDH patterns of cell cultures derived from a few animal species and suggested that isoenzyme analysis might provide a method for identifying cells from various species when grown in tissue culture. GARTLER in 1967 pointed out that electrophoretic variants of polymorphic enzymes might also serve as useful genetic markers for the detection of intraspecific contamination of cell cultures. More recently MONTES DE OCA et al. (1969) tested a total of 86 characterized animal cell lines certified by the Advisory Committee to the Animal Cell Culture Collection (1964). They observed that 20 out of 22 taxonomic groups can be easily distinguished from each other by comparison of their G6PD and LDH isoenzyme patterns.

Our studies demonstrated that isoenzyme analysis is also a valuable test for the interspecies identification of insect cells. Using isoenzyme patterns of glucose-6-phosphate dehydrogenase, lactate dehydrogenase, malate dehydrogenase, 6-phosphogluconate dehydrogenase and acid phosphatase, we could distinguish moth (*A. eucalypti*) and mosquito (*A. albopictus*) cell cultures from each other and from mammalian, fish and amphibian species.

Montes de Oca et al. (1969) have shown that once the electrophoretic enzyme pattern is determined for a large number of species and a "finger print" identification chart constructed, the species of an unknown cell line can be established. The technique which is relatively simple and rapid has proven of value for identifying species of mammalian, fish and amphibian cell cultures and our results indicate that the method will be equally valuable in confirming the species of insect cell cultures.

Acknowledgment. This investigation was supported by the Public Health Service Grant C-4953 from the National Cancer Institute and Damon Runyon Memorial Fund Grant DRG924A.

V. Cytogenetic Studies on Cells in Culture from the Class Insecta

Warren W. Nichols, Carole Bradt, and Willard Bowne

The class Insecta has provided some of the classical material for studies in genetics and cytogenetics. The availability of cells in culture from this class adds a new dimension to the possibilities available for study. The present paper describe various cytogenetic aspects of three cell lines derived from the class Insecta. One from the order Lepidoptera is from the moth, *Antheraea eucalypti;* the other two, from the order Diptera, were derived from two species of mosquito, *Aedes albopictus* and *A. aegypti*.

A. Materials and Methods

Cells. The cell lines used in these studies were obtained from the cell bank facility at the Institute for Medical Research, Camden, N. J., from Dr. Greene. The *A. eucalypti* cells were originally derived by Grace (1962) and modified by Yunker et al. (1967). The *A. albopictus* and *A. aegypti* were derived by Singh (1967). The culture methods used with these cells are described in another section of this chapter, Greene and Charney, p. 51.

Chromosome preparations. Since the *A. eucalypti* cells grow in suspension, the time of optimal numbers of mitoses was selected by making preparations each day after passing the cells. The greatest number of mitoses were found 6 to 8 days after passage or refeeding. The most satisfactory colchicine treatment was for 18 hrs at a concentration of 10^{-6} M.

The *A. albopictus* and *A. aegypti* cells grow as a sheet on the glass surface. Microscopic inspection of this sheet for metaphase plates indicated that the time of greatest mitotic activity was 3 to 5 days after passing the cells. For these cells, colchicine was used in a final concentration of 10^{-6} M for $1^1/_2$ hrs.

Metaphase preparations were made by subjecting the cells, suspended in media containing colchicine, to hypotonic expansion for 20 min. by adding 3 volumes of distilled water. They were then centrifuged, the supernatant fluid decanted, and

60% acetic acid—0.1 N hydrochloric acid fixative added to the cell pellet for 10 to 15 min. The fixative was then decanted and 2% orcein stain added. The cells were dispersed in the stain and squashes made on siliconized slides.

Anaphase preparations were made by obtaining cells in suspension, centrifugation, and adding fixative as described above with no pretreatment (no colchicine, no hypotonic expansion). Stain was added and slides made with no pressure applied to the coverslip.

B. Results

The cells from *A. eucalypti* are interesting from a cytogenetic standpoint, but preparations were obtained with a great deal of difficulty and the chromosome morphology was quite poor. In preparations both macro- and micro-chromosomes are present, and the micro-chromosomes make an accurate

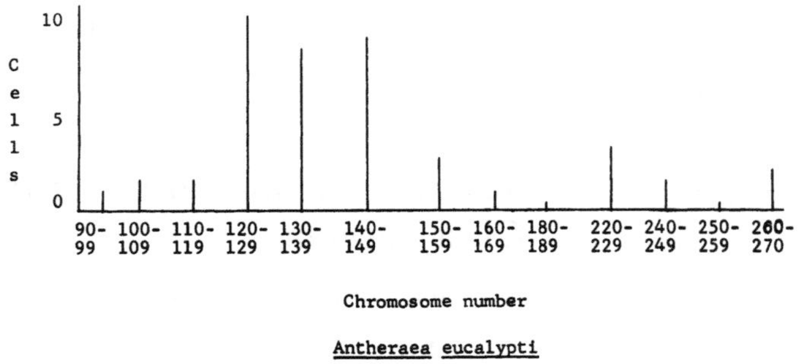

Fig. 53. Distribution of chromosome numbers from cell line derived from *Antheraea eucalypti*. No distinct stemline number is present but instead a range from 120 to 149

determination of chromosome number very difficult. The stemline chromosome number does not exhibit a sharp peak, but rather is seen as a range of chromosome numbers from 120 to 149 (Fig. 53). The diploid chromosome number of *A. eucalypti* has been reported to be 50 by Thomson and Grace (1963) on the basis of finding 25 bivalents in meiotic preparations. Examples of the morphology of the *Antheraea* chromosomes are seen in Fig. 54. The micro and macro nature of the chromosomes can be seen, and it is also noted that a distinct centromere is not seen in the macro-chromosomes, the two chromatids usually having the appearance of parallel rods.

Lepidoptera is one of the orders of insects that have been observed to possess diffuse centromeres or holocentromeres, rather than the more typical localized centromere. This means that either the entire chromosome acts as a centromere without any specifically differentiated structure, or that there are multiple small centromeres along the length of the chromosome. That this type of diffuse centromere is present can be determined by examining anaphase preparations which show that the chromosomes move to the poles without having a bending point, characteristic of a localized centromere. Also, when chromosomes with a diffuse centromere are fragmented by irradiation, each of the fragmented units moves to the pole, indicating centromeric activity over the entire chromosome.

In the metaphase preparations there are occasional constrictions that resemble a centromere, but the definitive determination of a diffuse versus a localized centromere is made on the basis of anaphase preparations. When anaphase preparations were made from these cells, individual chromosomes could not easily be distinguished, so that the point could not be completely

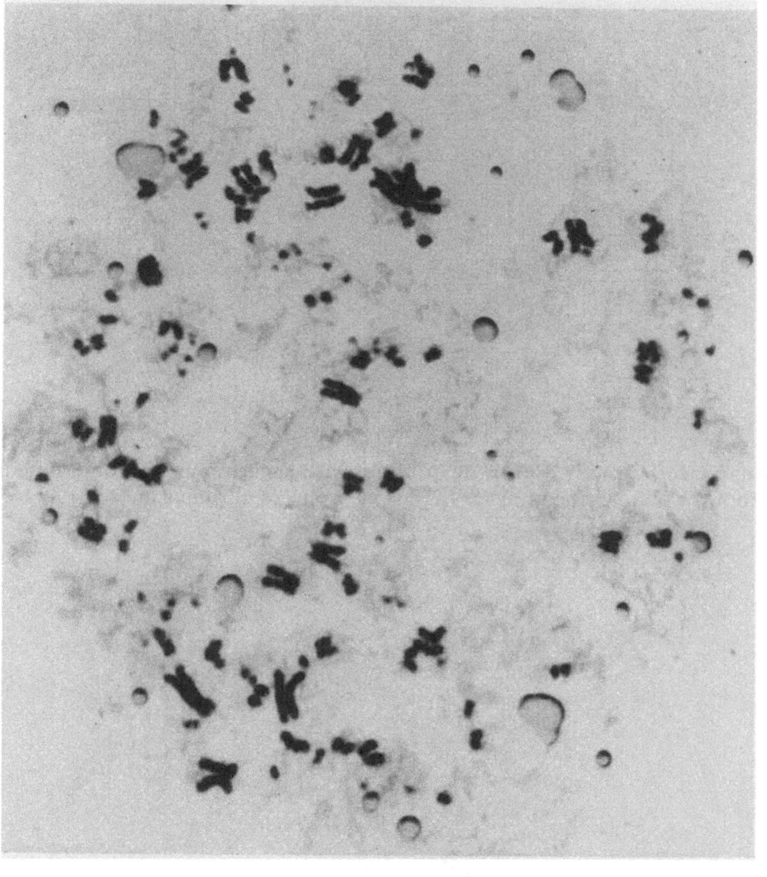

Fig. 54. Metaphase plates from cell line derived from *Antheraea eucalypti* demonstrating micro and macro-chromosomes and the lack of a localized centromere. 1935 ×

clarified; however, in some of the anaphase figures the chromosomes did appear to be moving without a point of bending (Fig. 55). Earlier anaphase figures should clarify this point further.

The anaphase preparations also demonstrated an exceedingly prominent spindle apparatus in these cells. Spindle fibers were easily seen in most anaphase figures, and it appeared that the continuous spindle fibers were quite persistent, even up to the stage of cytokinesis, resulting in the eccentric placement of the nuclei in the resultant daughter cells (Fig. 56a–c). These cells should offer excellent material for studies of spindle mechanisms.

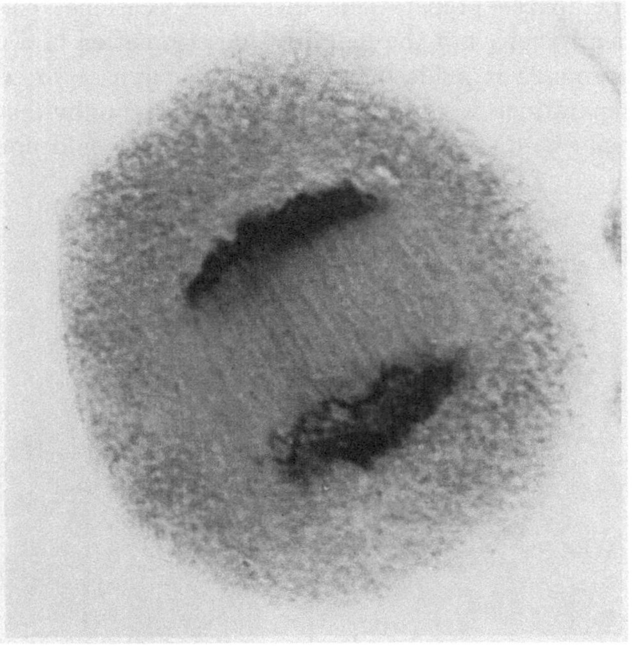

Fig. 55. Anaphase cell from *Antheraea eucalypti* with suggestion of chromosome migration without a point of bending in the lower group of chromosomes. 1400 ×

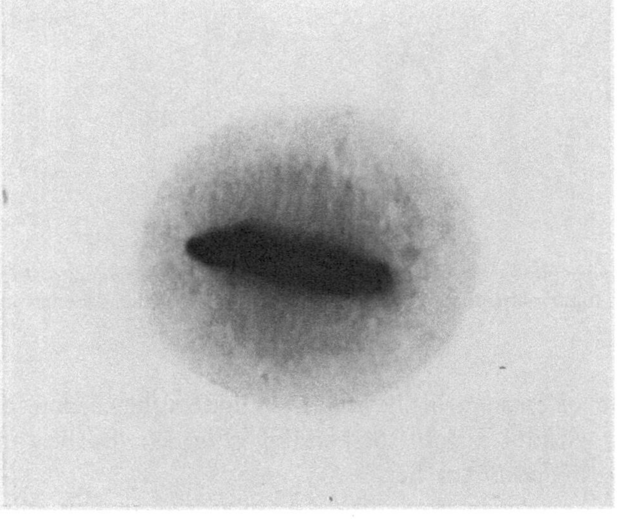

a

Fig. 56a–c. Preparations from *Antheraea eucalypti* line. a Metaphase plate demonstrating prominent spindle fibers, 2265 ×. b Late anaphase or early telophase demonstrating persistence of prominent spindle fibers producing eccentric nuclei at early stage of cytokinesis, 2165 ×. c Two daughter cells at completion of cytokinesis demonstrating eccentric nuclei, 1700 ×

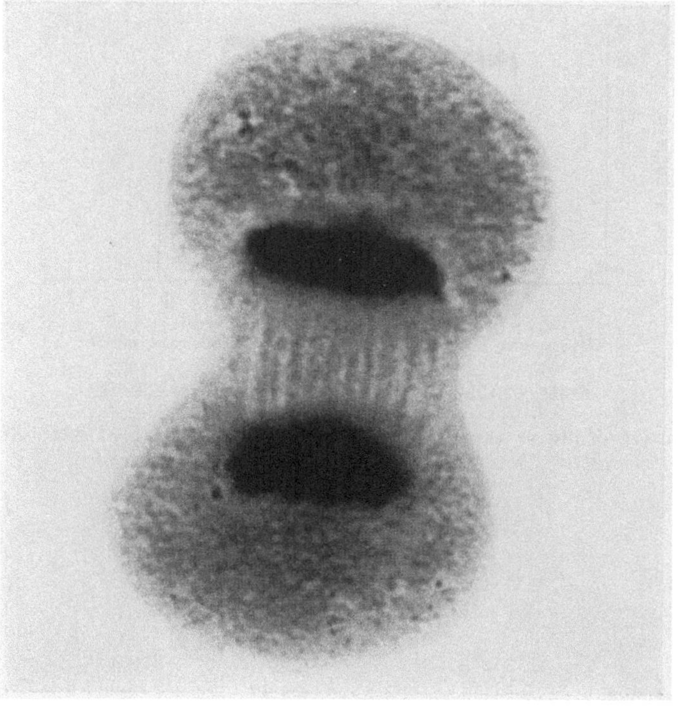

56b

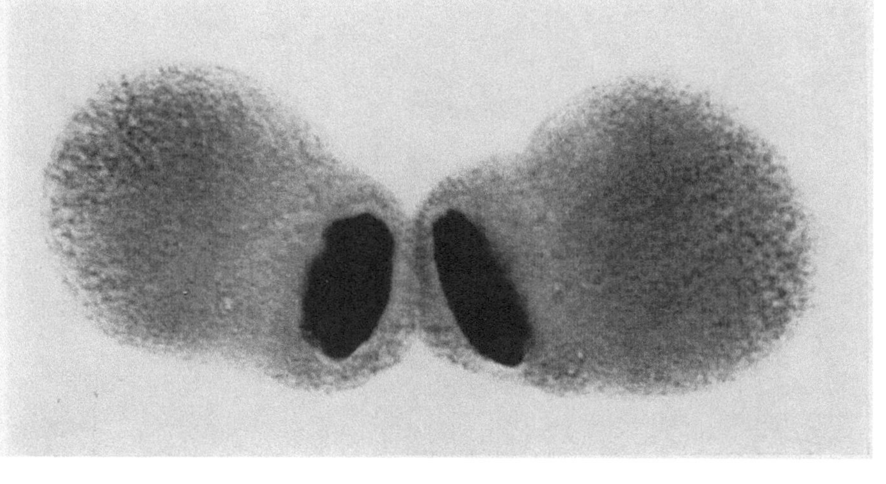

56c

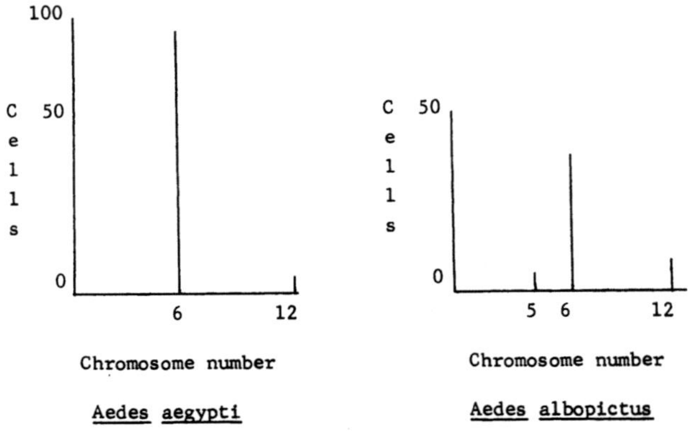

Fig. 57. Distribution of chromosome numbers from cell lines derived from *Aedes aegypti* and *Aedes albopictus*. Both show a distinct stemline number of 6

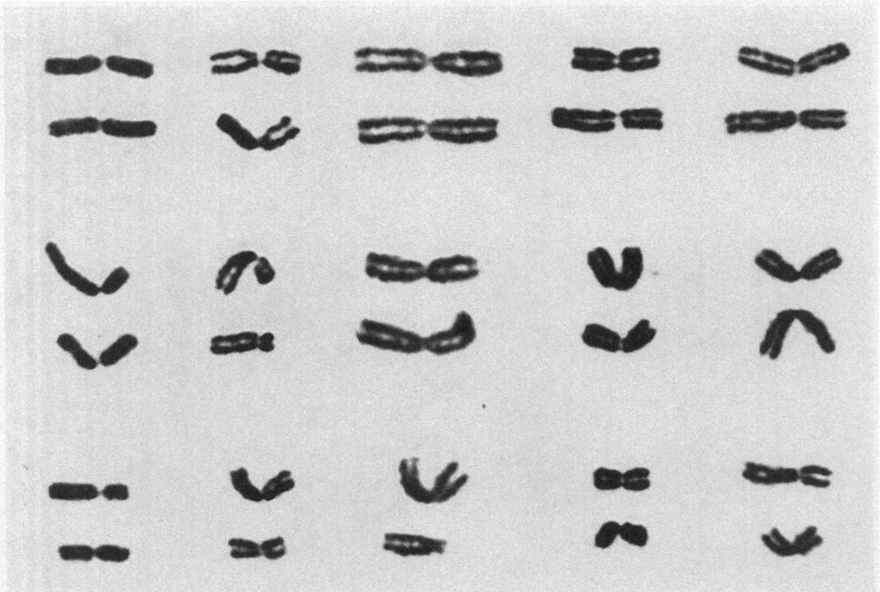

Fig. 58a. Five karyotypes from *Aedes albopictus*. Each has the stemline number of 6 but considerable variation in chromosome morphology exists from cell to cell. 1450×

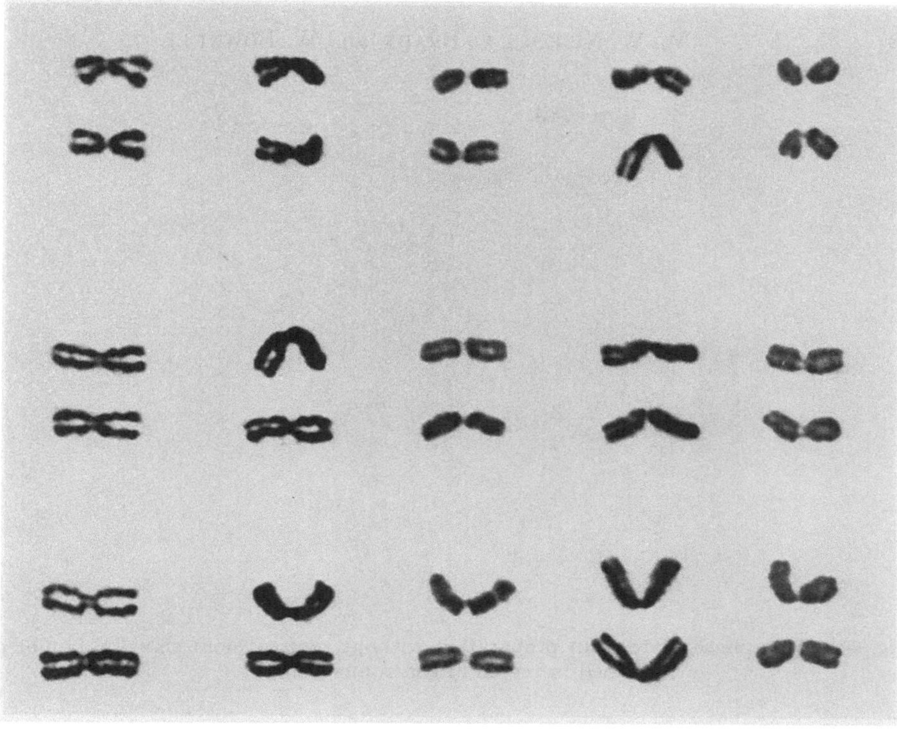

Fig. 58b. Five karyotypes from *Aedes aegypti* exhibiting little or no variation in chromosome morphology from cell to cell. 1965 ×

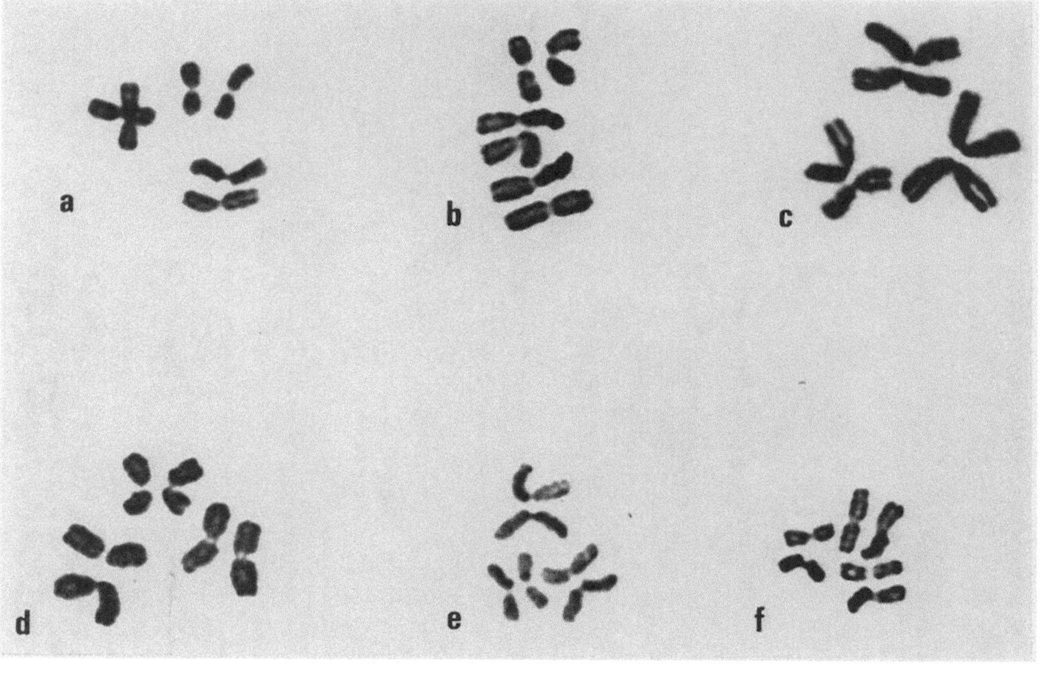

Fig. 59a–f. Six metaphase plates from *Aedes aegypti* exhibiting marked pairing of the homologous chromosomes of each pair. 1870 ×

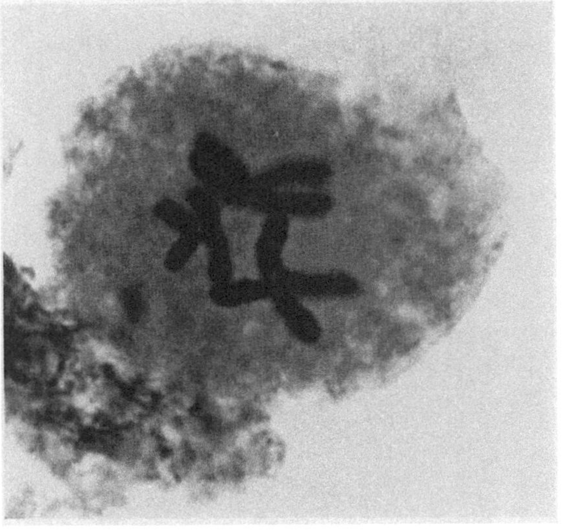

Fig. 60a. Metaphase plate from preparation without pretreatment showing homologues joined in region of centromere. 2035 ×

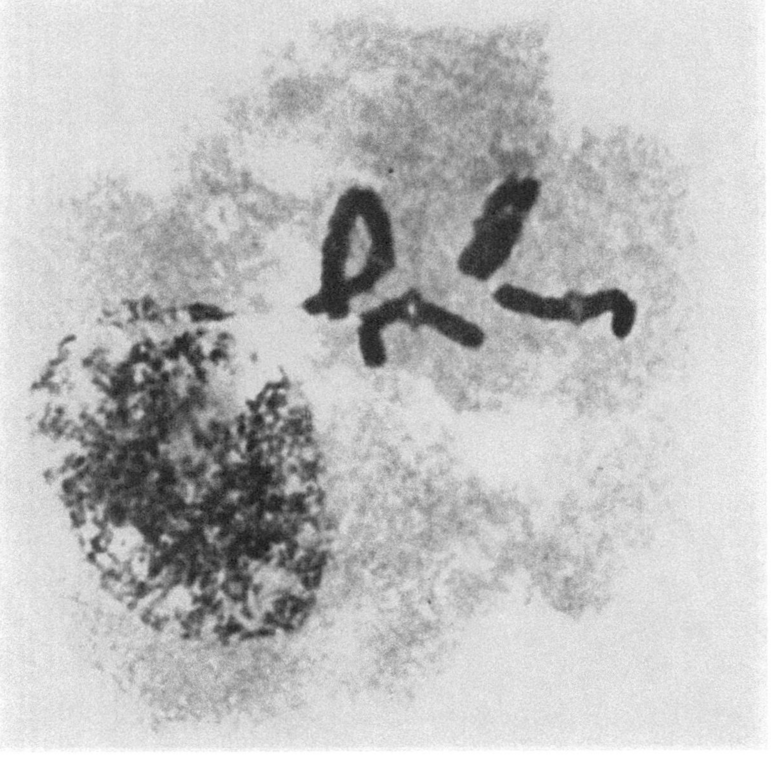

Fig. 60b. Early anaphase from same preparation with an appearance somewhat resembling synapsis. 1930 ×

The two cell lines from the other Diptera (the mosquitoes, *A. albopictus* and *A. aegypti*) have a very distinct stemline number. Their chromosomes are few and are easily studied morphologically. In all species of mosquitoes in which the chromosomes have been studied, 3 pairs of chromosomes were found, giving a diploid number of 6. Both of the cell lines described here have a stemline number of 6 (Fig. 57). There is, however, an interesting variation between these two lines. The *A. albopictus* has a great deal of variation in the chromosome morphology within the stemline number of 6. Fig. 58a shows karyotypes of 5 cells from the *albopictus* line. The difference in chromosome morphology from cell to cell is easily discerned, thus classifying this line as pseudodiploid. In contrast, the *A. aegypti* cells (Fig. 58b) exhibited very little morphologic variation from cell to cell within the line, indicating that these undergo much less variation and are true diploid cells.

An interesting phenomenon found in these lines of mosquito origin is somatic pairing. Somatic pairing and the possible resultant crossing over is often discussed for humans and many other species. However, the only animal life in which unequivocal somatic pairing is seen is in the order Diptera. A collection of 6 metaphase plates is seen in Fig. 59; all of these exhibit striking somatic pairing. When preparations were made from these mosquito cells without any pretreatment to avoid disruption of the pairing process, it can be seen that the homologs are joined in the region of the centromere (Fig. 60a). In Fig. 60b, which represents a very early anaphase, there is an appearance resembling synapsis, as is seen in meiotic chromosomes. These cells offer excellent material for cytogenetic studies from the standpoint of their chromosome morphology, their small chromosome number, and the interesting phenomenon of somatic pairing.

Acknowledgment. Supported by Research Career Award 5-k3-CA-16, 749 from the National Institutes of Health. We gratefully acknowledge the assistance of Miss Mary Federico, photography technician.

Chapter 3

Physiology of Cultivated Arthropod Cells

I. Introduction

T. D. C. GRACE

Physiology is the study of the specific functions carried out by the various organs and tissues which make up a multicellular organism. The organs and tissues themselves are the result of a number of cells adhering to one another and aggregating in specific patterns to form organized structures. The form these structures (organs and tissues) will take is determined partly by the genetic makeup of each cell and partly by the environmental factors acting upon the cells during embryonic development and throughout the life of the organism. In order, therefore, to gain some insight into the functions of cells and tissues, it is necessary to study their growth, the changes which occur in them during development and differentiation and the ways in which they convert available nutrients into energy and other products of metabolism. An obvious way to study these problems is to place tissues or organs in an environment which resembles, as closely as possible, the environment within the organism, but which can be varied or controlled as desired. When a piece of tissue or an organ is placed in a medium under *in vitro* conditions it will generally behave in one of two ways: It may (1) continue to grow, develop, differentiate and maintain fairly closely, normal physiological functions, much as it would have done in the whole animal, or (2) its cells may multiply rapidly, migrate away from the explant and exist as more or less independent units. In such cultures, the explant sooner or later loses its organization and the cells, after a time, adapt themselves to their new surroundings, lose many of their specific functions and become "dedifferentiated". Thus, in this chapter on the physiology of cultured arthropod cells we are concerned on the one hand with studies of organized tissues in which the cells have reattained their differentiated state and on the other in cells and cell populations in which most of the organization has been lost. The chapter can rather conveniently be divided into three sections. The first section will consist of papers which describe physiological studies on organized tissues, or their differentiation. SEECOF and TEPLITZ have studied the *in vitro* differentiation of neurons from *Drosophila* and MARKS describes the *in vitro* cultivation and activity of the endocrine glands from the cockroach *Leucophaea maderae*.

In the second section the paper by CONOVER, ZEPP, HIRSCHHORN, and HODES describes the production of human-mosquito somatic cell hybrids and their response to virus infection.

The third section is concerned with the functions of cells and their metabolism. VAUGHN, LOULOUDES and DOUGHERTY have studied the uptake of sterols, both free and serum-bound, by a line of moth cells, whereas JENKIN, TOWNSEND, MAKINO, and YANG have made a comparative study of the lipids in *Aedes aegypti* and monkey kidney cells (MK-2) cultivated *in vitro*. WHITE's paper describes a histochemical study of the cells from *Antheraea eucalypti*. The last paper by GRACE and MITSUHASHI describes the effect of insecticides on the growth of cells from *A. eucalypti* growing *in vitro*.

II. *Drosophila* Neuron Differentiation *in Vitro*

ROBERT L. SEECOF and RAYMOND L. TEPLITZ

A. Introduction

Many investigators have described the growth of axons *in vitro* from explants of embryonic mammalian or bird tissue (MURRAY, 1965). In all cases the explant contained cells from already differentiated nervous tissue and no claims were made for neuron differentiation from stem cells. NIU and TWITTY (1953) and BARTH and BARTH (1958) observed the development *in vitro* of long processes which presumably were axons, from neurons that had differentiated from fragments of amphibian gastrula ectoderm. Neuroblasts in explants taken from grasshopper embryos show repeated divisions *in vitro* (reviewed by CARLSON and GAULDEN, 1964). The cytoplasm is distributed unequally at each division to generate a small ganglion cell and a large cell that will divide again. Each neuroblast thus serves as a stem cell for several ganglion cells but no axons are produced *in vitro*.

The differentiation of the nervous elements of the *Drosophila melanogaster* ventral nervous system has been summarized by POULSON (1950). The information was gathered from cell counts made on sections from fixed whole embryos. Gastrulation commences about 3 hrs after fertilization. About 2 hrs later neuroblasts can be detected as large cells, about 12.5 μ in diameter, in the ventro-lateral ectoderm. Each neuroblast undergoes probably 8 unequal divisions, generating a small ganglion cell (diameter about 3.3 μ) each time. Each ganglion cell divides once to give a total of 16 ganglion cells from each neuroblast, with an estimated 26 to 32 min. between each division in the sequence. The ventral nervous system is formed by a condensation of these cells, and nerve fibers can first be seen clearly about 10 hrs after fertilization.

LESSEPS (1965) reported that cells with long processes appeared in cultures of dissociated *Drosophila melanogaster* embryos and he identified them as neurons by their morphology in the light microscope. Since the youngest embryos were about 6.5 hrs old when dissociated and axons had probably not developed, LESSEPS attributed their appearance to differentiation *in vitro*. This conclusion may not be warranted, however, because *Drosophila* sometimes lay partially developed eggs. Similar results were obtained by SEECOF and UNANUE (1968) who dissociated *D. melanogaster* embryos 4–7 hrs after they were laid and plated out the cells. They inspected the cells immediately but could not detect any morphological differentiation. After 7 days *in vitro*, however, cells with long processes that had the morphology of neurons were observed. It was concluded that differentiation had taken place and these cells were probably neurons, but it is possible that these cells, as in the studies of LESSEPS, could also have been already differentiated cells re-establishing their morphology *in vitro* rather than stem cells differentiating for the first time.

In the present study, individual cells from dissociated *Drosophila* embryos were monitored by light microscope. Many cells were observed to make unequal divisions and to generate axon-like processes. Inspection by electron microscopy demonstrated the presence of axons among the cultured cells.

B. Observations on Living Cells

Cells which were prepared as described previously by SEECOF and UNANUE (1968) were placed between the two coverslips of a Sykes-Moore chamber (SYKES and MOORE, 1960) at a concentration of 2×10^6 cells per ml. The cells were allowed to settle on one coverslip for a minute and then the chamber was turned over. This procedure resulted in a small number of cells attaching to the first coverslip and the great majority to the second.

The chamber was placed on the stage of a microscope so that the cells to be monitored were attached to the underside of the top coverslip. The cell layer on the bottom provided a milieu that sustained the isolated cells above them. Room temperature was 25 to 26° C while the cells were observed.

Immediately following preparation, cell aggregates and individual cells were found to be attached to the glass. Many cells were seen to make unequal divisions and generate long processes along the surface of the coverslip. Figs. 61 a–d illustrate a typical sequence of these events. A cell was chosen before division and photographed by phase contrast at intervals. Between observations the culture was not illuminated to avoid heating the cells. Fig. 61 a was taken 1 hr and 40 min. after the culture was initiated. The cell, labeled N, was rounded up at the start and had a diameter of about 13 μ. Three hours and ten minutes later this cell had divided to give a group containing at least six small cells and one large cell. Twelve hours and thirty minutes later a branched axon about 50 μ long had formed (Fig. 61 b). The camera was focused on the axon for this photograph and the neuron cell body with small daughter cells, and the adjacent cell that is seen in Fig. 61 a, are out of focus.

Fig. 61 c shows the cell 4.5 hrs later. Three processes emanate from the neuron cell body. It is not possible to determine whether these are one branched axon or three separate ones. Two small daughter cells, each 2–3 μ in diameter,

Fig. 61. Phase contrast photographs of living *Drosophila* neuroblast undergoing unequal divisions and producing a branched axon *in vitro*. a Two cells after 1 hr and 40 min. *in vitro*. The cell marked *N* is followed in the other figures. b After 17 hrs and 20 min. Cell *N* has produced an axon about 50 μ long. c After 21 hrs and 50 min. The axon is longer and branched. d After 44 hrs. One axon branch is 100 μ long and has a diameter of about 0.7 μ

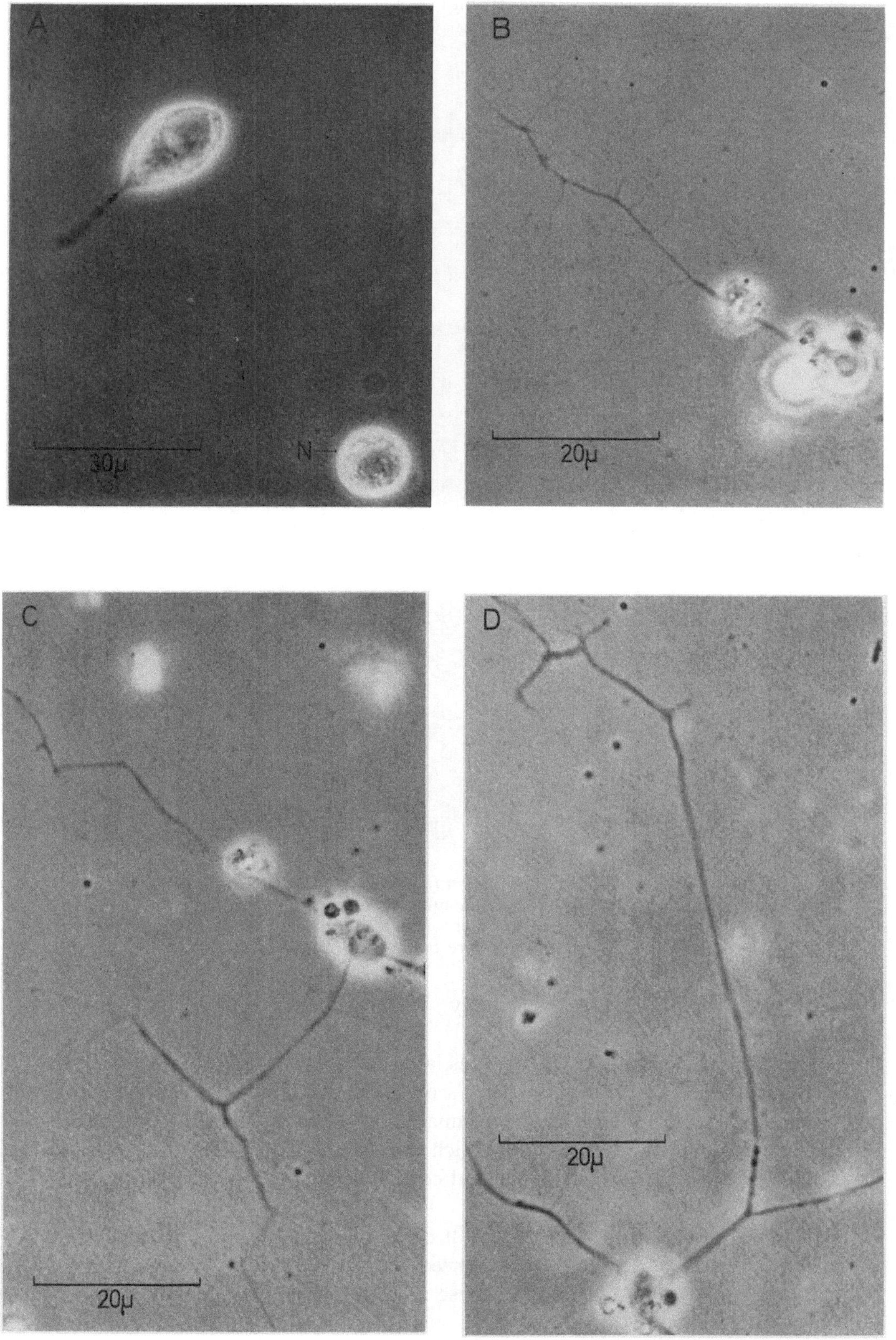
A
30μ
N
B
20μ
C
20μ
D
20μ
C

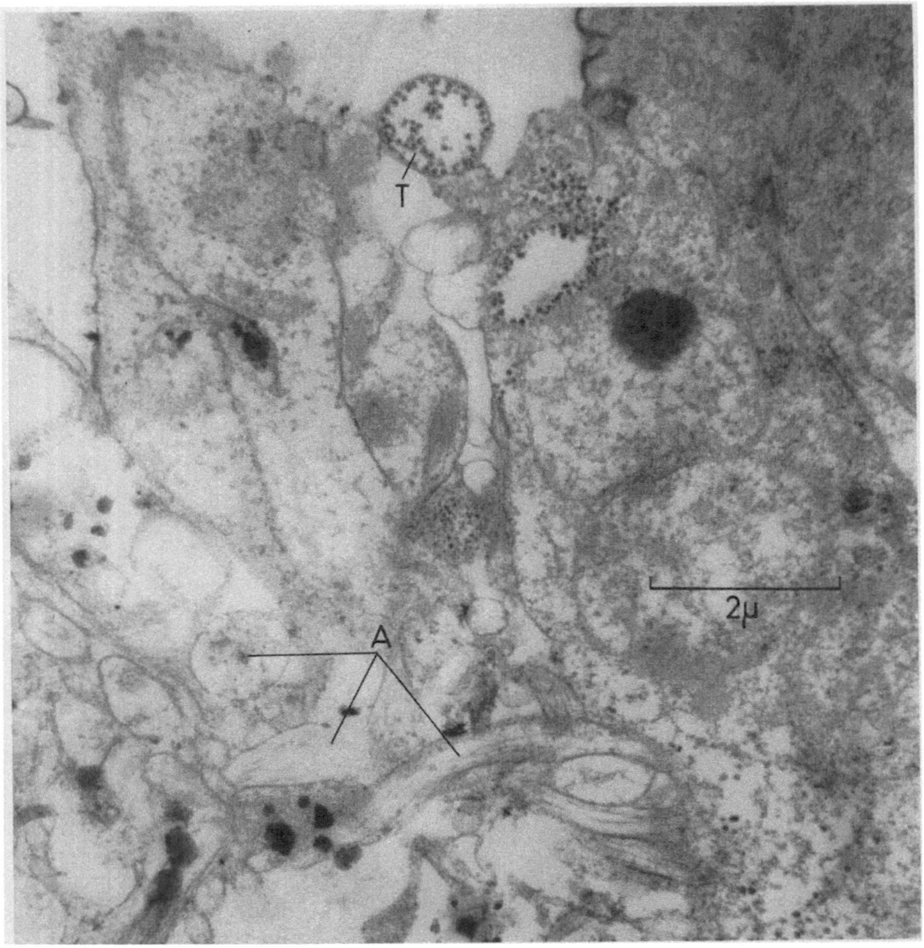

Fig. 62. A mixture of cell types after 48 hrs *in vitro*. Cross section of axon terminal (*T*) showing stained vesicles enclosed by a membrane. A second axon terminal is nearby with vesicles partially displaced by artifact. *A* sections through more proximal portions of axons. Opaque spots are precipitates from fixatives

are visible near the neuron cell body, the other small cells having already fallen away.

Twenty two hours and ten minutes later one process has attained a length of about 100 μ (Fig. 61 d). The axon's diameter is about 0.7 μ at this time. We have observed many cells, presumably neurons, generate an axon in a similar fashion. The rapidity with which the daughter cells are produced, and their small size relative to the original cell suggests there is little synthesis of new cytoplasm.

The figures indicate that the small daughter cells do not generate axons *in vitro*. Occasionally we have observed axonated cells unaccompanied by small cells but we do not know if these are a different cell type, or a type in which the production of ganglion cells has failed *in vitro*.

C. Observations on Fixed Material

To confirm the identity of the cell followed in Fig. 61 and to learn more about its ultrastructure, we examined cells from cultures similar to those used in the above study, by electron microscopy. The cultures were allowed to stand for 48 hrs without medium change and then fixed for 10 min. in a zinc-iodine-osmium (ZIO) solution which selectively stains the synaptic vesicles of axon terminals.

As technical limitations prevented us from processing single cells, we prepared whole cultures and scanned them for axon terminals. Fig. 62 shows two cross sections of cell processes about 1 μ in diameter bounded by a membrane about 30 nm thick. This is the expected dimension for an axon and corresponds in size to the axons we observed by the light microscope (Fig. 61). The cell processes contain dense bodies 50–65 nm in diameter. The cell processes and contained vesicles fulfill the dimensional and histochemical requisites for axon terminals.

D. Conclusions

Our observations show that individual cells undergo the sequence of events previously reported for neuroblasts differentiating *in vivo*. The unequal divisions and cell-process-generation described are typical for neuroblasts and unknown for other cell types, so the observed differentiation was almost certainly that of neuroblast proceeding to neuron and ganglion cells. The neural identity of these cells is supported by the finding of axon terminals in the cultures. These axons probably differentiated *in vitro*, but there is a small possibility they were formed *in vivo* in embryos of advanced age. Regardless of their origin, they show the approximate dimensions of the cell processes observed by the light microscope, thus supporting the axonal expectation.

Acknowledgment. Supported by National Institutes of Health grants A 105038 (R.L.S.), CA 08791 (R.L.T.), the Samuel A. Miller Research Fellowship (R.L.S.), and the Zachary Pitts Research Fellowship (R.L.T.).

III. Cultivation of Insect Endocrine Glands *in Vitro*

Edwin P. Marks

A. Introduction

The possibility that insect hormones can be used as agents for pest control and the abundant evidence that insect glands are active in *in vitro* preparations (Marks, 1970) have led to renewed interest in the question of what happens to insect endocrine glands when they are held *in vitro* for long periods.

Among the earliest studies of endocrine glands maintained *in vitro* are those of Horikawa and Sugahara (1960) who incubated brain-ring gland complexes from pupae of *Drosophila* in nutrient medium containing ^{14}C-thymidine. They found that 7.5 krad of X-rays not only stopped the incorporation of thymidine in those cells responsible for secretion of the molting hormone, but that the hormone effect produced when the ring gland was incubated was also destroyed. OBERLANDER et al. (1965) maintained prothoracic glands of *Samia cynthia ricini* (DONOVAN) with an active pupal brain, and the presence of the brain stimulated RNA synthesis in the cells of the gland. LELOUP and GIANFELICI (1966) and SCHALLER and MEUNIER (1967) used brains from the blowfly *Calliphora erythrocephala* MEIGEN, and from a dragon fly, *Aeschna cyanea* MULLER, respectively, and showed that the neurosecretory cells of brains maintained *in vitro* demonstrated the same staining reactions as those from normal brains. GIANFELICI (1968a, b) maintained the brain-ring complex of *Calliphora* for as long as three weeks *in vitro* and was able to follow the accumulation of granules in the median neurosecretory cells and their eventual storage in the corpus cardiacum. COURGEON (1969) studied the cephalic ganglia and ring gland of *Calliphora* and found that after 8 days *in vitro*, mitoses persisted in the cerebral hemispheres and that the peritracheal cells of the ring gland remained in good condition. The effects of endocrine glands (MARKS and REINECKE, 1965), endocrine gland incubates (MARKS, 1968), and purified hormones (MARKS and LEOPOLD, 1970) on the development of cockroach leg regenerates *in vitro* showed that the gland explants were not only viable but that they interacted and produced various effects on the target tissues.

It is apparent from these findings that insect endocrine glands maintained *in vitro* are able to retain not only their structural integrity but also some of their secretory activities. Therefore, a series of studies of the endocrine glands of the cockroach, *Leucophaea maderae* (FAB.) was undertaken to determine the culture conditions necessary for long-term maintenance, the morphogenetic changes in the explants, and the criteria for assaying the secretory activities that occur in these explants.

B. Methods

In the first of the two culture methods used, one or more explants were placed in Rose multipurpose tissue chambers under dialysis strip, examined by phase-contrast microscopy, and then recorded photographically. In the second method, the explants were placed in culture tubes containing 1 ml of M-18 medium, allowed to sink, and then incubated for various periods. Following incubation, the tissues were removed and examined as whole mounts according to the method of DOGRA and TANDAN (1964).

The M-18 medium has been used successfully to culture nerve tissues (MARKS et al., 1968) and epithelial tissues (MARKS and LEOPOLD, 1970). It supported normal activity in these tissues for as long as 90 days and has maintained cultures of dispersed embryo cells for more than two years. Fetal calf serum was added (5%) and the pH was adjusted to 7.5. The medium was changed whenever the phenol red indicated an acid shift in pH.

C. Results

1. Prothoracic Glands

When single pairs of prothoracic glands were placed in Rose chambers, in nearly every instance the tissue died within a few weeks without any visible sign of activity. When the number of glands was increased to 8 or 10 pairs

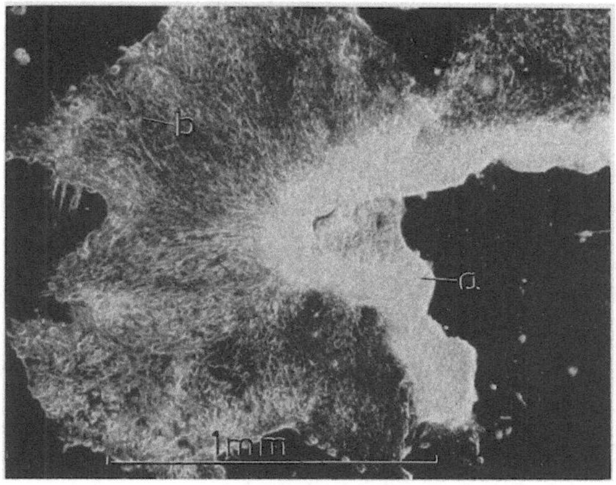

Fig. 63. Outgrowth of cells from a prothoracic gland (a) has formed a monolayer (b); 40 days *in vitro*

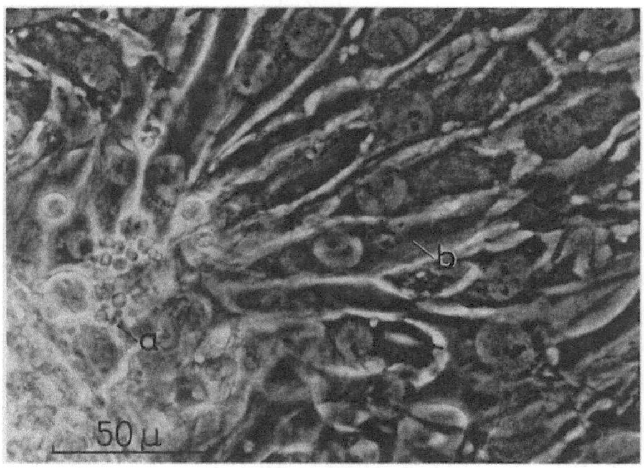

Fig. 64. Migrating cells (b) pour out of prothoracic gland (a) onto the coverslip; 28 days *in vitro*

per chamber, cell migration and monolayer formation with mitotic activity persisted for several weeks. Since the survival of the glands seemed to depend upon a "tissue mass effect", other tissues such as blood clots, brains, and immature leg regenerates were added to single gland cultures; all were effective maintaining mitosis and cell migration in the gland. The most successful preparations consisted of one leg regenerate and two pairs of prothoracic glands placed on either side of the chamber at such a distance that the cell migrants from the various explants did not come into contact. Nearly 100% of these cultures survived. Cell migration usually began about 10 days after explantation, either at the cut ends of the glands or from breaks in the sheath (Fig. 63). The cells, because they lacked phagocytic vacuoles and possessed finely stippled cytoplasm, could be distinguished from the blood cells that occasionally clung to the explant (Fig. 64). In well-defined sheets, the cell boundaries were irregular, and in isolated cells, the cytoplasm was often extended in broad, thin sheets and occasional cells had phase-dark vesicles around the nucleus.

As the cells formed into monolayers, a second type of cell migrated from the gland. This type of cell could be distinguished from the others by the large nuclei and the system of very fine phase-negative tubules that extended throughout the cytoplasm. Perhaps the most striking feature of the cells was their high rate of mitotic activity. In the explants cultured 8 days after molting, this mitotic behavior was apparent as soon as monolayers appeared around the edges of the gland (usually between 14–21 days *in vitro*) and continued for 30–60 days. By selecting an area in the monolayer that showed

Table 9. *Average rate of mitoses and duration of single mitosis in cultures of prothoracic glands*

Days *in vitro*	Cells/field	Average mitoses/ cell hr	Average duration single mitosis (hr)
25	98	1/554	30
35	108	1/333	22
39	120	1/181	49
42	15	1/140	50

mitoses and taking time-lapse photographs of it, it was possible to study the mitotic behavior with some accuracy. The number of mitoses per cell per hour and the average duration of a single mitotic event were measured. The results for four specimens are given in Table 9. Both the mitotic rate and the average duration of a single mitosis varied widely, and there seemed to be no correlation between them. It was apparent from the time-lapse sequences that not all the cells in any one field were involved in mitosis over the period observed. Therefore, the mitotic rate involved not only the rate of division of the active cells but also the relative number of active cells in a given field. Since the average duration of a single mitosis does not seem to vary with age *in vitro*, it may be assumed that the increase in mitotic rate is the result of an increase in the number of dividing cells.

The monolayer around the explants grew rapidly by cell division and by migration from the explant for up to 60 days, but by 90 days, numerous dead and dying cells were apparent in the cultures, and no cultures survived more than four months.

2. Corpus Allatum-Cardiacum Complex

The small size of the corpora allata of *L. maderae* made it impossible to separate them cleanly from the corpora cardiaca to which they were attached and to place enough glands in a chamber to yield a sufficient tissue mass for culturing. Therefore, the cultures were prepared by placing a single allatum-cardiacum complex in a Rose chamber together with an 8-day leg regenerate at a distance sufficient to prevent intermingling of the cell migrants; under these conditions, the cultures survived quite well for up to 90 days. Cells migrated from both the allatum and cardiacum portions of the explant (Fig. 65). The cells migrating from the allatum had a striking resemblance to those migrating from the prothoracic gland, and the cells from the allatum showed the same nuclear dimorphism and frequent mitoses. Pinocytic vacuoles were scattered throughout the cells, and many phase-dark vacuoles were found in

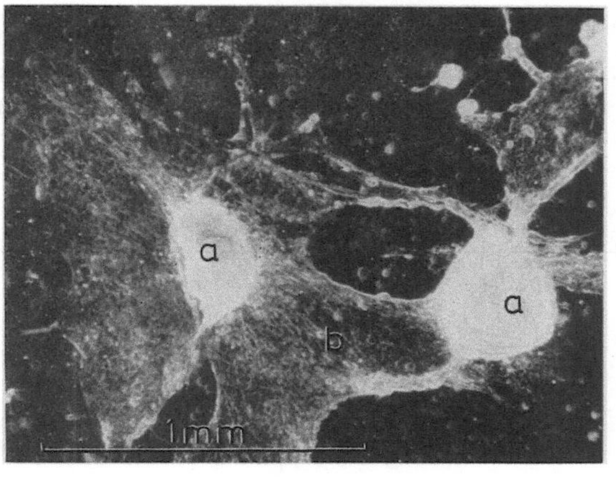

Fig. 65. Two corpus cardiacum-allatum explants (a) are surrounded by monolayer of cells (b) migrating from the corpus allatum; 40 days *in vitro*

clusters around the nucleus and at the periphery of the cell. These differed from the phase-dark vacuoles found in the cells of the prothoracic gland since they were surrounded by a well defined vacuolar membrane. Mitotic activity occurred as early as 25 days *in vitro* (when the cell migrants started to form monolayers) and persisted for 20 to 30 additional days.

Two types of cells emerged from the cardiacum portion of the explant, and these resembled the neurons and glial cells that were reported by Marks et al. (1968) from cultures of the prothoracic ganglion. In older cultures, the migration of these nerve cells was extensive, and frequent contacts occurred between these neurons and the migrating gland cells.

3. Brain-Corpus Cardiacum Complex

When brain explants were placed in Rose chambers, considerable cell migration occurred from the bases of the optic nerves and other areas where the brain sheath was ruptured. Such migrants included glial cells, perineuria, and neurons (Marks et al., 1968). The thickness of the explant and the heavy sheath surrounding it made phase-contrast visualization of the neurosecretory cells and tracts impossible, and any attempt to remove the sheath and flatten the explant resulted in disruption of the nerve tracts and the death of the cells. Therefore, the process of neurosecretion *in vitro* was studied by first culturing brains in tubes and then at intervals fixing and staining them following the method of Dogra and Tandan (1964). In these preparations, the neurosecretory cells and tracts remained visible because of the bright blue stained neurosecretory granules that they contained.

The types of explants treated in this manner were those with corpora cardiaca and allata still attached and those with these organs removed. When the brain, corpora cardiaca, and corpora allata were removed from late-instar nymphs 25 days after molting and stained, the median neurosecretory cells and associated axon tracts were almost devoid of neurosecretory material (Fig. 66), while the fiber tracts nearest the median line of the corpora cardiaca were filled with stainable material (Fig. 67). The corpora allata showed only a few granules around the periphery, and these appeared to lie between the sheath and the cells of the gland. When brains with the corpus cardiacum-allatum complexes removed were incubated for 10 days, considerable neurosecretory material was found in both the median neurosecretory cells and the axon tracts of these cells, which were sharply outlined by the blue-stained granules (Fig. 68). In these preparations, it was apparent that the neurosecretory material was being elaborated by the neurosecretory cells but stored in the fiber tracts and cell bodies since the normal storage depots had been removed. When brains with the corpora cardiaca-allata still attached were treated in the same manner, very little material was present in the neurosecretory cells and fiber tracts, and only a small quantity was found in the corpora cardiaca (Fig. 69). The neurosecretory material that was elaborated by the cells was apparently being moved along the fiber tracts in the corpora cardiaca and then released into the nutrient medium. These results are summarized in Table 10.

Table 10. *The presence of stainable material in brains stained by the method of* Dogra *and* Tandan *(1964)*

Preparation	Days *in vitro*	Amount of stainable material in		
		median neurosecretory cells	fiber tracts	corpora cardiaca
Brain + corpus cardiacum + corpus allatum	0	none	none	heavy
Brain alone	10	heavy	moderate	
Brain + corpus cardiacum + corpus allatum	10	trace	trace	moderate

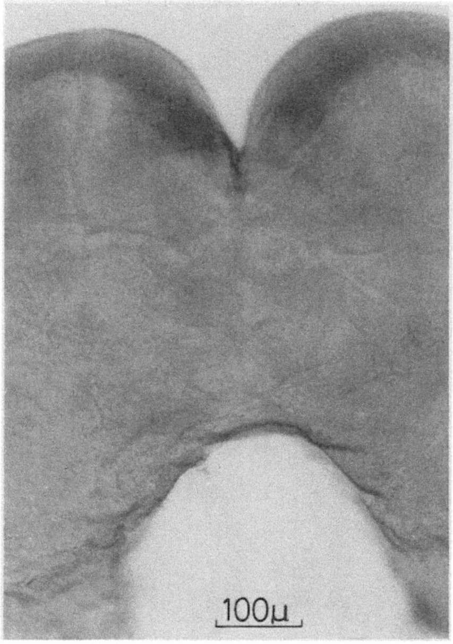

Fig. 66. Mid-frontal area of brain of 25-day nymph prepared by the method of DOGRA and TANDAN (1964). Median neurosecretory cells and fiber tracts are empty of neurosecretory material; 0 days *in vitro*

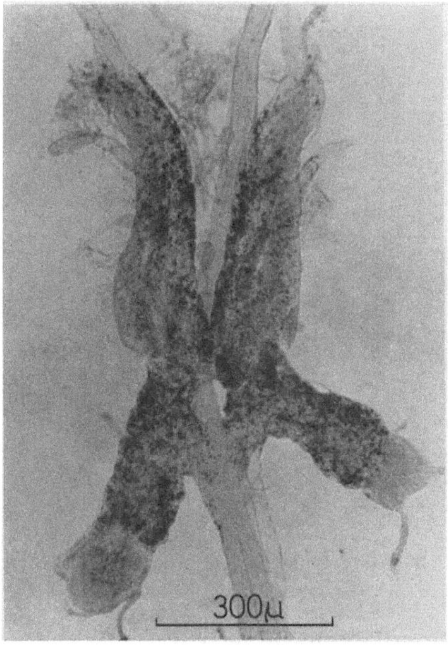

Fig. 67. Corpus cardiacum-allatum complex of 25-day nymph prepared by the method of DOGRA and TANDAN (1964). Stainable neurosecretory material fills the fiber tracts of the corpus cardiacum but is absent from the corpora allata; 0 days *in vitro*

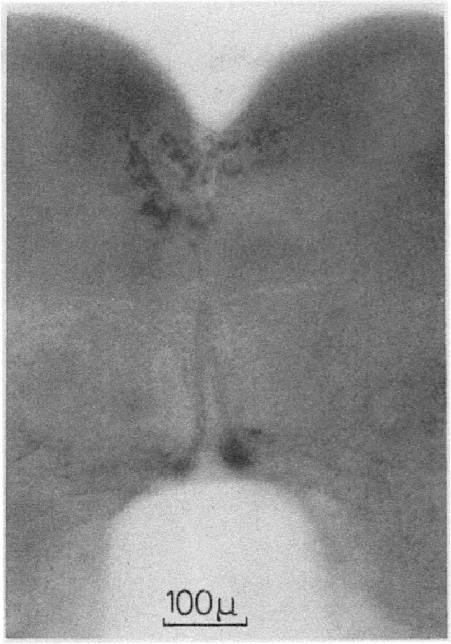

Fig. 68. After maintenance *in vitro* for 10 days, a brain from which the allatum-cardiacum complex had been previously removed shows accumulation of stainable neurosecretory material in the fiber tracts and cell bodies; 10 days *in vitro*

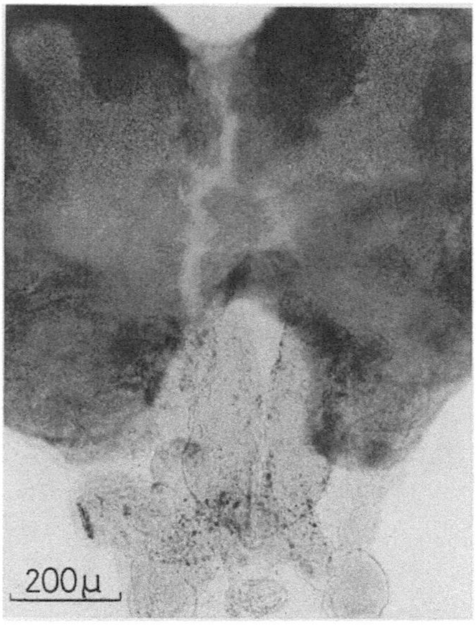

Fig. 69. Brain with corpus cardiacum-allatum attached after 10 days *in vitro*. The neurosecretory cells and fiber tracts are barely visible, and the corpus cardiacum remains partly filled with stainable material

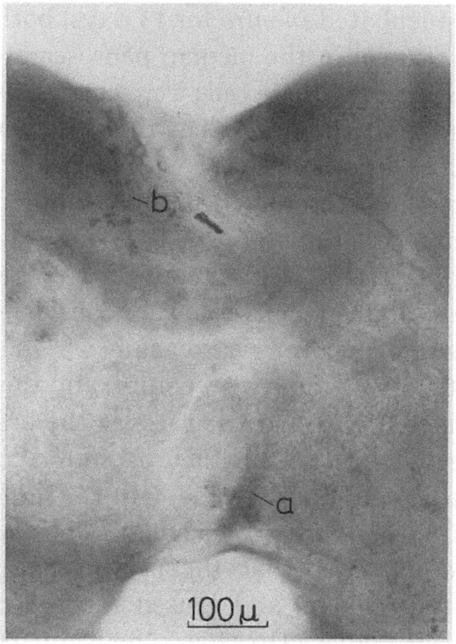

Fig. 70. Brain to which the corpus allatum-cardiacum (Fig. 71) remained attached. Damage to one of the nervi corporis cardiaci resulted in accumulation of stainable material in one fiber tract (a) but not in the other. Associated cell bodies (b) on opposite side (see text) also show increased accumulation of stainable material; 13 days *in vitro*

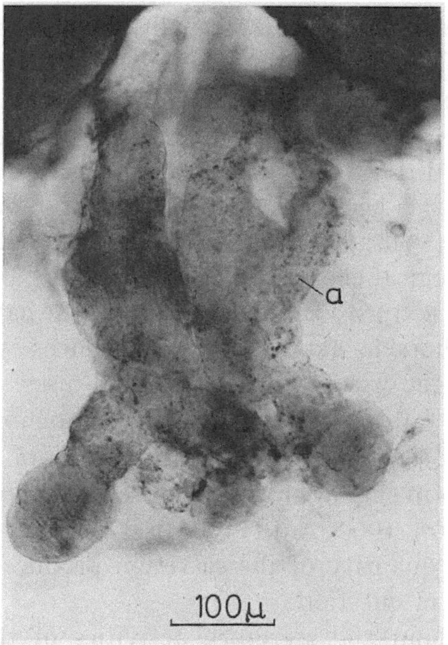

Fig. 71. Corpus allatum cardiacum of brain shown in Fig. 70. Note that the side with the partially blocked fiber tract (a) contains little stainable material; *in vitro* 13 days

When two preparations that included the brain and corpus cardiacum-allatum complex were incubated *in vitro* for 13 days, both showed considerably more neurosecretory material in the median neurosecretory cells than did the 10-day brains. In one specimen, these cells were evenly distributed with 10 cells visible on one side and 8 on the other. The fiber tracts within the brain were all but empty, and the corpora cardiaca were partially filled. In the second specimen, the median neurosecretory cells with stained granules were more numerous and unevenly distributed with 15 on one side and 31 on the other (Fig. 70). When the fiber tracts were examined, the tract on the side of the brain with the fewest stained cells showed considerable accumulation of granules, and the associated corpus cardiacum was nearly empty (Fig. 71). On the opposite side, the fiber tracts were empty, and the associated cardiacum was partially filled. Since the fiber tracts cross in the brain (SCHARRER, 1952), the side with the fewest stained median neurosecretory cells is associated with the partially filled cardiacum, and the side with the largest number of stained cells is associated with the empty cardiacum. This fact is consistent with previous findings and indicates that the nervous corporis cardiaci on one side were injured during removal and caused the neurosecretory material to back up toward the cell bodies. However, they apparently remained intact on the other side, and the neurosecretory material passed into the corpus cardiacum.

D. Discussion

It is apparent that all three of the major endocrine glands of *Leucophaea* survive and continue to function *in vitro* for considerable periods. It is more difficult to determine if anything is actually being secreted by the glands. CLARK and LANGLEY (1963) attempted to link mitotic activity with secretion in the prothoracic gland *in vitro*. HERMAN (1967) concluded that while mitosis and secretory activity may be coincident in some insects, they are probably not directly linked, and thus the presence of mitosis in a gland does not necessarily indicate that it has been activated to secrete. Similar conclusions were reached by SCHARRER (1948; 1964a, b), who found no evidence to link secretory activity with morphological changes in the cells.

Studies (SCHNEIDER, 1964; BURDETT et al., 1968) have indicated that the presence of the prothoracic gland *in vitro* produces well defined morphogenetic changes in target organs such as imaginal discs. However, the deposition of cuticle by the target tissues as a result of incubation in the presence of the gland has not been reported. The stimulation of cuticle deposition by ecdysone or its analogs in similar cultures has been reported by several workers (SENGEL and MANDARON, 1969; AGUI et al., 1969; MARKS and LEOPOLD, 1970). Thus, the nature and quantity of the secretion produced by the prothoracic gland *in vitro* remains in question.

Little is known about the secretory activities of the prothoracic gland *in vitro*, but somewhat more is known about those of the corpus allatum. Studies of several insects (LENDER and LAVERDURE, 1967; ITTYCHERIAH and

STEPHANOS, 1968) indicated that vitello-genesis in cultured insect ovaries can be induced either by adding explants of the brain, corpus cardiacum, and corpus allatum to the culture or by adding synthetic juvenile hormone. Furthermore, RÖLLER (personal communication) recovered and identified juvenile hormone from culture medium in which corpora allata from *Hyalophora cecropia* (L.) had been incubated. It is apparent that corpora allata continue to secrete juvenile hormone *in vitro* at least for short periods.

The evidence that cultured brains continue to supply secretory products rests primarily on the assumption that the neurosecretory granules found in these cells actually represent hormone material. The present study of *Leucophaea* and the work of GIANFELICI (1968a, b) with *Calliphora* show that protocerebral neurosecretory material moves along the axons of the median neurosecretory cells into the corpus cardiacum from which it eventually disappears. Studies using gland combinations with target tissues suggest that at least some of the neurosecretory material actually enters the culture medium (OBERLANDER et al., 1965; MARKS, 1968) and activates the prothoracic glands. The problem of relating the stainable neurosecretory material seen in cultured brains with specific hormones has not yet been solved.

Acknowledgment. The author acknowledges the contribution to this work made by Mr. P. I. ITTYCHERIAH who stained and mounted the cockroach brain preparations.

IV. Production of Human-Mosquito Somatic Cell Hybrids and their Response to Virus Infection

JAMES H. CONOVER, HELEN D. ZEPP, KURT HIRSCHHORN, and HORACE L. HODES

A. Introduction

Somatic cell hybridization techniques have given cell biologists a unique opportunity to obtain specific information concerning cellular function and gene expression. Intra- and inter-species hybrids have been produced by means of biochemical selection (LITTLEFIELD, 1964), viral induction (HARRIS and WATKINS, 1965), or a combination of these methods (MIGGIANO et al., 1969).

In the investigation of specific cellular traits, inter-species hybrids are most frequently chosen because homologous traits in such hybrids can be distinguished more easily from each other, and this possibility becomes greater the more genetically distant are the cell sources. The potential usefulness of species differences was recently demonstrated by the development of immunity against Ehrlich ascites tumor cells by the use of an experimentally produced hamster-Ehrlich ascites hybrid cell line (WATKINS and CHEN, 1969). As another example, GUGGENHEIM et al. (1969) have shown that chick red blood cell—human fibroblast heterokaryons were more effectively protected against vesicular stomatitis virus destruction by both human and chick interferons together than by either interferon alone. These heterokaryons were also much less protected by human interferon than were human fibroblast without chick nuclei.

Investigations concerning immunological factors on tumor regression and the interaction of interferon and viral infection seem to be most advantageously approached through hybrid cells, when the "parent" cells are more phylogenetically remote. Thus, an interphylum hybrid would perhaps offer a better system for such studies than any interspecies hybrid yet available.

B. The Production of Interphylum Hybrids: Cultural Morphology and Evidence of Hybrid Production

We have recently achieved interphylum heterokaryon and hybrid formation from human (HeLa, clone 5 of S3; LEIDY et al., 1959) and mosquito (*Aedes aegypti* L.; SUITOR et al., 1966a) cell lines using UV-inactivated Sendai virus as the fusion-inducing agent.

HeLa cells were maintained in monolayer culture at 37° C in Eagle's Minimum Essential Medium (MEM) (LOCKART and EAGLE, 1959), supplemented with 10% fetal bovine serum (FBS) and antibiotics (100 units/ml penicillin, 100 μg/ml streptomycin). The mosquito cell line was propagated in suspension culture at 26–28° C in GRACE's (1966) Insect Tissue Culture (GITC) medium, supplemented with 10% FBS and antibiotics (Grand Island Biological Co.). In selecting a common culture environment for the eventual co-cultivation of these cell types, parameters of nutritional requirements and temperature were primarily investigated. The insect culture is characterized predominantly by its sucrose content whereas mammalian nutrient media contain hexoses. The human cell line exhibited a moderate to slow growth when cultivated in GITC medium at 37° C. However, little or no growth was obtained when HeLa cells were incubated at 26–28° C in this medium. The mosquito cells could not be maintained in MEM at any of the three temperatures. When equal parts of complete MEM and GITC media were mixed (EPM), mosquito cells displayed moderate growth at 26–28° C and HeLa cells grew at 37° C.

Sendai virus, obtained from Microbiological Associates, was UV-inactivated and the hemagglutinating titer determined by standard techniques (DAVIS et al., 1968). Infectivity titers before and after inactivation were assayed in WI-38 cells.

Fusion of cells was produced in the following manner: HeLa cells were preincubated with 2 μC/ml of tritiated thymidine (H^3T, 1.9 C/mM) for 24 hrs before use, which resulted in the uniform labeling of the HeLa cell nuclei. UV-inactivated Sendai virus in various titers was added to mixtures of mosquito and labeled or unlabeled HeLa cells. Confluent monolayers of HeLa cells were prepared in 30 ml Falcon flasks. Mosquito cells were cultured in the same type of flask and averaged 4×10^5 cells per flask. The HeLa cultures were trypsinized (0.25% trypsin solution, Grand Island Biological Co.) and centrifuged at 250 × g, and the pellet was resuspended in a volume of virus containing the required number of hemagglutinating units. After thorough mixing, the suspension was used to resuspend a mosquito cell pellet of approximately equal volume to the HeLa cell pellet. These pellets were obtained from 2 flasks of HeLa cells and 6–8 flasks of mosquito cells. The mixtures were placed in a 4° C water bath on an automatic shaker and shaken at low speed for 20 min. and then shaken for 60 min. at 37° C. Direct cell smears of virus-treated and untreated (control) cell suspensions were made immediately after shaking at 37° C and cells from H^3T treated cultures were coated with Kodak stripping film for subsequent autoradiographic detection of nuclear H^3T in the fusion products.

Autoradiography of direct smears made immediately after viral treatment and shaking provided conclusive evidence of human-mosquito heterokaryon formation (Figs. 72a and b). Heterokaryons were detected only in the Sendai-treated cell suspensions. Evidence of heterokaryon formation was also obtained directly without the use of H^3T, since the nuclei of both "parents" are morphologically distinguishable. The mosquito cell line is composed of morphologically different cell types, the most predominant being a crescent-shaped cell with very scanty cytoplasm. At either end of these cells are several filamentous processes, which are probably of cytoplasmic origin, and it appears that these processes are the focal point for the virus-mediated cell to cell interaction with the HeLa cells. Initially, in the virus-treated cell mixtures, the mosquito cells appeared to attach to the HeLa plasmalemma by the filamentous processes (Fig. 72c). A second step in the fusion process seemed to involve the apposition of the concave surface of the crescent-shaped mosquito cell to the surface of the round HeLa cell. The last step appears to be the breaking down of intercellular barriers, producing a heterokaryon state.

Each of the Sendai-treated cell suspensions was divided into two equal parts, which were then diluted 25 : 1 with the appropriate culture medium. Half of each suspension was diluted with MEM for culture at 26°, 32°, and 37° C, and the other half was diluted with EPM for culture at the same three temperatures. Since plating efficiency and growth characteristics proved inadequate at 26° and 32°, these temperatures were abandoned. Although cells formed monolayers within one day at 37° C, an additional 3–4 days were required before sufficient cells existed to warrant subculture in either type of culture medium. In general, cells grown in EPM appeared more irregular than those grown in MEM. The gross cellular morphology in each medium, however, readily distinguished the hybrid cells from their "parents". Ten passages of the mixed (hybrid cells + residual HeLa cells) cell populations have been made and the differences in morphology between the MEM and EPM replicate cultures have become more pronounced with each passage.

The line of HeLa cells used is characterized by a range of 40 to 46 metacentric and 5 to 10 acrocentric chromosomes. The *Aedes aegypti* cell line, said to be 32n (n = 3) (GRACE, 1966), was characterized by approximately 96 "micro-chromosomes". Using an improved method of chromosome preparation it was found that the cells possess about 130 micro-chromosomes (Fig. 73b). Most of the mosquito chromosomes give the appearance of having a long and a short arm, connected by a thin chromatin strand (detailed description: AKSTEIN, 1962). Evidence of interphylum hybridization was obtained from chromosome analysis of 36–40 hrs first passage cultures. In Fig. 73a the presence of both human and mosquito chromosomes, which apparently share the same spindle apparatus, is seen. However, incomplete parental complements were consistently observed. In fact, in no instance were complete summation complements found.

Chromosomal analysis of the initially fused cells obtained by adding a mitotic arrester (Velban, Eli Lilly Co.) to one of the replicate primary cultures

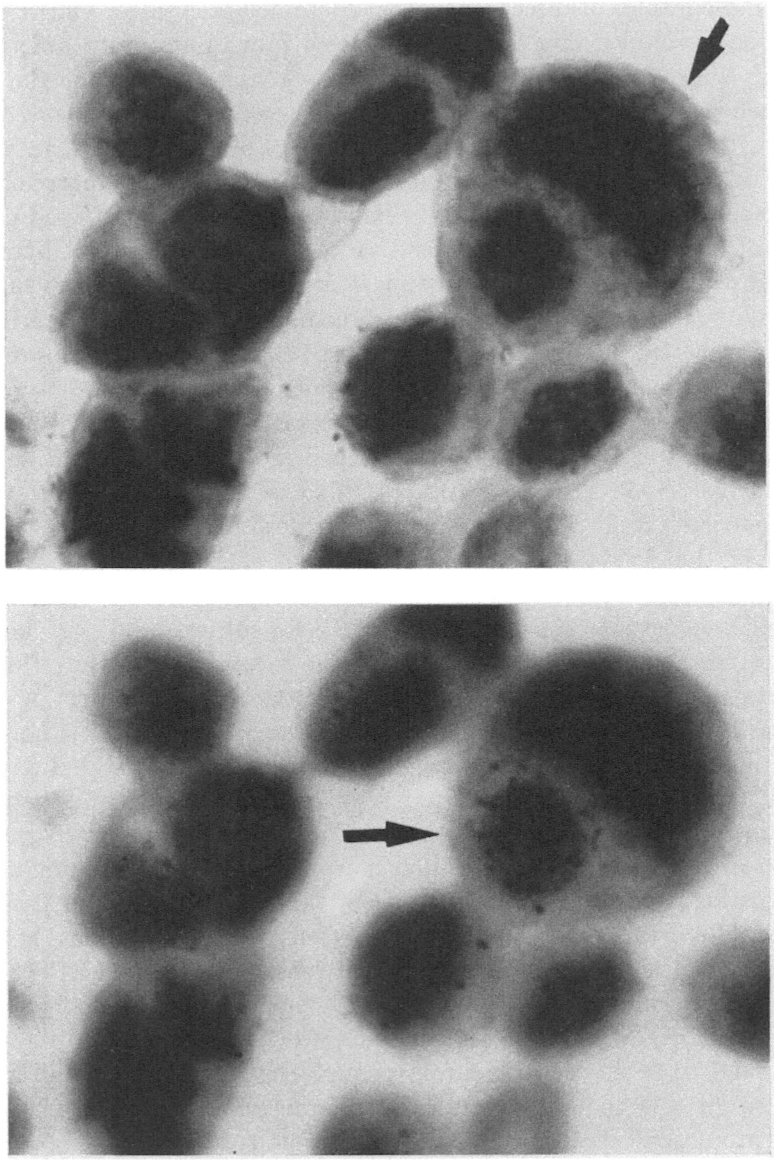

72a and b

at a time immediately following fusion, revealed the presence of both parental types of chromosome. Subsequent chromosomal analysis at each passage also showed both parental chromosome types present within single cells grown in both types of medium. In general, chromosomal analysis at each passage always demonstrated a predominance of human chromosomes, with progressively smaller numbers of insect chromosomes. Also, the number of obvious human-insect hybrids which could be chromosomally identified were much less predominant in the MEM cultures than in the EPM cultures, particularly

Fig. 72. a Human-mosquito heterokaryon—the elongated, crescent-shaped nucleus is characteristic of a mosquito cell. Arrow denotes hybrid cell. b The same field after autoradiography. Note that only the HeLa nucleus is labeled (arrow). c One of the stages in the cell fusion process, showing apparent attachment of crescent-shaped mosquito cell to HeLa plasmalemma via elongated filamentous process. Note that the round HeLa cell is labeled with H^3T, but not the mosquito cell

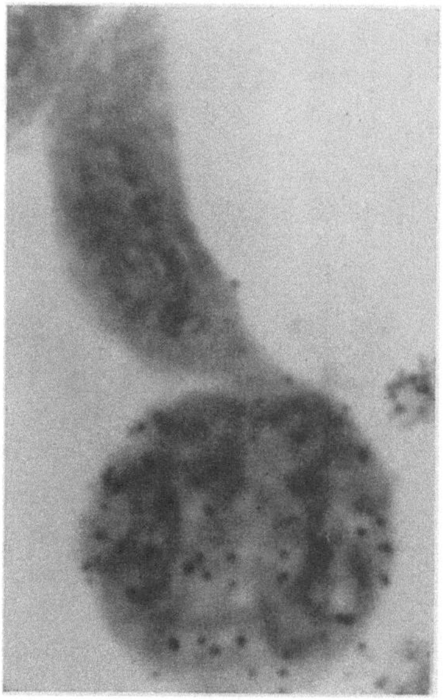

72c

Table 11. *Effect of virus concentration on the character of the heterokaryon produced*

Counts made on autoradiographs of direct smears after viral treatment	Concentration of virus used after ultraviolet inactivation			
	640 HAU[a]	1600 HAU	3200 HAU	4800 HAU
Percent of cells showing heterokaryon formation	10	25	25	—
Percent of binucleate cells containing one HeLa and one insect nucleus	50	55	56	syncytia
Percent of multinucleate cells which were not heterokaryons	15	16	24	syncytia
Percent of multinucleate cells which were heterokaryons	35	29	20	syncytia

[a] HAU = hemagglutinating units.

after the second passage. This trend may be attributed to the selective factors operating in the different culture media. Since *Aedes aegypti* cells do not survive at 37° C or in MEM, the hybrids surviving in MEM must be able to withstand nutrient and thermal selection against them. In addition, residual HeLa cells in MEM tended to overgrow the existing hybrid cells. HeLa cells grow optimally in 37° C in MEM, but are only barely sustained in GITC medium at the same temperature. This is due to the fact that the GITC medium contains sucrose as its primary carbohydrate source—a sugar that

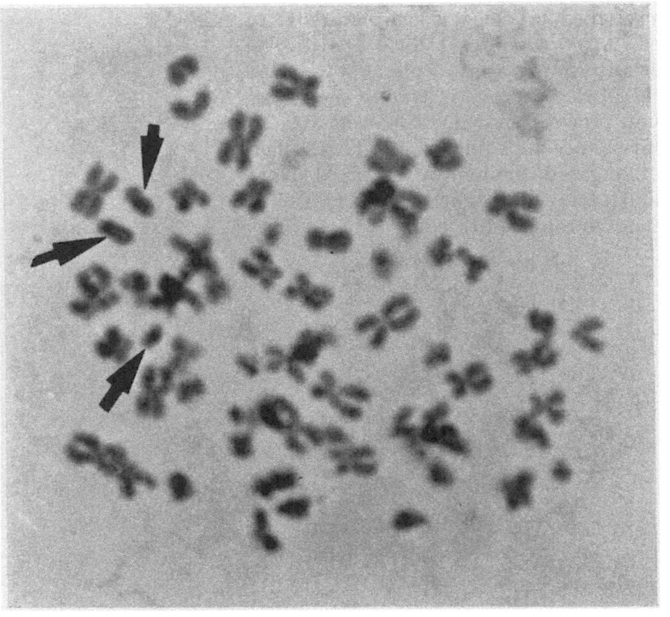

a

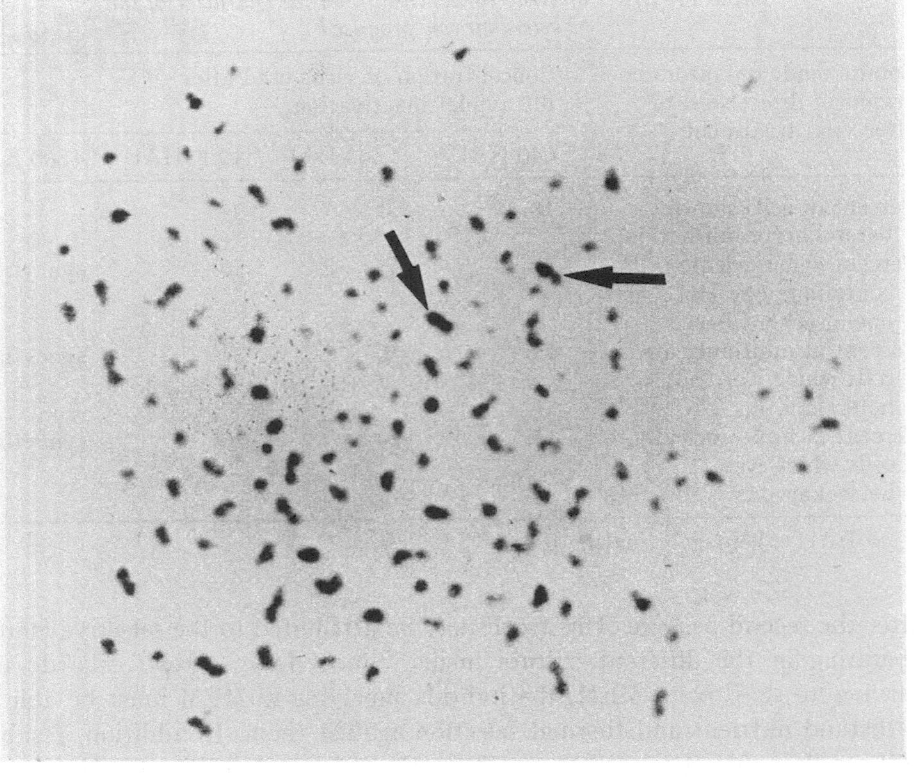

b

Fig. 73. a Chromosome preparation of human-mosquito hybrid cell from a 36–40 hrs first passage culture. Typical mosquito chromosomes are denoted by arrows. b Chromosome preparation of cell line *Aedes aegypti* (L.) which displays approximately 130 microchromosomes. Chromosomes indicated by arrows resemble the insect chromosomes in (a) cannot be metabolized by human cells because they lack the enzyme sucrase. Therefore, the hybrid cells may be selected because of the presence of GITC in EPM.

Heterokaryon formations were consistently obtained when the amount of virus used contained 1600, 3200, or 4800 hemagglutinating units (HAU). Table 11 shows the effect of virus concentration on the character of the heterokaryons produced. Note that the lower three doses seem equally effective, but that 4800 HAU elicited the formation of syncytia. The lower 640 HAU dose, in addition, showed less than a 10% incidence of heterokaryon formation, as compared with 25% for the 1600 and 3200 HAU doses. An occasional heterokaryon was obtained when lesser amounts of the virus were used.

C. Virus Infection of Interphylum Hybrids

Experiments were designed to determine whether the hybrid cells would provide a system for investigation of virus growth and interferon activity. The Mahoney strain of type 1 polio virus was used as the challenging virus. This virus had a titer of $10^{7.5}$ ID_{50}/ml and each of the hybrids, either in MEM or EPM, was challenged with 0.1 ml of 10^{-5} dilution of the virus. As a control, a HeLa cell culture grown in MEM received the same amount of virus. The control HeLa cells were destroyed completely on the third post-inoculum day, whereas the hybrids were not destroyed until day 10. Table 12 shows that the HeLa cell controls consistently produced more virus than did the hybrid. It is possible that the higher titers obtained from the HeLa cells reflected a difference in the number of HeLa cells present in each preparation. Alternatively, the hybrid cells may have been more resistant to virus infection, resulting in longer survival and lower titers.

To determine whether the hybrid cultures produced an interfering agent, an experiment was done as outlined in Table 13. The supernatant media, in which virus had been demonstrated, were heated at 56° C for 75 min. to destroy the infectivity of the virus. None of these fluids showed infectivity or toxicity for HeLa cells. These virus-inactivated supernatant fluids were each allowed to adsorb on six tubes of HeLa cells (after first decanting the culture medium) for 3–4 hrs at 37° C. The fluids were then removed, fresh culture medium added and three of each of the six replicates were challenged with 0.1 ml of a 10^{-5} or 10^{-6} dilution of polio virus. Table 13 shows that the heated supernatant fluids protected the HeLa cells against subsequent challenges with polio virus. The protection produced by media from hybrid culture was more effective than that provided by medium from HeLa cells alone. In fact, the hybrid media provided full protection even if obtained as early as the third day of culture while HeLa medium did this only after 8 days of culture. The interfering agent produced by the hybrid cells has not been characterized.

Investigations of the viral effects on the cultures and the characterization of the nature of the interfering agent are in progress. Pure clones of the human-

Table 12. *Titer of polio virus in the fluid supernatants of HeLa and hybrid cells*

Day after inoculation[a]	HeLa[b]	Hybrid (MEM)[b]	Hybrid (EPM)[b]
3	6.6	5.5	4.7
6	6.6	4.0	2.36
8	7.0	—	—
10	—	3.3	1.5

[a] With 0.1 ml of a 10^{-5} dilution of the virus.
[b] Log_{10} titers of the virus as assayed in replicate HeLa tubes.

Table 13. *Protection of HeLa cells by interfering factors in culture media*[a]

Dilution of virus	Pretreatment of cells with heat-inactivated medium from									
	no pre-treat-ment	HeLa			MEM			EPM		
		day 3	day 6	day 8	day 3	day 6	day 10	day 3	day 6	day 10
10^{-5}	2/3[a]	3/3	2/3	0/3	0/3	0/3	0/3	0/3	0/3	0/3
10^{-6}	3/3	3/3	0/3	0/3	0/3	0/3	0/3	0/3	0/3	0/3
No virus	—	0/3	0/3	0/3	0/3	0/3	0/3	0/3	0/3	0/3

[a] No. HeLa tubes showing cytopathic effect/No. HeLa tubes tested.

mosquito cells would be useful for several types of virus studies, particularly those involving insect-borne viruses affecting humans. The experiments imply that the protective effect (interferon?) is not due simply to contamination of the hybrid cultures by HeLa cells. This finding justifies the belief that the longer survival and lower viral titers in hybrid cultures as compared to HeLa cells alone are due to greater resistance to polio infection of the hybrid cells.

Acknowledgment. This work was supported in part by NIH Grant Nos. 5-R01AI07131 and HD-02552.

V. The Uptake of Free and Serum-Bound Sterols by Insect Cells *in Vitro*

J. L. VAUGHN, S. J. LOULOUDES, and K. DOUGHERTY

A. Introduction

The importance of sterols in insect nutrition has been accepted for some time, and all synthetic diets for insects contain one or more sources of these compounds (SMITH, 1966). The sterols are necessary because, unlike other animals, insects are incapable of synthesizing them from simpler, precursor compounds (CLARK and BLOCK, 1959). It therefore seems likely that insect cells grown in culture also cannot synthesize these vital materials. STANLEY and VAUGHN (1967) considered a marked sensitivity of insect cells to the antifungal agent amphotericin B circumstantial evidence that the cells were dependent on exogenous cholesterol because amphotericin B apparently binds the sterols of the cell membrane and thus alters the physical structure of the membrane (LAMPEN et al., 1965). If the cells had been capable of synthesizing cholesterol the removal of exogenous sterol could have been overcome.

Despite the evidence for the importance of the sterols in insect nutrition, they have received almost no consideration in the development of media for the culture of insect cells. Cholesterol has been added to one medium by GRACE (1958) but he reported that it had no effect on cell growth or survival. In the medium used by HIRUMI and MARAMOROSCH (1964a) for the culture of leafhopper tissue, cholesterol was present in trace amounts as a result of the incorporation of Medium 199 in their formula. Most other minimal insect media contain no cholesterol, and no studies of the utilization of sterols by insect cells *in vitro* have been made.

In contrast, the use of sterols by vertebrate cells has received considerable study. This work has been extensively reviewed by BAILEY (1967) and by ROTHBLAT et al. (1967), and only a brief review will be given here. The source of the sterols in most of the media used for culturing vertebrate cells has been the serum added as a medium supplement. BAILEY (1961) demonstrated that the amount of cholesterol incorporated into cells was related to the cell type and to the source of this serum. For example, cells grown in rabbit serum had a higher cholesterol content than cells grown in adult human serum even though the cholesterol content of the rabbit serum was much less than that of human serum.

Cholesterol is also present in serum as cholesteryl esters that are taken into the cells and hydrolyzed by cellular esterases to free cholesterol (BAILEY, 1967). The amount of esters hydrolyzed varies with the cell type, but generally the uptake of the esters was at a rate considerably lower than that for free cholesterol. If the hydrolyzed esters released more cellular cholesterol than required by the cell, the excess was excreted into the medium (ROTHBLAT and KRITCHEVSKY, 1968).

When relatively small amounts of free cholesterol (non-protein bound) was added to the medium, large increases in cell cholesterol resulted (BAILEY, 1961). Thus, when both serum cholesterol and free cholesterol were present, the free cholesterol was taken up preferentially, but not to the exclusion of serum cholesterol.

In addition, ROTHBLAT and KRITCHEVSKY (1968) reported that the uptake of cholesterol was influenced by factors other than the concentration and the form present in the medium. For example, the presence of various phospholipids markedly reduced the uptake of cholesterol by mouse lymphoblasts. The efficiency of inhibition varied with the phospholipid tested, sphingomyelin was the most efficient followed by lecithin, phosphatidyl ethanolamine, and phosphatidyl serine.

Thus, the use of cholesterol by vertebrate cells *in vitro* appears to be a complex but well regulated system. This complex system occurs in cells which if deprived of an exogenous source of sterols are perfectly able to synthesize all that they require. Since insect cell lines were derived from animals without the capability for sterol synthesis they must either have acquired this ability as a result of mutation or they must depend entirely upon the serum-bound cholesterol. Because of our general interest in improving the media for the culture of insect cells and because of our specific interest in the metabolism of sterols by insects, we undertook a study of the utilization of sterols by insects *in vitro*.

B. Sterol Uptake by Insect Cells

1. Sterol Content of the Medium

The minimal medium used was that formulated by Grace (1962) for his line of *Antheraea* cells and contained no sterols. It was assumed that the cell requirements for sterols had to be filled by the sterols present in the serum supplements. The

Table 14. *Level of sterols present in serum supplements*

Supplement	Sterol content (μg/ml)	ml supplement to 100 ml minimal medium	Sterol level in complete medium (μg/ml)
Fetal bovine serum	305	7	17.7
Bovine serum albumin (35% aqueous)	30.6	10	2.5
Egg ultrafiltrate	0	3	0
Total			20.2

sterol content of each supplement was therefore analyzed separately as follows: The lipids were extracted from each serum with 10 ml of a 3 : 2 mixture of chloroform-methanol (acidified with 0.5% acetic acid, v/v methanol) for each 1 ml of serum. The extracts were then analyzed for sterols on a Barber-Coleman Gas-Liquid Chromatograph[1] (GLC) using three phases, SE-30, QF-1, and NGS (Vanden Heuvel et al., 1961). Quantitation was accomplished on the GLC by comparison with known samples of cholesterol.

The results are shown in Table 14. The principal source of sterol in the medium was the fetal bovine serum; some sterol was found in the bovine serum albumin, but none was found in the whole egg ultrafiltrate. From these data it was calculated that the tissue culture medium contained 20.2 μg/ml of sterol. The level of serum cholesterol (305 μg/ml or 30.5 mg-%) in fetal calf serum was unusually low compared with other animal sera. Bailey (1961) reported that human adult serum contained 218.7 ± 49.6 mg-%, human cord serum 97.4 ± 35.7 mg-%, and adult rabbit serum 85.8 ± 29.5 mg-%.

---

1 Mention of a proprietary item in this paper does not constitute an endorsement of this product by the U.S. Department of Agriculture.

2. Uptake of Serum Sterols by Insect Cells.

The cell line used for the study of serum uptake was the RML-10 strain of the GRACE *Antheraea* line, adapted to grow in hemolymph-free medium (YUNKER et al., 1967). The complete medium consisted of 100 parts of GRACE minimal medium, 7 parts fetal bovine serum, 10 parts whole egg ultrafiltrate and 3 parts 35% aqueous solution of bovine serum albumin. The cultures were grown in Falcon plastic tissue culture flasks. In each experiment, twelve 18 ml cultures were seeded with 100,000 cells/ml and incubated at 26° C for 8 days. The contents of six flasks were pooled to give two replicate samples, and the cell number was determined by counting the cells with a Coulter Counter, Model B. There was an average 10-fold increase in cell number after 8 days.

To assay for sterols, the cells were removed from the growth medium by centrifugation and washed in minimal medium. After the last centrifugation, the cells were resuspended in 1 ml of distilled water, 2 ml methanol was added, and any remaining whole cells were broken up in a Potter-Elvehjem tissue grinder. The cell

Table 15. *Uptake of serum sterols during growth of RML-10 cells*

Fraction	Sterol recovered (μg)			
	0 days	%	8 days	%
Medium	1912.00	99.5	93.1	52.7
Cells	7.84	0.5	83.3	47.3
Total	1919.84	100	176.4[a]	100

[a] Represents 9.1% of sterol present in medium at 0 days.

debris was removed by centrifugation, and the supernatant was analyzed for sterol content on the GLC. The cell-free spent medium was also analyzed for the remaining sterols on the GLC.

The results are shown in Table 15, along with the results of the assay of unused medium and cells equivalent in number to the original inoculum. The cells incorporated an average of only 4.39% of the medium sterols during growth but only 12.45% of original sterols could be recovered from the medium leaving 83.16% unaccountable. The data in Table 15 show an apparent increase in the level of sterol per cell after growth. However, the range of the measurements of the cell inoculum was 0.2 to 0.8 μg/10^6 cells, which approaches the level of sterol per cell found after multiplication. Therefore the increase may not be real.

3. Uptake of 4-^{14}C-Labeled Cholesterol

To follow the sterol metabolism more closely and to determine what happened to the missing serum sterols, we studied the fate of ^{14}C-cholesterol in our cultures.

Because of the difficulty of incorporating the water insoluble cholesterol into the medium in a biologically active form the following technique was used: 4-^{14}C-cholesterol, obtained from New England Nuclear and purified on a Unisil column (DUTKY et al., 1963) to a purity of 99.9% with a specific activity of 288,348 cpm/μg was used. The desired amount of cholesterol was dissolved in a small amount of benzene and coated onto the inside of a test tube. The benzene was evaporated off with nitrogen, and the coated test tube was autoclaved to sterilize the cholesterol. When the tube had cooled, a small amount (1–5 ml) of fetal bovine serum was added to the tube and shaken for 5 to 10 min. on a wrist action shaker. The serum was then transferred to the medium, and the tube was rinsed 2 or 3 times with small amounts of medium. The amount of cholesterol actually transferred to the medium was determined by assaying the radioactivity recovered from the test tube by benzene extraction on a Tri Carb Liquid Scintillation Counter, Model 578 (Packard Instrument Co.). The scintillation solvent was 12.0 g of 2,5-diphenyl-oxazole and 200 mg 1,4-bis (4-methyl-5-phenyl-2-oxazolyl) benzene in 3 liters of toluene. All samples were counted for a sufficient time to give standard errors of less than $\pm 5\%$. When applicable, the counting rates were corrected for self-quenching.

This method generally results in the successful transfer to the medium of 75% or better of the labeled cholesterol, but has several drawbacks (ROTHBLAT and KRITCHEVSKY, 1968), since it may lead to a nonproportionate distribution of the

Table 16. *Uptake of ^{14}C-cholesterol during growth of RML-10 cells*

Component assayed	Radioactivity (cpm/ml)	Cholesterol recovered (μg/ml)	Percent of total cholesterol recovered
Control[a]:			
Medium	58,092	0.2014	100
Test culture after 8 days:			
Medium	46,768	0.1622	83.3
Cells	9,394	0.0325	16.7
		0.1947	100

[a] The control was medium only incubated under the same conditions as the test culture.

sterol among the serum proteins. If this happens and the cells use one of the serum fractions preferentially, the uptake of labeled sterol is affected. These factors must be kept in mind when interpreting the data obtained by this technique.

The results of one such experiment are shown in Table 16. The amount of labeled cholesterol added was 1/100 ($\frac{0.21}{20.20}$ μg/ml) of the amount of serumbound sterol estimated to be in the complete medium (Table 14), and with the assay procedures used, slightly more than 90% of the ^{14}C-cholesterol was accountable. The ratio of labeled cholesterol to unlabeled cholesterol (Table 15) in the cells was 1 : 25 ($\frac{0.0397}{0.9190}$ μg) indicating a preferential uptake of the added cholesterol. To more accurately identify the location of the label the chloroform-methanol extract was fractionated on a Unisil column (DUTKY et al., 1963) to separate cholesterol from its esters and/or its derivatives. The activity of each of these fractions was then determined on the scintillation counter. The results showed that 6.5% of the label was still in the cholesterol and the remainder was in fractions containing more polar compounds. Of the label remaining in the medium, only 12.5% was in cholesterol. Thus, only 10% of the recovered

^{14}C-cholesterol was found in the cells. These findings indicate that considerable alteration of the cholesterol molecule had occurred. The nature of the alteration is currently under study.

C. Discussion

The studies show that insect tissue culture media do contain some sources of sterols which are used by the cells. However, when additional cholesterol was present in the medium as *in vitro* serum-bound sterol this cholesterol was preferentially taken up by the cells. In view of BAILEY's (1961) findings, this may indicate that not all of our cholesterol was bound to serum. We have not as yet demonstrated that the cells cannot synthesize sufficient sterols to meet their needs, but we feel that we have shown that sterol metabolism in insect cells *in vitro* is as complex and as interesting as that already described for vertebrate cells.

VI. Comparative Lipid Analysis of *Aedes aegypti* and Monkey Kidney Cells (MK-2) Cultivated *in Vitro*

HOWARD JENKIN, DEWAYNE TOWNSEND, SHIGERU MAKINO, and TZE-KEN YANG

A. Introduction

Lipid composition of several types of mammalian cells has been reported, such as that of the mouse "L" cell (ANDERSON et al., 1969; BAILEY and MENTER, 1967) and the monkey kidney cell (MK-2) (MAKINO et al., 1970), but the arthropods (BARROSO et al., 1969; AGARWAL and RAO, 1969; FAST, 1966) or the established insect cell lines have not been fully analyzed. Arthropods have been reported as having an active lipid metabolism and appear to depend on fatty acids rather than carbohydrates as a main energy source (O'BRIEN and WOLFE, 1964). However, arthropods do require a source of dietary sterols since they are unable to synthesize the sterol ring from acetate or mevalonate, but are capable of converting sterol to cholesterol (O'BRIEN and WOLFE, 1964).

The purpose of this study was to determine the lipid composition of *Aedes aegypti* cells and to compare their lipids to those of monkey kidney MK-2 cells (MAKINO et al., 1970). Both cell types were cultivated *in vitro* in defined media supplemented with calf serum. The calf serum was the sole source of lipid for both types of cells; therefore, the lipids of calf serum were analyzed as a control and were compared with the lipids of *Aedes aegypti* and MK-2 cells.

B. Methods

Cloned cells of *Aedes aegypti* (SUITOR et al., 1966a) were cultivated in 250 ml Bellco spinner flasks containing 100 ml of GRACE's medium (GRACE, 1962) supplemented with 10% fetal calf serum. The cells were grown for 10 days at 28° C, washed 5 times in Hanks' balanced salt solution (HANKS and WALLACE, 1949) and once in distilled water. In each of two experiments the cells were pooled from 34 spinner flasks.

Monkeys kidney cells (MK-2) (HULL et al., 1962) were cultivated in monolayers using 32 oz prescription bottles containing 35 ml of minimum essential medium (EAGLE, 1959) supplemented with 5% fetal calf serum. MK-2 cells were prepared from 200 bottles.

The lipid methodology employed in this study has been reported by MAKINO et al. (1970).

The methyl esters were analyzed in a Victoreen Model 4000 gas chromatograph (Cleveland, Ohio) using a 10 ft column containing 15% ethylene glycol succinate plus 2% phosphoric acid on Gas Chrom P, 80–100 mesh, with a flame ionization detector. The methyl esters were separated isothermally at 180° C. Weights of lipids were determined with the aid of an electronic Cahn balance, Model G (Paramount, California). All results were reported as the average of three determinations and have a variation of less than 5% from the mean.

C. Results

The lipid content of the two types of cells and calf serum are presented in Table 17. *Aedes aegypti* had a high lipid content (44%) compared to that of MK-2 cells (15%) and of calf serum (4.5%). Glycolipid could be detected only in MK-2 cells (Table 17). The observations from thin-layer chromatography suggested that the glycolipids contain a glycerol backbone rather than a sphingosine base usually found in mammalian glycolipids (MARTENSSON, 1970). Characterization of these glycolipids are in progress. The *Aedes aegypti* cells had a neutral lipid to phospholipid ratio of 1 : 1 which was similar to that of calf serum, but quite different from that of MK-2 cells (1 : 3).

Fatty acids of the total lipids of *Aedes aegypti*, MK-2 cells and calf serum are shown in Table 18. The fatty acid composition of the total lipid of *Aedes aegypti* showed a high percentage of monoenes (61%) and a low percentage of dienes (2%) when compared to MK-2 cells (37% and 15%, respectively) or calf serum (16% and 43%, respectively).

The high amount of linoleic acid (18 : 2, 42.9%) in the total lipid of calf serum (Table 18) was associated with the sterol esters. The sterol esters of calf serum contained 70% 18 : 2 fatty acid (MAKINO et al., 1970). *Aedes aegypti* cells had a high content of free fatty acids (Table 19). Differences were found in the distribution of the neutral lipids (Table 19) and in the composition of the constituent fatty acids of the total neutral lipids (Table 20) from *Aedes aegypti*, MK-2 cells and calf serum. The calf serum differed from *Aedes aegypti* and MK-2 cells in the percentage of triglyceride, sterol esters and sterols.

Table 20 shows the fatty acid composition of the total neutral lipids from *Aedes aegypti*, MK-2 cells and calf serum. The fatty acid composition of the total neutral lipids and the total lipids were similar for each of the samples. The percentage of arachidonic acid (20 : 4) in the total neutral lipid of *Aedes aegypti* (Table 20) was higher than in the total lipid (Table 18) in contrast to MK-2 cells and calf serum both of which contain nearly the same percentage of 20 : 4 fatty acid. *Aedes aegypti* characteristically contained a larger number

Table 17. *Lipid content of Aedes aegypti cells, monkey kidney cells (MK-2) and calf serum*

Lipids	*Aedes aegypti*	MK-2	Calf serum
Total lipids (mg)	90.8	76.3	170.3
Total lipids (% of dry weight)	44.0	14.8	4.5
Neutral lipids (% of total lipids)	49.3	24.1	52.5
Phospholipids (% of total lipids)	50.7	68.2	47.5
Glycolipids (% of total lipids)	0.0	7.7	0.0

Table 18. *Constituent fatty acids from the total lipids*

Fatty acid	*Aedes aegypti* cells	MK-2 cells	Calf serum
14:0[a]	0.1[b]	1.2	0.5
15:0	tr	1.1	
16:0	*4.7*	*15.3*	*12.5*
16:1	*13.5*	*6.3*	*2.0*
18:0	*10.9*	*17.2*	14.0
18:1	*48.0*	*30.5*	*13.8*
18:2	*2.1*	*14.6*	*42.9*
18:3	1.5	1.7	
20:0	2.2	tr	1.1
20:3	1.3		
20:4	6.7	*11.0*	*9.6*
22:0			*3.7*
22:3	1.4	0.4	
22:5	*2.6*	*0.6*	
22:6	*5.3*		

[a] Number of carbon atoms in acid: number of double bonds.

[b] Percentage of fatty acid. Italicized numbers emphasize the differences in composition of the fatty acids.

Table 19. *Distribution of neutral lipids*

Neutral lipid classes	Percentage of neutral lipid		
	Aedes aegypti cells	MK-2 cells	Calf serum
Monoglycerides	*13.6*[a]	4.9	6.6
Diglycerides	3.2	5.4	4.4
Steroids	2.1	*55.5*	16.8
Free fatty acids	*49.5*	4.2	8.6
Triglycerides	20.5	19.2	4.6
Sterol esters	11.3	10.1	56.7

[a] Italicized numbers emphasize the differences in the distribution of the neutral lipids.

of C_{22} fatty acids (10%) whereas the MK-2 cells contained C_{24} fatty acids (5%) in the neutral lipid fraction (Table 20).

The phospholipid composition of *Aedes aegypti* and MK-2 cells (Table 21) was similar in most respects; however, there were two major differences. *Aedes aegypti* contained lysophosphatidylcholine not found in MK-2 cells and MK-2 cells contained phosphatidylserine, cardiolipin, and a glycolipid not found in *Aedes aegypti*. The second major difference between the cells was the

Table 20. *Constituent fatty acids of the total neutral lipids*

Fatty acid	*Aedes aegypti* cells	MK-2 cells	Calf serum
14:0[a]		1.9[b]	1.0
16:0	*5.6*	*17.6*	11.2
16:1	*12.2*	*5.9*	*2.4*
17:0	1.0	*1.7*	
17:1		1.5	
18:0	*9.0*	*14.8*	*6.5*
18:1	*45.6*	*26.2*	*13.0*
18:2	*1.9*	*8.4*	*53.8*
18:3	1.7	0.6	
20:0	2.4	0.8	1.3
20:1			*3.1*
20:3	2.5	1.2	
20:4	7.3	9.7	6.3
22:0	0.5		1.5
22:2		*3.4*	
22:3	1.2	1.0	
22:5	*2.8*		
22:6	*4.2*		
24:0	1.0	1.4	
24:1		1.6	
24:3		2.4	

[a] Number of carbon atoms in acid:number of double bonds.

[b] Percentage of fatty acid. Italicized numbers emphasize differences in composition of the fatty acids.

presence of plasmalogen in the phospholipids of MK-2 cells, but plasmalogens could not be found in *Aedes aegypti* (TOWNSEND, unpublished data).

The fatty acid profile of the phospholipid fraction of *Aedes aegypti*, MK-2 and calf serum (Table 22) were similar to that found in the total lipid except the lower percentage of 18 : 2 fatty acid found in the calf serum phospholipids. The difference in the relative concentration of 18 : 2 fatty acid could be attributed to the high percentage of 18 : 2 acids found in the sterol esters of calf serum. A high percentage of monoenes was found in the phospholipid fraction of *Aedes aegypti*. A larger amount of 16 : 0 and 18 : 2 fatty acids was found in MK-2 cells and calf serum compared to *Aedes aegypti*.

Table 21. *Distribution of phospholipids*

Phospholipid classes	Percentage of total phospholipid		
	Aedes aegypti cells	MK-2 cells	Calf serum
Lysophosphatidylcholine	*7.7*[a]	*0.0*	8.2
Sphingomyelin	15.2	21.4	26.5
Phosphatidylcholine	42.0	36.1	*57.9*
Phosphatidylserine	0.0	*2.5*	0.0
Phosphatidylinositol	4.0	*7.2*	tr
Phosphatidylethanolamine	*31.1*	24.9	*4.7*
Phosphatidylglycerol	0.0	0.0	*2.6*
Cardiolipin	0.0	*7.9*	0.0

[a] Italicized numbers emphasize the differences in distribution of phospholipids.

Table 22. *The constituent fatty acids of total phospholipids*

Fatty acid	*Aedes aegypti* cells	MK-2 cells	Calf serum
14:0[a]		1.3[b]	
16:0	4.8	*16.6*	*16.6*
16:1	*21.0*	*4.5*	*1.0*
17:0		1.2	1.0
18:0	*20.3*	*17.7*	*26.4*
18:1	*33.1*	*28.5*	*15.5*
18:2	*2.4*	*14.2*	*14.1*
18:3	2.2	1.0	
20:0	*2.3*	*0.5*	
20:1			11.9
20:4	*5.0*	*9.3*	*8.6*
22:0	0.9	1.4	*4.9*
22:1	1.0		
22:2	1.8		
22:3	2.2	2.2	
22:5	1.4	0.8	
22:6	1.5	0.8	

[a] Number of carbon atoms in acid: number of double bonds.
[b] Percentage of fatty acids. Italicized numbers emphasize the differences in composition of the fatty acids.

D. Discussion

The comparison of the lipid composition of *Aedes aegypti* and MK-2 cells has revealed that cells grown in similar media can have very different lipid profiles. While this result was to be expected in view of the phylogenetic differences in the cells and the differences in temperature of cultivation, the types of lipids in the medium were expected to be reflected in the lipid composition of the cells. The lipids of calf serum are the sole source of exogenous lipids for both cell types, yet very few similarities were found in the lipid composition between either cell type or the cells and calf serum. The results of these data suggest that very little, if any, intact ester lipid from the medium are incorporated into both types of cells. Most likely the cells may have very active lipases to catabolize complex lipids of the serum. The analyses of the individual calf serum lipids show that it is not selective absorption or selective hydrolysis of a specific lipid class that supplies the fatty acids for the MK-2 cells (MAKINO et al., 1970).

The results of this investigation are in agreement with the hypothesis in the report by O'BRIEN and WOLFE (1964). The high percentage of free fatty acid in *Aedes aegypti* cells (Table 19) support the suggestion that arthropods have a high fatty acid requirement. The low concentration of free sterol in the *Aedes aegypti* (Table 19) probably reflects the inability of arthropods to synthesize the sterol ring. These results also suggest that sterol ester hydrolysis is minimal.

The importance of phospholipids in membrane structure is well documented (VANDENHEUVEL, 1965; BENSON, 1966; GREEN and TZAGOLOFF, 1966). The occurrence of plasmalogens which are usually found in membranes warrants their further study in arbovirus infected cells. MK-2 cells are destroyed during infection with cell lysis, but *Aedes aegypti* cells can maintain a chronic infection. This observation may be related to the absence of plasmalogens in *Aedes aegypti* but their presence in MK-2 cells.

Acknowledgment. The authors wish to thank Dr. E. C. SUITOR, R. GRAYS and F. PAUL for supplying the mosquito cells for the analysis and SAMUEL YU for his excellent technical assistance. This investigation was supported in part by the Public Health Service Research Grant No. HE 08214 from the Program Projects Branch, Extramural Programs, National Institutes of Health and The Hormel Foundation.

VII. Studies Attempting to Elucidate the *in Vivo* Function(s) of Grace's *Antheraea eucalypti* Cell Strains

JOAN F. WHITE

Diapause ovarioles of giant silkmoth pupae contain cells that continuously multiply *in vitro* in GRACE's lines and clones (GRACE, 1962; 1968). Histochemical evidence for the thesis that the continuously multiplying cells originate from a so-called intermediate layer is based mainly on the presence of periodic acid Schiff reaction positive (PAS+) and paraldehyde-fuchsin reaction positive (PAF+) granules found in the migrating cells in primary culture (WHITE, 1965; WHITE and LARSEN, 1966). Such granules are not found consistently in other diapause ovariole cell types in the intact silkmoth ovary. TRAGER (1935), JONES and CUNNINGHAM (1961), MARTIGNONI (1963), KRAUSE et al. (1966) and STANLEY and VAUGHN (1968) have also reported different types of evidence for the intermediate layer being the source of migrating cells in cultures of both pupal and late larval silkworm ovaries.

As previously described for *Samia cynthia* and substantiated for *Hyalophora cecropia* and *Antheraea eucalypti* (WHITE and LARSEN, 1966; SASTRODIHARDJO, 1965, 1967; WHITE, 1969) the diapause ovariole is a tubule invested by a sheath of relatively non-differentiated cells arranged in circular layers between two basal lamina. Beneath the inner basal lamina is a space containing loose, round to amorphous cells arranged irregularly and enclosed centrally by a thick basal lamina surrounding the follicular cells of the developing oocyte. These loose cells form the intermediate layer, and it is the purpose of this paper to present evidence that will suggest that they may perform a variety of functions *in vivo*, particularly during metamorphosis.

Some histochemical evidence has already been given relating to possible functions of the intermediate cells in *Samia cynthia* (WHITE, 1969), and here we will reinforce that evidence with further studies of the histochemical and ultrastructural nature of the intermediate cells of *A. eucalypti* ovarioles *in vivo* and of the *A. eucalypti* cell lines *in vitro*. The presence and changing distribution of PAS+ granules in the intermediate cells during metamorphosis strongly suggested that lysosomes played a primary role during this developmental process. The granules take up neutral red both *in vivo* and *in vitro*, and they contain acid phosphatase in primary culture *in vitro*. These are all criteria used in defining lysosomes (DE DUVE and WATTIAUX, 1966; GAHAN, 1967; STRAUS, 1967; LANE, 1968), and indeed, when we look at the intermediate cells at the ultrastructural level, we see the typical fine structure of the lysosome, a vesicle limited by a unit membrane and containing electron opaque inclusions.

The histochemistry, size and ultrastructure of these granules also change during the metamorphosis of the moth, and this suggests that during adult development lysosomes perform a variety of functions. In the diapause intermediate cell, only a few pale PAS+ granules are found; as development proceeds they become more numerous and larger and more intensely PAS+, and when the adult emerges the intermediate cells themselves dwindle in numbers or entirely disappear. Those that remain contain small, pale or no PAS+ granules (WHITE and LARSEN, 1966). Also, at the ultrastructural level, both intermediate cells and lysosomal numbers decrease and the size and structure of the lysosomes change drastically.

What are the functional possibilities of these intermediate cells *in vivo*? With respect to their lysosomal content, a number of functions can be proposed. DE DUVE and WATTIAUX (1966) have outlined how lysosomes, in the economy of the cell may perform a variety of functions. Those of a heterophagic or heterolytic nature include intracellular digestion, defense against micro-organisms and lysis of obstructing structures. Autophagic or autolytic functions include self-nutrition under unfavorable conditions by means of piecemeal self-digestion, cellular differentiation and metamorphosis, intracellular scavenging as part of self-rejuvenation of long-lived cells, a possible involvement in mitotic divisions and self-clearance of dead cells.

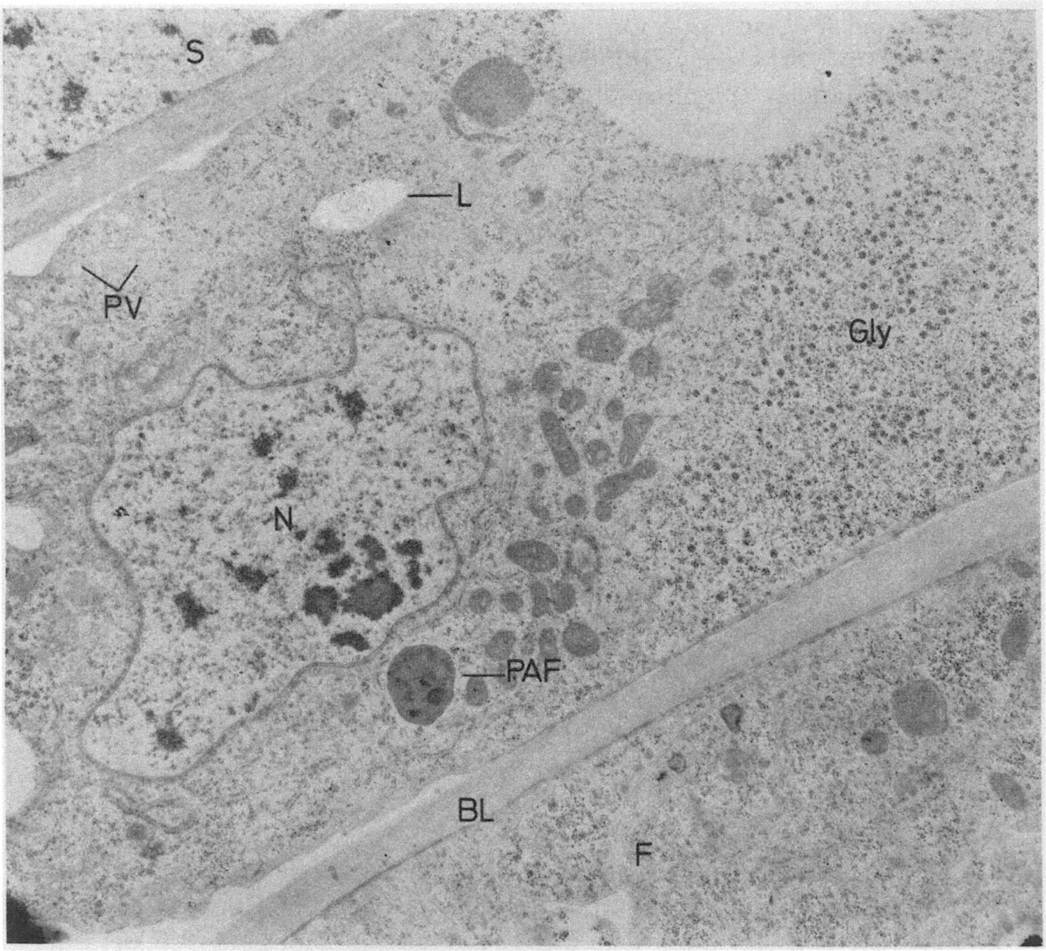

Fig. 74. Sheath, intermediate cell, and follicle cells of *A. eucalypti* diapause ovary. 12000×. *S* ovariole sheath; *L* lysosome; *PV* pinocytic vesicle; *PV'* forming pinocytic vesicle; *N* nucleus of intermediate cell; *Gly* glycogen; *PAF* granule that stains with paraldehyde-fuchsin reagent; *BL* basal lamina of follicle; *F* follicle cell

Let us relate these functions to the intermediate cell in terms of its lysosomal content. In the diapause cell, the number of lysosomes is low, but some are present. Lysosomal enzymes may be stored and then act on exogenous supplies of nutrients brought in by endocytosis (Straus, 1967) from the sheath side of the ovariole (lying in the fatbody surrounded by hemolymph), and pass the transformed products to the follicular side where they leave by exocytosis. Glycogen is present in these diapause cells (Fig. 74) and certainly is processed or exported during metamorphosis. Pinocytic vesicles (Figs. 74 and 75) are seen on the sheath side of the intermediate cells, and vesicles enclosed by membranes and containing flocculent or granular material seem to be moving inward. On the other side of the cell are similar structures, with

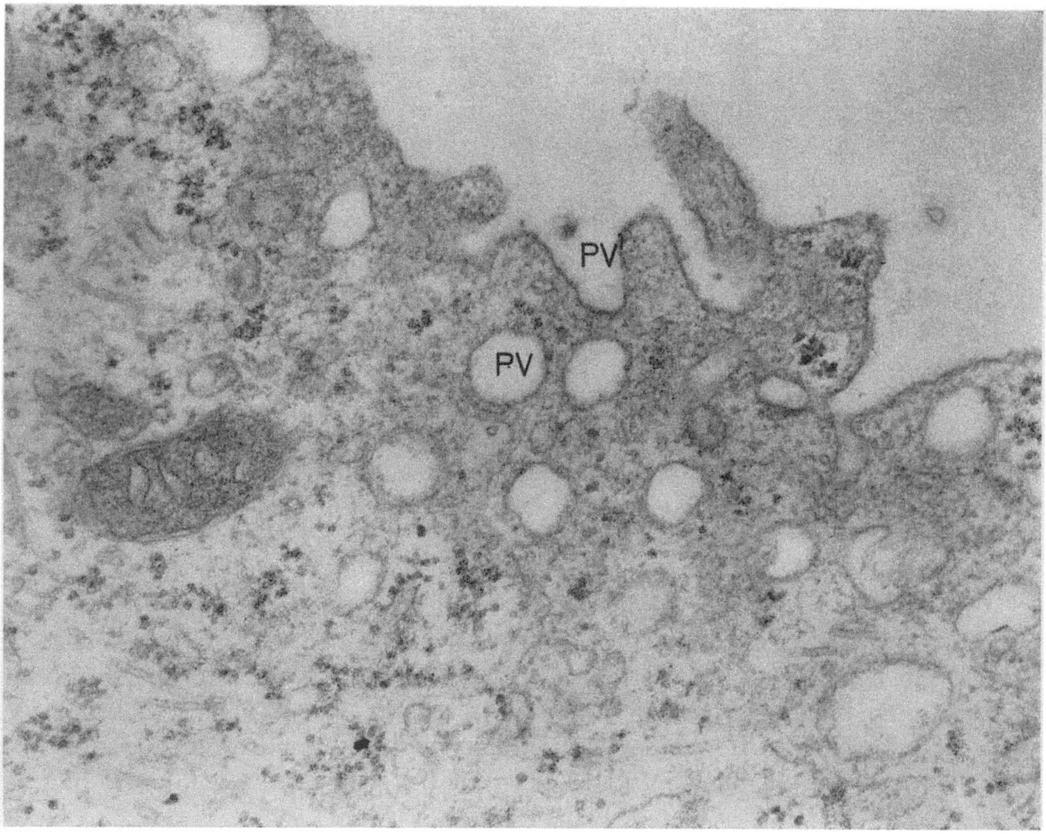

Fig. 75. Higher magnification of region in *A. eucalypti* intermediate cell. 60000 ×. See Fig. 14

both particulate and amorphous material between the cell and the follicular basal lamina. A case for the reverse process of transport can also be made—wastes might be passed out of the oocyte to the intermediate cell for processing.

That the intermediate cell defends itself against virus infections is suggested by the fact that GRACE's cell lines have a very low infectivity (GRACE, personal communication; SUITOR, 1966b); they are loaded with lysosomes which contain enzymes capable of acting on viral coats and nucleic acids (DAVID-FERREIRA and MANAKER, 1965). Viral particles are found in very few intermediate cells *in vivo*.

Self-nutrition or storage is evident from electron micrographs showing mitochondria and remnants of other cells organelles in lysosomes, where the state of breakdown varies (Fig. 74).

Preparation for differentiation and metamorphosis *per se* is not likely in the case of the diapause intermediate cell, for most of them show post-lysosomes or residual bodies and eventually disintegrate. They may, by exocytosis, pass on active enzyme molecules to other differentiating systems in either direction through extracellular spaces. Differentiation of both muscle cells in the ovariole sheath (WHITE and LARSEN, 1970) and vitellogenesis in the oocyte (ROTH and PORTER, 1964; STAY, 1965) proceed during metamorphosis, and the intermediate cell may play a part in these processes.

Intracellular scavenging as part of the self-rejuvenation of long-lived cells is probable and is suggested by the fact that the intermediate cells are long-lived in culture. After the cells leave the ovariole, some "transform" and become mitotically active. These may be the ones which contain larger populations of lysosomes at the time, of the type that store and release the enzymes known to be implicated in cell division (DE DUVE and WATTIAUX, 1966; DAVIES and WEISSMANN, 1969) and initiating the mitotic cycle.

In the economy of the organism we have discussed some functions of lysosomes concerned with autonomous cell life. Similar functions concerned with autonomous cell life are repeated in the economy of the whole organism. Programmed cell breakdown during metamorphosis occurs in many tissues (LOCKSHIN, 1964, 1965, 1969; SCHIN and CLEVER, 1965), and during vitellogenesis (STAY, 1965) and the thinning of the sheath, the intermediate cell may play a part. The breakdown of secreted macromolecules, for example, those in the basal laminae of the ovariole, is known to occur during metamorphosis (STANLEY and VAUGHN, 1968). The fact that the intermediate cell can migrate *in vitro* through two basal laminae composed of muco-polysaccharide is evidence that lysosomal enzymes (glycosidases) may aid the cell in invading the PAS+ material.

Conferring immunity, scavenging and detoxication are also possibilities. In the latter case, it is of interest that GRACE and MITSUHASHI (p. 108) noticed marked differences between clones in their toxicity to rotenone; and I have found, in preliminary observations, that the clones show marked differences histochemically in their lysosomal populations.

In primary cultures of *A. eucalypti* ovarioles grown on Araldite, pinocytic vesicles were found which may form lysosomes and further aid in the "transformation" of some of the intermediate cells.

The formation and "activation" of lysosomes in the intermediate layer of the ovariole could conceivably be stimulated by ecdyson, for in diapause intermediate cells numbers of PAF+ granules occur, and their relative numbers and staining intensity go through cyclic changes during adult development (WHITE and LARSEN, 1966). In injured diapause silkmoths, the intermediate cells show increased numbers of PAF+ granules (WHITE, unpublished) which suggests that injury activates hormone uptake. Investigations are in progress to ascertain whether injury affects the lysosomal population of the intermediate cells, for preliminary observations suggest that there is an increase in the number of mitoses in this cell after injury. In electron micrographs the hormone material must be masked, but some cells show vesicles and granules (Fig. 74) of the size and density of the neurosecretion in other cells of the moth; and its initial presence and ultimate disappearance from the intermediate cells during metamorphosis suggests that it is being used as development proceeds. The titer of ecdyson during metamorphosis correlates with the number and staining intensity of the PAF+ granules *in vivo* during development (VAN DER KLOOT, 1955; KARLSON and SHAAYA, 1965; WILLIAMS, 1968) and the PAS+ granule population also increases steadily.

In primary cultures of ovarioles, the PAF+ material is found in migrating cells in *small* granules which, over a few days of culture, become paler and eventually the cells contain only PAS+ granules which contain acid phosphatase and take up neutral red. In the cell lines of GRACE, there seem to be two types of granules also, but the smaller granules' intensity of staining with the PAF+ reagent is very much diminished, and in some cells there are no PAF+ granules at all, only those which are PAS+, acid phosphatase+ and which take up neutral red.

In the intermediate cells we may have a system which responds to hormonal control through the formation of lysosomes, which then, in turn, respond to different environments in a variety of ways. Briefly outlined, the hormone could effect the formation of lysosomes by entering the cell, changing the permeability or pore size of the nuclear membrane (ITO and LOEWENSTEIN, 1965) to act on the genetic material. The GERL apparatus would then operate to form lysosomes containing enzymes specific for these cells. Release or storage of enzymes would subsequently depend on the cell's environment.

WIGGLESWORTH (1963) observed during insect metamorphosis the enlargement of epidermal and ventral muscle nucleoli and the appearance of RNA in the cytoplasm, and he proposed that ecdyson "activated" mitosis. The biochemical effects of ecdyson on RNA synthesis have long been known as a concomitant of metamorphosis (KARLSON and PETERS, 1965). Injury to diapause insect is thought to release a hormone which acts at the cellular level to cause an increase in RNA synthesis (BARTH et al., 1964; BERRY et al., 1967; CONGOTE et al., 1969). Whether the intermediate cell *in vivo* is capable of producing a hormone is not known, but the diapause cell *in vivo* may be a very sensitive responsive target to ecdyson and *in vitro* descendents of these cells may respond to conditions which can be brought about by hormones.

In conclusion, it is proposed that the intermediate cell of the silkmoth ovary is very versatile, having the ability to function in a number of ways, depending on the milieu in which it finds itself, and reacting to that environment by virtue of its multi-functional lysosomal content.

Acknowledgment. Some of this work (with *Cynthia* and *Cecropia*) was performed in collaboration with Dr. J. R. LARSEN (Department of Entomology, University of Illinois, Urbana) and Dr. S. SASTRODIHARDJO (Department of Biology, Bandung Technological Institute, Bandung, Java, Indonesia) as well as with Dr. GRACE and Dr. B. FILSHIE of the Division of Entomology, C.S.I.R.O., Australia.

VIII. The Effects of Insecticides on Insect Cells Grown *in Vitro*

T. D. C. GRACE and J. MITSUHASHI

A. Introduction

In 1958, one of us (GRACE) placed some DDT crystals of high purity into several primary cultures of insect cells and was amazed to see that the cells multiplied and the tissues remained healthy, despite the fact that many of them were lying on or alongside the DDT crystals. The medium was removed after about 6 days, a series of dilutions made and six mosquito larvae were added to each dilution. At a dilution of 1/1 000 the larvae died within 17 hrs and at dilutions below about 1/250 the larvae were killed in 20 min. to 1 hr. There was no doubt, therefore, that the medium contained DDT in solution and there was every reason to believe also that the DDT would have been taken in by the cells. Although the results were not very encouraging, the opportunity came in 1968 to investigate the effect of 20 insecticides on the growth of cells from the established cell lines of the moth *Antheraea eucalypti* and the mosquito *Aedes aegypti*. No attempt was made to determine the mode of action of the insecticides.

B. Materials and Methods

In most instances the insecticides, which were of high purity, were dissolved in ethyl alcohol and a very fine dispersion produced by adding 1.0 ml of sterile insecticide to 9.0 ml of insect culture medium. The insecticides isolan, dimetilan, sodium arsenate, 4,6-dinitro-o-cresol (DNOC), nicotine sulphate and lead arsenate were dissolved directly into culture medium and sterilized through a 0.45 μ pore size filter (Millipore Corp., Bedford, Mass.). All the dispersions were prepared just before use because most produced deposits if they were kept for long. All dilutions were made with insect culture medium.

The effect of the insecticide on the cells was measured by determining whether the growth rate of the cells was changed. The cells were cultured in petri dishes and three replicates were made of each dilution tested. The cell density in each culture was calculated from the recorded number of cells, the area of the bottom of the culture dish and the amount of fluid, and was expressed as number of cells per ml. The cells were maintained for either 4 or 6 days without a change of medium and counts were made every 2 days. When each experiment had been completed the culture medium was tested for insecticidal activity. Final instar larvae of *Aedes aegypti* were used as test animals. The media were diluted with distilled water and 10 larvae placed in each experimental group. The mortality after 24 hrs was recorded. Freshly prepared insecticide solutions were also tested in the same manner to examine whether or not the culture medium had altered the activity of the insecticides during the experiment.

In the experiments where both the mosquito and moth cells were tested, the results were similar.

The effect of ethyl alcohol on cell growth was determined before the experiments were done. The highest concentration used—1%—had no effect on growth, and a slightly stimulative effect was obtained with 0.5% aqueous ethyl alcohol. Above 2% cell growth was impaired and all the cells were killed within a short period at 4.0% or higher.

C. Results

1. Effect of Insecticides on Insect Cells

Organophosphates. The cells were killed by diazinon and malathion at 112 µg and 123 µg/ml, respectively, and at higher concentrations. In the media containing either of these insecticides at these concentrations droplets were deposited and the cells, which did not adhere to the glass, had many droplets on their surfaces. The cell growth in the medium containing malathion at 12.3 µg/ml was not affected until the 4th day, after which the growth rate was markedly decreased. This drop in cell growth after the 4th day was examined more closely by growing cells in media containing malathion at concentrations between 12.3 µg and 123 µg/ml. The higher the concentration of malathion the larger was the drop in the number of live cells after the 4th day, which suggested that malathion or its toxic metabolites may have accumulated in the cells but killed them only after the concentration in the cells had reached a critical level.

Carbamates. Isolan and dimetilan had some effect on cell growth at concentrations between 10^2 and 10^4 µg/ml. Nevertheless the cells continued to multiply in media containing 10^2 isolan and 10^3 dimetilan. The cells became granulated after 4 days and the cell population decreased slightly in the medium containing isolan at 10^3 µg/ml. In media containing either insecticide at 10^4 µg/ml, some of the attached cells became detached after the 2nd day and only a few surviving cells were found by the 6th day.

Carbaryl killed the cells at 10^2 µg/ml. It is interesting that in the presence of this insecticide at 10^2 µg/ml, inhibition and death of the cells was quite sudden, whereas at 10 µg/ml or less, there was no effect.

DDT and its derivatives. At the concentration of 1 µg/ml neither pp'-DDT, pp'-DDD, pp'-DDE, nor methoxychlor had any effect on cell growth. Some cells however became granulated. At 10 µg/ml most of the cells became granulated and growth was slightly impaired. Although methoxychlor had less effect than the others, the decrease it produced in the growth rate was significant ($P=1\%$). At 10^2 µg/ml concentrations of these insecticides, the media in each instance were turbid, especially those containing DDT, in which needle-like crystals were deposited. At this concentration the cells died in all media.

Cyclodienes. Aldrin and dieldrin had no effect on cell growth at the concentration of 1 µg/ml, but at 10 µg/ml of aldin the cells did not adhere to the glass and started to degenerate soon after the experiment began. Dieldrin at this concentration, however, did not prevent the cells from multiplying even up to the 6th day. However, most of the cells became granulated, and some cells produced blebs. The media containing either aldrin or dieldrin at 10^2 µg/ml were turbid and deposited crystals, and the cells died very quickly.

At concentrations of 1 µg and 10 µg/ml of lindane, there was no effect on the growth rate of the cells. The medium was turbid and some crystals were formed at 10^2 µg/ml, but some cells survived the whole experimental period.

Arsenicals. Sodium arsenate decreased the population density at concentrations of 10^2 μg/ml or higher. At 1 μg/ml and 10 μg/ml the cells multiplied until the 4th day, after which the growth rate decreased markedly.

Lead arsenate, which is virtually insoluble in both water and alcohol, had no effect on the cells at a concentration of 10^2 μg/ml, although particles of lead arsenate covered the bottom of the vessels. The cells remained apparently healthy although many of them had obviously phagocytosed the particles.

DNOC. Up to a concentration of 10 μg/ml there was no decrease in growth. At 10^2 μg/ml the medium was quite yellow and cell growth was retarded, but some cells survived for up to 4 days.

Rotenone. Rotenone had the most effect on the growth of the cells. At a concentration of 10^{-3} μg/ml there was some suppression of growth and granulation of the cells. At 10^{-2} μg/ml and higher concentrations there was increasingly significant suppression of growth.

Nicotine. Nicotine sulphate had very little effect on the growth of the cells. In medium containing 1 mg/ml in which the smell of nicotine was quite noticeable, the growth rate was suppressed to a small extent, but the cells continued to multiply.

Pyrethrins. Media containing 10^2 μg/ml of either pyrethrins or allethrin were turbid and killed all the cells in a short time. These insecticides impaired growth slightly at 10 μg/ml, but neither insecticide had an inhibitory effect on cell growth at a concentration of 1 μg/ml or lower.

Piperonyl butoxide. This substance showed an inhibitory effect on cell growth at concentrations of 40 μg/ml or higher. To examine the synergistic action of pyrethrins and piperonyl butoxide on the growth rate the following experiment was done: Media were prepared which contained (1) pyrethrins 2.5 μg/ml and piperonyl butoxide 10 μg/ml; (2) pyrethrins 5 μg/ml and piperonyl butoxide 20 μg/ml; and (3) pyrethrins 10 μg/ml and piperonyl butoxide 40 μg/ml. In medium (1) no inhibitory action would have been expected unless there was synergistic action. Without synergistic action the effects of media (2) and (3) should have been about the same as those of pyrethrins at 10 μg and 20 μg/ml or those of piperonyl butoxide at 40 μg and 80 μg/ml. The results showed that the effect of the presence of both chemicals was more than additive in medium (1). However, the growth rate in medium (2) did not differ significantly from that of cells in either 10 μg/ml of pyrethrins or 40 μg/ml of piperonyl butoxide. Similarly, no significant difference was obtained in the growth rate of the cells between medium (3) and either the media containing 20 μg/ml pyrethrins or 80 μg/ml piperonyl butoxide. The results, in effect, show that synergism occurred at lower concentrations of both chemicals, but not at higher concentrations.

2. Relative Susceptibilities of Cells and Mosquito Larvae

With 13 of 19 insecticides tested the cells were able to survive in concentrations 100 to 10000 times higher than those killing last instar *A. aegypti* larvae. In medium containing sodium arsenate or DNOC, the toxicity was higher for the cells by a factor of 10 than for the larvae. The toxicity of rotenone was much higher, by a factor of 10000 for the cells than for the larvae. Piperonyl butoxide showed a higher toxicity for mosquito larvae than might have been expected as it is not, by itself, used as an insecticide.

Five insecticides, malathion, carbaryl, pp'-DDT, pp'-DDD, pp'-DDE and lindane showed some interaction with the medium during the six days incubation of the cells, which resulted in a reduction of toxicity compared with freshly prepared dispersions. Most of the insecticides were, however, stable in the medium.

D. Discussion

Most of the insecticides had relatively little effect on the survival and multiplication of insect cells grown *in vitro* in this and related studies (MITSUHASHI et al., 1970). The principal exception was rotenone in which the effective inhibiting concentration was comparable with that of a biologically active substance such as colchicine. The differences between rotenone and the other insecticides may have been due to its highly specific mode of action. Rotenone acts on the point of coupled oxidation of reduced nicotinamide adenine dinucleotide and flavoprotein (FUKAMI, 1961) and if this system is blocked a critically important energy source is eliminated. Although this system has not been demonstrated in cultured insect cells, its existence is a possibility.

The organophosphates and carbamates are known to be cholinesterase inhibitors. No information is available on the presence of cholinesterase in the insect cell lines, but the moth cells were derived from ovary tissue and cholinesterase is mostly distributed in nervous tissue and it is unlikely that the enzyme is present in the cells, or, if present, that the cells depend upon cholinesterase for any vital function. In either instance, the ineffectiveness of the organophosphates and carbamates is understandable and their inhibitory action at high concentrations may be due to a mechanism other than cholinesterase inhibition.

pp'-DDT, lindane, the cyclodienes, nicotine and the pyrethrins are said to be neurotoxins, although their mode of action is not thoroughly understood. Their effects on cell growth at high concentrations should probably be considered as different from their mode of action in intact insects, where they act at much lower concentrations on specific nerve sites.

Arsenicals and DNOC are believed to act mainly by uncoupling oxidative phosphorylation, and their mode of action is somewhat similar to that of rotenone. If the arsenicals or DNOC were taken into the cells and acted by inhibiting the respiratory system, the energy source should have been eliminated. In the present study, they were ineffective compared with rotenone and their mode of action cannot be explained on this basis. In the case of arsenate poisoning the toxic effect is believed to be due to the reduction of arsenate to arsenite. If this is the case, the rate of reduction in the cells must be very slow, and perhaps this is the explanation why the toxic effects in the sodium arsenate medium became evident only after 4 days.

Some insecticides have little toxicity until they are converted to highly toxic materials after entry into the body. The possibility of explaining the low toxicity of the insecticides on the cultured cells from this point of view should be kept in mind. Malathion, for example, is said to be a rather poor inhibitor of cholinesterase *in vitro*, but is converted to malaoxon which is an active anticholinesterase. In the present study, high doses of malathion inhibited cell growth, but it is not known whether malathion was converted to malaoxon in the cells.

The fact that the mosquito larvae were generally much more sensitive to the insecticides than the cells seems curious at first, but it is less surprising if the mode of action of insecticides is considered. To interfere lethally with any one of a large number of vital processes in a complex, highly integrated organism is one matter and to kill one of its component cell types is another. Although the mode of action of most insecticides is still not well understood, it is possible to generalize that most insecticides appear to interfere with the function of highly organized tissues and organs more readily than they act on the basic functions of most of the general somatic cells.

Chapter 4

Arthropod Tissue Culture in the Study of Arboviruses and Rickettsiae: A Review

CONRAD E. YUNKER

In 1959 GRACE noted a decided time lag between early attempts to culture insect tissues *in vitro* and the development of techniques that would permit long term survival or growth of insect cells and tissues under these conditions. Lack of these techniques was undoubtedly responsible for the nearly complete absence of studies on multiplication of arthropod-borne vertebrate viruses and rickettsiae in arthropod tissue and cell cultures during the first four decades of insect tissue culture, 1916–1956.

A. Arbovirus Studies[1]

1. Primary Arthropod Tissue and Cell Cultures

(Table 24)

A noteworthy exception to this inactivity was TRAGER's (1938) demonstration of the multiplication of WEE virus in surviving tissues of *Aedes aegypti*. After 28 days in tissue culture the virus had multiplied 100,000 times, providing rare evidence that a vector-borne arbovirus is capable of multiplying in cells of its arthropod host.

Despite the significance of this experiment similar studies on arboviruses in surviving insect tissues were not reported again until 1956, when PRICE observed that JBE virus lost its infectivity for a time when grown in mosquito tissue culture. This finding supported the hypothesis that, in the mosquito vector, the virus undergoes an eclipse phase similar to that described for many animal viruses or bacterial viruses in their respective host cells. Unfortunately, details of methods employed or the species of mosquito used were not given, so the significance of this work remains moot.

In 1959, HAINES reported a large series of attempts to cultivate tissues from *Aedes aegypti in vitro*. When larval midgut cells were exposed to EEE virus, enhanced virus survival over a 5 day period could not be demonstrated, although cell survival and some degree of development were seen. Subsequently, JOHNSON (1969) showed that if primary tissue cultures are prepared from minced larvae, EEE virus, as well as VEE virus, will readily grow in cells of this species.

1 See Table 23 for key to virus names.

Table 23. *Key to viruses discussed in this review*[a]

Abbre-viations	Virus	Group[b]	See Table
AHS	African Horse Sickness	AHS	27
BAND	Bandia	QAL	26
BEB	Bebaru	A	26
BUN	Bunyamwera	BUN	25
CE	California BFS-283	CAL	25
CHP	Chandipura	PIR	26, 27
CNU	Chenuda	KEM	25, 26
CHIK	Chikungunya	A	25—27
CHIT	Chittoor (Batai)	BUN	26, 27
COX	Coxsackie B5	(+)	26–27
CR	Cowbone Ridge	B	26
CTF	Colorado Tick Fever	(—)	24—27
CV	Cache Valley	BUN	25
C5581	(Unnamed; tickborne)	(—)	26
DEN-1-4	Dengue-1, 2, 3 and 4	B	26, 27
DHO	Dhori	(—)	26, 27
EEE	Eastern Equine Encephalitis	A	24–27
EH	Edge Hill	B	26
EMC	Encephalomyocarditis	(+)	24—27
FAR	Farallon	HUG	26
GAN	Ganjam	(—)	26, 27
MHE	Mouse Hepatoencephalomyelitis	(+)	26, 27
HS	Herpes Simplex	(+)	25
ITQ	Itaqui	C	25
JA	Johnston Atoll	QRF	26
JBE	Japanese B Encephalitis	B	24—27
JUN	Junin	TAC	25
KFD	Kyasanur Forest Disease	B	24, 26, 27
KOK	Kokoberra	B	26
KSO	Kaisodi	KAI	26, 27
KUN	Kunjin	B	26
LGT	Langat	B	24—27
LCM	Lymphocytic Choriomeningitis	(+)	24
LI	Louping Ill	B	24
MAC	Machupo	TAC	25
MTB	Marituba	C	25
MET	Metucare	(—)	25
MOD	Modoc	B	25, 26
MVE	Murray Valley Encephalitis	B	26
ND	Newcastle's Disease	(+)	24
NYM	Nyamanini	(—)	25
OMSK	Omsk Hemorrhagic Fever	B	24
PHL	Phlebotomus Fever	PHL	25
POL	Poliovirus	(+)	24, 26, 27
POW	Powassan	B	24, 25, 27
PSU	Pseudorabies	(+)	24, 25

[a] Adapted from McLean (1969).

[b] (+) = Not an arbovirus; (—) = Ungrouped arbovirus; HUG = Hughes group; KEM = Kemerovo group; PIR = Piry group; SIM = Simbu group.

Table 23 (Continued)

Abbre-viations	Virus	Group	See Table
QAL	Qalyub	QAL	26
QRF	Quaranfil	QRF	25
RB	Rio Bravo	B	25
RSSE	Russian Spring-Summer Enceph.	B	24
SAT	Sathuperi	SIM	26
SF	Semliki Forest	A	24, 26, 27
SFN	Naples Sandfly Fever	PHL	26
SFS	Sicilian Sandfly Fever	PHL	26
SH	Snowshoe Hare	CAL	25
SIL	Silverwater	KAI	26
SIN	Sindbis	A	24, 26, 27
SLE	St. Louis Encephalitis	B	24—27
TAH	Tahyna	CAL	25
TBE	Tickborne Encephalitis	B	24
TCR	Tacaribe	TAC	25
TRB	Tribec	KEM	24
TUR	Turlock	TUR	25
URB	(Unnamed; tickborne)	(−)	26
VAC	Vaccinia	(+)	24
VEE	Venezuelan Equine Encephalitis	A	24, 26, 27
VSI	Vesicular Stomatitis (Indiana)	VS	24, 26
VSNJ	Vesicular Stomatitis (New Jersey)	VS	26
WEE	Western Equine Encephalitis	A	24, 25
WN	West Nile	B	24, 26, 27
YF	Yellow Fever	B	24—27

Cultures of surviving, explanted tissues from the ticks *Dermacentor pictus* and *Ixodes ricinus* were found suitable for substantial growth of EEE virus (REHACEK and PESEK, 1960). REHACEK (1965a, b) observed that primary cultures of growing, dispersed tick cells were satisfactory systems for the propagation of numerous viruses. In these studies and another (REHACEK, 1963) viruses of the serological subgroup A and most tick-borne members of subgroup B (except LGT) readily multiplied in tick cells *in vitro*. Mosquito-borne viruses of the B subgroup also multiplied, but not to the extent to which tick-borne entities did. Most non-arboviruses failed to multiply in tick cell culture, but an exception was LCM virus, which multiplied well. None of the viruses produced obvious cytopathic effects (CPE) in the host cells. It was repeatedly noted that growth of some viruses was obtained from very dilute virus suspensions, in some cases so dilute that corresponding inocula failed to infect intracerebrally inoculated suckling mice or HeLa cell cultures. The last observation was substantiated by the demonstration that primary tissue cultures of *Hyalomma asiaticum* ticks were at least 6 times more susceptible than chick embryo cells to infection by small amounts of TBE virus (REHACEK and KOZUCH, 1964). Because of these findings, these same authors (1969) used

Table 24. *Growth of viruses in primary tissue and cell*

Investigator	Arthropod species	Type of culture
TRAGER (1938)	*Aedes aegypti*	Surviving larval and pupal tissues
PRICE (1956)	"Mosquito"	"Tissue culture"
HAINES (1959)	*Aedes aegypti*	Larval midgut cultures
REHACEK and PESEK (1960)	*Ixodes ricinus* and *Dermacentor pictus*	Fat body, hypodermis tissue cultures
REHACEK and PESEK (1960)	*D. marginatus*	Fat body, hypodermis tissue cultures
REHACEK and PESEK (1960)	*D. marginatus* and *I. ricinus*	Fat body, hypodermis tissue cultures
REHACEK and PESEK (1960)	*Periplaneta americana*	Fat body, hypodermis tissue cultures
REHACEK (1963)	*D. marginatus*, *Hyalomma asiaticum*	Dispersed cell cultures
PELEG and TRAGER (1963a and b)	*A. aegypti*	Imaginal discs in hanging drops
REHACEK and KOZUCH (1964)	*H. asiaticum*	Dispersed cell cultures
REHACEK (1965a and b)	*H. dromedarii*	Dispersed cell cultures
REHACEK (1965a and b)	*H. dromedarii*	Dispersed cell cultures
REHACEK (1965a and b)	*H. dromedarii*	Dispersed cell cultures
YUNKER and CORY (1967)	*D. andersoni*	Primary tissue cultures
PELEG (1968a)	*A. aegypti*	Dispersed embryonic cells
FUJITA et al. (1968)	*Culex molestus*	Primary ovarian cultures
REHACEK and KOZUCH (1969)	*H. dromedarii*	Dispersed cell cultures
REHACEK et al. (1969)	*H. dromedarii*	Dispersed cell cultures
JOHNSON (1969)	*A. aegypti*	Minced larval cultures
JOHNSON (1969)	*A. triseriatus*	Minced larval cultures

[a] See Table 23 for key to virus names.

primary tick cell cultures to isolate TBE virus from samples of rodent bloods and ticks collected in natural foci of the virus. Five viral strains were isolated in tick tissue cultures from 187 samples tested. Identical results were obtained with chick embryo cultures, however, which probably means that the infected samples contained significant amounts of virus. The authors believed that the use of tick cell cultures to isolate viruses is at least equivalent to the use of chick embryo cell cultures (until that time the most satisfactory method). Tribec virus, which has serological affinities to the Kemerovo subgroup of *cultures from arthropods*

Virus[a]	Results
WEE	5 dex virus increase
JBE	Loss of infectivity
EEE	Survival; no growth
EEE	4 dex increase
EEE	Survival; no growth
ND	Enhanced survival; no growth
EEE	Virus inactivation
TBE	1-4 dex virus increase
WN	Sustained virus growth
TBE	Minute amounts of virus detected
EEE, WEE, SF, SIN, TBE, LI, RSSE, LCM	4–5 dex virus increase
JBE, SLE, WN, YF, KFD, LGT, LI, OMSK, POW	2–3 dex virus increase
EMC, POL, VSI, VAC, ND, PSU	No multiplication
CTF	5 dex virus increase; sustained virus proliferation
EE, SF, WN	Sustained virus growth
JBE	Virus growth
TBE	Virus isolated from nature in tick cell culture
TRB	4 dex virus increase; growth in cytoplasm
VEE, EEE	Virus growth
VEE	Virus growth

arboviruses, multiplied well in primary, dispersed cell cultures of *Hyalomma dromedarii*, providing additional evidence of its status as an arbovirus (REHACEK, RAJCANI and GRESIKOVA, 1969). Viral antigen was confined to the cytoplasm of infected cells, as shown by the fluorescent antibody technique.

In this decade also, PELEG and TRAGER (1963a, b) demonstrated that WN virus will proliferate markedly in primary cultures of imaginal discs of *Aedes aegypti*. Cultures of *Philosamia cynthia* gonads, however, failed to support WN virus growth. Later, PELEG (1968a) reported that SF and EEE multiplied in primary cultures of *A. aegypti* embryos. No indications of CPE were seen in any of these experiments with WN virus. Although occasional alterations of embryonic cells after infection with SF or EEE were noted, the authors did not attribute these alterations to infection by the viruses. A titrable EEE hemagglutinin was detected.

CTF virus multiplied markedly in primary cultures of metamorphosing nymphal tissues of its vector, the tick *Dermacentor andersoni* (YUNKER and CORY, 1967). A $4^1/_2$ dex[2] increase was detected by the fourth or fifth week after cultures had been inoculated. In cultures prepared from ticks that had been infected by feeding, maximum virus production (a 5 dex increase) occurred on the second week after preparation of the cultures. Virus titers of ticks which imbibed the same infected blood meal and were held alive as controls remained relatively constant from feeding until molting, illustrating a significant difference in virus supporting ability between intact vectors and primary cultures of their tissues. (A similar quantitative difference between virus growth in mosquitoes and mosquito tissue cultures had already been noted by TRAGER [1938] for WEE virus.) Because these tick tissue cultures continued to yield CTF virus for extremely long periods, an effort was made to determine if some strain characteristics were altered by prolonged growth in tick cells. However, after 98 and 124 days in tick cells, the Florio strain of CTF virus retained its virulence for adult mice inoculated intracerebrally and average survival time of infected suckling mice remained unchanged.

Primary ovarian cultures from *Culex molestus* mosquitoes were used to propagate two strains of JBE virus (FUJITA et al., 1968). Multiplication was demonstrated by viral growth curves, successful serial subcultivation in *C. molestus* cells, and specific immunofluorescence in infected cells. As previously noted by REHACEK (1965a) for TBE virus grown in tick cells, JBE virus could only be detected in the cytoplasm of the mosquito cells. No differences in viral growth patterns were detectable between a strain that had undergone 200 mouse brain passages, and a low passage strain. It was suggested that the virus' affinity for insect tissues is not altered by adaptation to the rodent host.

2. The *Antheraea eucalypti* Cell Line

(Table 25)

In 1966, SUITOR demonstrated that GRACE's *Antheraea eucalypti* cells would support the growth of JBE virus. Virus titers increased 10–100 times over starting amounts after 3 or 4 weeks incubation. The cells used were of a substrain capable of being maintained on a medium containing hemolymph from a distantly related species of moth. However, virus growth occurred only when congeneric hemolymph was added to the culture medium, leading SUITOR to believe that the physiological state of the cells determined their capacity to support JBE virus growth.

2 Dex: an order of magnitude expressing a number as its logarithm to the base 10 (HALDANE, 1960).

In 1967, CONVERSE and NAGLE recovered YF virus from GRACE's lines of *Antheraea eucalypti* and *Aedes aegypti* cells for 8 and 5 days, respectively, after inoculations. Because inocula contained larger amounts of virus than those recovered, no net virus increase was demonstrated. However, virus decreased exponentially in cell-free controls and, after inoculation of the cultures, an initial drop in titer was followed by 2–3 dex higher virus yields, indicating that virus growth probably did occur in these cells.

Table 25. *Virus studies in* GRACE's *Antheraea eucalypti cell line*[a]

Investigator	Virus	Results[b]
SUITOR (1966b)	JBE	1–2 dex virus increase; no CPE
CONVERSE and NAGLE (1967)	YF	2–3 dex virus increase; no CPE
YUNKER and CORY (1968)	CV, JBE, SLE, VSI, YF	3.7–6.0 dex virus increase; no CPE
YUNKER and CORY (1968)	BUN, CE, SH, TAH	Enhanced virus survival or some growth; no CPE
YUNKER and CORY (1968)	CNU, CHIK, CTF, EEE, EMC, HS, ITQ, JUN, LGT, MTB, MET, MOD, NYM, PHL, POW, PSU, QRF, RB, TCR, TUR, WEE	No virus growth evident
JUSTINES (unpubl.)	MAC	No virus growth evident
YANG et al. (1969)	VSI	3 dex virus increase; no CPE; virus penetrates cell, cytoplasm, replicates near plasma membrane

[a] See Table 23 for key to virus names.
[b] CPE = Cytopathic effects.

At that time we were investigating the ability of the *Antheraea* cell line to support the growth of a number of arboviruses, including YF, and observed a 6 dex increase of that virus in these cells (YUNKER and CORY, 1968a). As did CONVERSE and NAGLE, we used a subline of *Antheraea* cells that had been adapted to grow in the absence of hemolymph (YUNKER, VAUGHN and CORY, 1967) and peak titers that we obtained were about equal to those observed by CONVERSE and NAGLE.

In our study also, SUITOR's demonstration of JBE virus growth in *Antheraea* cells was confirmed. Three other arboviruses also multiplied to relatively high titers (3–6 dex) in these cells: SLE, CV, and VSI. A lower but significant increase in titer was seen for BUN and SH viruses. The latter, a member of the California subgroup, showed a 3.4 dex increase after an initial drop in titer. Two related viruses, TAH and the BFS 283 strain of CE, were sustained by the *Antheraea* cells, but failed to show a significant increase in titer over the amounts inoculated. Viruses that failed to propagate or to survive longer than corresponding controls included all members tested of subgroups A, C and Tacaribe, members of subgroup B that are tick-borne or of unknown vector affinities, eight ungrouped arboviruses and three non-arboviruses. With regard to the negative results obtained here with two members of the Tacaribe subgroup, it has recently been shown that a third member, MAC, which is the agent of Bolivian hemorrhagic fever, also failed to grow in a hemolymph-free subline of *Antheraea* cells (G. JUSTINES, personal communication). Thus, it appears that the *Antheraea* cell line is a selective system for the growth of viruses and may aid in the identification and classification of arboviruses.

YANG, STOLZ and PREVIC (1969) grew VSI in a hemolymph-free subline of *Antheraea eucalypti* cells and studied virus-cell dynamics by means of plaque assay in L cells, immunofluorescence, and electron microscopy. By use of specific antiserum, they demonstrated that presence of infectious virus in these cultures was due to synthesis of new virus and not to spontaneous release of the infective virus from cellular receptor sites to which virus had attached during adsorption. This work also showed that the virus was able to penetrate host cells during the one-hour adsorption period. Immunofluorescence reactions were used by YANG et al. (1969) to study a previously suggested hypothesis (YUNKER and CORY, 1968a) that morphologically different cell types of the *Antheraea* cell line may have different capabilities for supporting the growth of virus. Virus antigen was synthesized in only a portion of the cell population, but no great differences among morphological types of cells in their ability to synthesize this antigen were detected. As noted for TBE and JBE viruses in arthropod cell cultures (REHACEK, 1965a; FUJITA et al., 1968), VSI virus antigen was confined exclusively to the cell cytoplasm. Electron-microscopic examinations suggested that virus formation takes place at the host-cell plasma membrane or, less frequently, at intracytoplasmic vesicles. Marked CPE were not seen.

3. Mosquito Cell Lines

(Tables 26 and 27)

A new era for vector tissue culture research began in 1966, when GRACE reported the establishment of a stable cell line from larval *Aedes aegypti*. The following year development of additional lines of mosquito cells from larval *A. aegypti* and *A. albopictus* was announced (SINGH, 1967). Subsequently, PELEG (1968b) established in continuous culture a third line of embryonic *A. aegypti* cells. These mosquito cell lines were immediately used in arbovirus studies, those of SINGH finding widespread popularity because of independence from any insect hemolymph requirement.

Aedes aegypti cell lines (Table 26). As noted above, CONVERSE and NAGLE (1967) reported a 2 dex increase in titer of YF virus in GRACE's cell line after an initial drop in titer following inoculation of large amounts of the virus.

Table 26. *Growth of viruses in Aedes aegypti cell lines*[a]

Investigator	Growth detected	Growth not detected	Remarks[b]
CONVERSE and NAGLE (1967)	YF		
REHACEK (1967, 1968a)	JBE, KUN, MVE, WN	BEB, EH, KOK, SF, SIN	
REHACEK (1968b)	JBE, MVE		Persistent infection; loss of virulence
REHACEK (1968b)	JBE, KUN, MVE		Superinfection
FILSHIE and REHACEK (1968)	JBE, MVE		Virus in ER; CPE with JBE
SINGH and PAUL (1968a and b)	CHP, CHIK, WN	CHIT, COX, DEN, DHO, EMC, GAN, MHE, JBE, KFD, KSO, POL, SAT, SIN	No CPE
BUCKLEY (1969a)	CHIK, EEE, SF, SLE, VEE, VSI, VSNJ, WN, YF	BAND, CNU, CR, CTF, C5581, FAR, JA, LGT, MOD, QAL, SFN, SFS, SIL, URB	No CPE
PELEG (1968b)	EEE, SF, WN		No CPE
PELEG (1969a)	SF, WN	EMC, POL	Arbovirus RNA, but not picornavirus RNA, infective
PELEG (1969b)	SF		Persistent infection; only few cells infected

[a] See Table 23 for key to virus names.
[b] CPE = Cytopathic effects; ER = endoplasmic reticulum.

The following year, REHACEK (1968a, b) reported that GRACE's *A. aegypti* cells supported growth of four B subgroup arboviruses, JBE, WN, MVE, and KUN. All of these, presumably, are mosquito-borne. Two additional viruses of the B subgroup and three of the A subgroup, all also presumably mosquito-borne, failed to grow in these cells. With one exception, no adverse effects of the viruses on the cells were apparent; virus particles were present in the endoplasmic reticulum of degenerating cells infected with JBE virus (FILSHIE and REHACEK, 1968). Later, REHACEK (1968c) described the course of persistent infections of *A. aegypti* cells with MVE and JBE viruses. After 210 days and 21 *in vitro* passages these viruses became less virulent for subcutaneously inoculated mice. In addition, cultures persistently infected with these viruses could be superinfected with KUN virus.

SINGH's *A. aegypti* cell line supported the growth of WN, CHIK, and CHP viruses (SINGH and PAUL, 1968a, b) but was refractory to a number of other viruses, including some that grew well in *A. albopictus* cells. This refractoriness was considered remarkable in view of the wide range of viruses that *A. aegypti* mosquitoes are capable of transmitting (SINGH and PAUL, 1968b). However, CHP virus grew markedly better in *A. aegypti* cells than in *A. albopictus* cells.

Recently, BUCKLEY (1969a) reported attempts to infect SINGH's *Aedes* cells with 23 arboviruses. Seven mosquito-borne viruses of subgroups A and B and both VSI and VSNJ viruses grew in *A. aegypti* cells. VSI multiplied to higher titers here than in *A. albopictus* cells. Thirteen other viruses of various groups, including tick-borne entities, failed to grow in *A. aegypti* cells. CPE were not seen.

PELEG (1968b) reported marked growth, in his line of embryonic *A. aegypti* cells, of 2 arboviruses of the A subgroup, EEE and SF, and one of the B subgroup, WN. Thus, this cell line differs from GRACE's line of larval cells from the same species in its ability to support growth of SF virus and from SINGH's line in its susceptibility to EEE virus. Here, also, CPE were not evident in infected cells. Later, PELEG (1969a) attempted to infect his *A. aegypti* cells with RNA from EMC and POL, two picornaviruses to which the cells were known to be refractory. For comparison, infectious RNA from SF and WN were also tested. Whereas the cells inoculated with the arbovirus RNA yielded viable virus progenies identical to the donors of the RNA, the cells failed to produce viruses after inoculation with RNA derived from the picornaviruses. Finally, PELEG (1969b) reported long-term, inapparent infection of the *A. aegypti* cell line with SF virus. It was demonstrated that only a small proportion of the total cell population was involved in virus production. Although gross CPE were not apparent, a decline in virus-producing cells concomitant with a drop in virus titer suggested destruction of the infected cells. The involvement in virus production of only a small proportion of cells has also been noted for MVE and JBE in *A. aegypti* cells by FILSHIE and REHACEK (1968) and for VSI in *Antheraea eucalypti* cells by YANG et al. (1969).

Aedes albopictus cell line (Table 27). In the last few years, the contributions of SINGH and co-workers to today's topic have been impressive. Not only was GRACE's unique success in establishing mosquito cell lines quickly repeated, but this feat was performed without using insect hemolymph (SINGH, 1967). Subsequently, the growth of certain culicine-borne, as opposed to tick-borne, arboviruses was reported (SINGH and PAUL, 1968a and b), CPE and a carrier state for some of these viruses in the *A. albopictus* cell line were described (SINGH and PAUL, 1968a and b; BANERJEE and SINGH, 1968; PAUL et al., 1969), and the sensitivity of *A. albopictus* cells to infection with CHIK, WN, JBE and DEN-2 viruses was demonstrated (PAUL and SINGH, 1969). Finally, employing these findings, it was shown that DEN viruses of all 4 serotypes may be isolated from human sera or mosquitoes in *A. albopictus* cell cultures, that the type of CPE produced afforded a rapid means of preliminary differentiation of the serotype and that although neutralization tests in mosquito cell cultures were of no use in identifying isolates, complement-fixation (CF)

Table 27. *Growth of viruses in* Singh's *Aedes albopictus cell line*[a]

Investigator	Growth detected	Growth not detected	Remarks[b]
Singh and Paul (1968a and b)	AHS, CHP, CHIK, CHIT, DEN-1, 2, 3, 4, GAN, JBE, SAT, SIN, WN	COX, DHO, EMC, MHE, KFD, KSO, POL	CPE with JBE, WN, DEN-1, 2, and 4
Banerjee and Singh (1968)	CHIK, JBE, WN		Initial CPE with JBE and WN; otherwise inapparent infection
Paul and Singh (1968)	CHIK, JBE, WN, DEN-2		Cells highly sensitive to infection
Paul et al. (1969)	CHIK, JBE, WN, DEN-2		CPE, including syncytia formation, described
Singh and Paul (1969)	DEN-1, 2, 3, 4		Isolated from blood and mosquitoes; identified by CF from cell culture fluids
Pavri and Ghosh (1969)	DEN-1		Isolated from acute human blood, identified by CF from culture fluid
Yunker and Cory (1968b)	JBE, SLE, WN	LGT, POW	CPE not seen
Yunker and Cory (1969)	CTF		CPE not seen
Buckley (1969a)	CHIK, CTF, EEE, SF, SLE, VEE, VSI, VSNJ, WN, YF	BAND, CNU, CR, C5581, FAR, JA, LGT, MOD, QAL, SFN, SFS, SIL, URB	CPE with WN only
Suitor (1969)	DEN-2		Syncytial formation enhanced in plastic
Suitor and Paul (1969)	JBE		Plaques produced

[a] See Table 23 for key to virus names.
[b] CPE = Cytopathic effects; CF = Complement-fixation.

tests, with cell culture fluids as antigens clearly identified the isolates (Singh and Paul, 1969; Pavri and Ghosh, 1969).

In 1968, we reported results of inoculation of Singh's *A. albopictus* cells with certain B subgroup arboviruses (Yunker and Cory, 1968b). Previous experiences with Grace's *Antheraea* cells had demonstrated their apparent predilection for culicine-borne viruses, especially of the B subgroup, as opposed to tick-borne viruses (Yunker and Cory, 1968a). We found a similar pattern occurring in the *A. albopictus* cells: SLE, JBE and WN viruses (B subgroup; mosquito-borne) propagated therein but LGT and POW (B subgroup; tick-borne) did not. As noted above, Singh and Paul (1968b) reached similar conclusions with different viruses. Later, growth of CTF virus, which is not a member of the B subgroup, in *A. albopictus* cell cultures was reported (Yunker and Cory, 1969). CPE were not seen in our infected cells.

Buckley (1969a) reported growth of 7 mosquito-borne members of subgroups A and B, VSI and VSNJ viruses and CTF virus in *Aedes albopictus* cells. Thirteen other arboviruses, including LGT, a tick-borne subgroup B virus, failed to infect *A. albopictus* cells. Only WN virus produced CPE in these cells.

The CPE noted by Singh and Paul (1968a) when *A. albopictus* cells were infected with JBE virus were utilized by Suitor (1969) in developing a plaquing method for this virus-cell system. Although virus plaque titers were approximately 1 dex less than the TCD_{50} of either hamster-kidney or *A. albopictus* cells in liquid medium, they were reproducible, directly proportional to the virus content and reducible by prior neutralization of the virus with specific immune ascitic fluid. Hence, if this technique can be extended to other strains of JBE virus or to other viruses, a valuable tool for virus quantitation will be available.

Also in 1969, Suitor and Paul described altered morphology of *A. albopictus* cell monolayers grown in plastic rather than glass vessels. Further, they found that whereas the CPE produced by DEN-2 virus in these cells grown on glass surfaces were not easily definable, if cells were infected after transfer to plastic, cytoplasmic fusion was increased until the entire monolayer often became one large syncytial mass. These authors, and Paul et al. (1969), noted that arboviruses have not been known to induce syncytial formation in vertebrate cells. Investigation of this basic difference between responses of arthropod and mammalian cells to infection with arboviruses of the B subgroup is worthy of further study, since it may aid in our understanding of a recurring question in pathogen-vector relationships: Why is the arthropod host cell unharmed by an organism that is destined to destroy its mammalian host cell?

B. Rickettsial Studies

Nauck and Weyer (1941) and Weyer (1947; 1952) reported multiplication of *Rickettsia prowazeki*, *R. mooseri* and *R. quintana* on surviving tissues of insects, particularly the gut of lice. In the last named study, Weyer also performed what I believe to be the earliest recorded tick tissue culture experiment. Explants of louse-guts infected with *R. prowazeki* and *R. mooseri* were combined with tissues of *Rhipicephalus bursa* in culture vessels. Rickettsiae were observed in smears of this material for eight days, but transfer of the rickettsiae from louse tissues to tick tissues was not observed.

REHACEK and colleagues demonstrated that *Coxiella burneti* will propagate in explanted organs and primary tissue cultures of ticks, *Hyalomma asiaticum* (REHACEK and BREZINA, 1964) and *H. inirmis*, *Ixodes ricinus* and *Dermacentor marginatus* (KORDOVA and REHACEK, 1965). Later, it was shown that primary, dispersed cell cultures of *H. dromedarii* were highly sensitive to infection with *R. conori* and *R. akari* and also supported growth of *R. prowazeki* and *R. mooseri* (REHACEK, BREZINA and MAJERSKA, 1968). The first two species occasionally grew intranuclearly and the last two underwent a pronounced change in morphology in tick-cell cytoplasm.

Recently we tested GRACE's *Antheraea* and SINGH's *A. albopictus* cell lines for ability to support growth of various pathogenic rickettsiae. Two strains of *Coxiella burneti*, Nine Mile and Australian QD, grew well in established cell lines from the insects *Antheraea eucalypti* and *Aedes albopictus* (YUNKER et al., 1970). Rickettsial growth curves in terms of infection rate of stationary cells, exhibited a prolonged lag phase followed by a period of exponential growth and, finally, infection of nearly all cells of the cultures. After 23 weeks and a total cell-dilution of 3×10^{32}, titer of *C. burneti* (Nine Mile strain) in an *Antheraea* cell culture was 1.1×10^7 $GPID_{50}$/ml, indicating a very substantial multiplication of the rickettsia during this period. Infection of *Aedes* cells with both strains of *C. burneti* was accelerated when roller cultures were employed. Effects of rickettsia on *Antheraea* cells were limited to enlargement, rounding and fragility of heavily infected cells. Infection of *Aedes* cells, on the other hand, was accompanied by progressive decline in cell number and occasional loss of the culture through complete destruction of cells. Use of a lactalbumin-hydrolysate medium containing 0.5 % bovine plasma albumin in lieu of fetal calf serum for infected cultures of *Aedes* cells resulted in production of highly satisfactory complement-fixing antigen.

Subsequently, we observed substantial growth of two *Rickettsia prowazeki* strains in the *Antheraea* cells, as judged by infection rate of cells (YUNKER, BURGDORFER and CORY, in preparation). Evidence of CPE was seen. A strain of *R. rickettsi* showed no growth in these cells until seven weeks after inoculation, when specific immunofluorescent reactions were observed. These reactions became more intense in succeeding weeks, indicating growth undetectable by Giménez staining. Infection of *Aedes* cells by either *R. prowazeki* or *R. rickettsi* could not be demonstrated by Giménez staining, but specific immunofluorescent reactions were observed for as long as eleven weeks after inoculation, indicating prolonged survival or some growth of rickettsia.

C. Concluding Remarks

In concluding this review, it may be worth observing that approximately a third of all publications dealing with applications of arthropod tissue culture to studies of arboviruses appeared last year. Another third appeared in 1968; the remainder were published at intervals during the preceding three decades. Obviously, this area is just beginning a "logarithmic" phase of growth, hence generalizations must await the test of time. However, some salient points and recurrent observations may be noted:

1. Viable arthropod tissues and cells in primary cultures will support growth of viruses that the arthropod, or a related species, is known to transmit. (Exception: LGT virus has not yet been grown in tick tissue cultures.) Certain alien arboviruses may grow in these cultures but, generally, to a relatively lesser degree.

2. Arboviruses may propagate to a higher degree in primary cultures of vector tissues or cells than in the intact arthropod.

3. Tissues or cells from some organs or developmental stages of arthropods may be more suitable for virus growth than are those from other organs or stages.

4. Primary tick tissue cultures and established insect cell lines are extremely sensitive to certain arboviruses and rickettsiae. In some cases they may be used to detect amounts of the organism that are not detectable by standard animal, egg or vertebrate tissue culture methods.

5. Established insect cell lines, regardless of vector status, show an affinity for culicine-borne viruses (especially of subgroup B) as opposed to tick-borne viruses. (Exception: CTF virus grows in mosquito cell cultures.) They are refractory to infection with non-arboviruses thus far tested. In one instance these cells discriminated between naked infectious RNA of arbo- and non-arboviruses.

6. Persistent inapparent infection is characteristic of certain arbovirus-arthropod tissue culture systems and only a small percentage of cells are involved in virus production.

7. Gross CPE in arthropod cells *in vitro* have been observed only for five B subgroup, mosquito-borne arboviruses, and only in the *A. albopictus* cell line. Some rickettsiae do cause gross CPE in insect cell lines.

8. The kind of CPE produced by various DEN virus serotypes is characteristic and may be used in the direct assay and identification of these serotypes.

9. Induction of syncytia by certain B subgroup arboviruses is apparently unique for arthropod, as opposed to vertebrate, cell cultures.

10. Infected arthropod cell culture materials may be used as CF antigens for identification of arboviruses or rickettsiae, and at least one arbovirus produces a titrable hemagglutinin in mosquito cell cultures.

The significance of this body of information for arbovirology and epidemiology and control of infectious diseases is largely yet to be determined. However, a promising note is the appearance of unpublished reports of the successful use of mosquito cell cultures in isolating and identifying dengue viruses from human sera or mosquitoes during outbreaks in the Philippines (J. SUMPAICO, personal communication) and Puerto Rico (W. A. CHAPPELL, personal communication).

Chapter 5

Propagation of Arboviruses in Singh's *Aedes* Cell Lines

I. Growth of Arboviruses in *Aedes albopictus* and *A. aegypti* Cell Lines[1]

K. R. P. Singh

A. Susceptibility of Mosquito Cell Lines to Viruses

The susceptibility of *A. albopictus* and *A. aegypti* cell lines to infection with 22 arboviruses and 4 other viruses is listed in Table 28. *A. albopictus* was susceptible to infection with all the mosquito-borne arboviruses tested, but was generally resistant, with certain exceptions, to infection with members of the tick-borne groups. *A. aegypti* was much more restricted in its susceptibility and could support the growth of only three arboviruses. Attempts to infect both cell lines with other species of animal viruses were not successful.

A. albopictus did support the growth of Ganjam (GAM) virus, an ungrouped tick-borne virus, and Yunker et al. (1969) reported that this cell line could be infected with Colorado tick fever (CTF) virus. With the exception of these viruses, only those that are normally transmitted by mosquitoes or can experimentally infect mosquitoes can multiply in one or both insect cell lines.

GAM virus was selected for further study because of its capacity to grow in *A. albopictus* cells. This virus multiplied in ticks but not mosquitoes when inoculated intrathoracically. An experiment was performed to determine if GAM virus passed serially in *A. albopictus* cells would acquire the capacity to infect adult mosquitoes. A strain of GAM virus passed several times in *A. albopictus* cells and the mouse adapted strain were inoculated intrathoracically in *A. aegypti* females. A mouse-adapted strain of GAM virus served as the control virus. Virus multiplication was clearly demonstrated in the mosquitoes inoculated with virus adapted to *A. albopictus* cells, but only a trace of virus could be detected in mosquitoes receiving the mouse-adapted strain. These results demonstrate that a change in the host range properties of the virus occurred as the result of serial passage in the *A. albopictus* cell line. It will be of interest to see, if CTF, like GAM virus, after passage in *A. albopictus* cells can also multiply in mosquitoes.

---

1 This is a review of work carried out in our laboratory.

Table 28. *Susceptibility of Aedes albopictus and Aedes aegypti cell lines to various arbo and other viruses*

Virus	Cell line		Virus	Cell line	
	A. albopictus	*A. aegypti*		*A. albopictus*	*A. aegypti*
Mosquito-borne			*Tick-borne*		
Group A:			*Group B:*		
Chikungunya	+	+	Kyasanur Forest Disease	0	0
Sindbis	+	0	*Kaisodi:*		
Group B:			Kaisodi	0	0
Japanese encephalitis	+	0	*Ungrouped:*		
West Nile	+	+	Ganjam	+	0
Dengue 1	+	0	Dhori	0	0
Dengue 2	+	0	Wanowrie	0	NT
Dengue 3	+	0	*Ungrouped:*		
Dengue 4	+	0	African horse sickness	+	0
Bunyamwera:			*Other viruses:*		
Batai	+	0	Polio 1	0	0
Piry:			Coxsackie B5	0	0
Chandipura	+	+	Encephalomyocarditis	0	0
Simbu:			Hepatoencephalomyelitis	0	0
Sathuperi	+	0			
Palyam:					
Palyam	+	NT			
VRC No. 15534	+	NT			
VRC No. 68886	+	NT			
Ungrouped:					
Minnal	+	NT			
Arkonam	+	NT			

+ = Susceptible; 0 = not susceptible; NT = not tested.

B. Growth of Arboviruses in Mosquito Cell Lines

The growth response of several arboviruses in *A. albopictus* and *A. aegypti* cell lines was compared with that obtained in Vero cells and in infant mice (Table 29) (PAUL and SINGH, 1969). *A. albopictus* cells supported equally as good or greater virus yields for CHIK, JBE, WN, and DEN-2 viruses[2] compared to those observed in Vero cells or mice, but lower titers were obtained for Batai and Chandipura viruses. *A. aegypti* cells were tested only with CHIK, WN, and Chandipura viruses. Virus yields in this cell line were consistently lower than those achieved in Vero cells or mice.

2 Abbreviations for arboviruses are those used in Chapter 4, Table 23, p. 114.

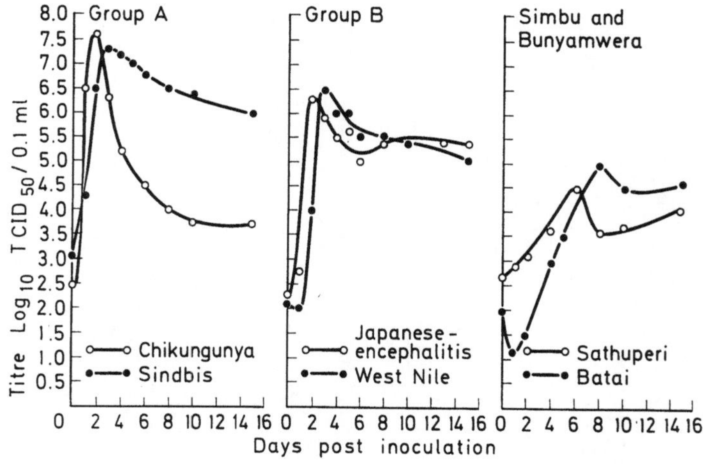

Fig. 76. Growth of arboviruses in *Aedes albopictus* cell cultures

Table 29. *Titers*[a] *of viral infectivity in various cell lines and infant mice*

Virus	Infant mice	Cell lines		
		Vero	*A. albopictus*	*A. aegypti*
Chikungunya	8.2	8.3	8.7	6.6
West Nile	8.6	9.0	9.0	7.0
Japanese encephalitis	8.4	9.0	9.5	NS
Dengue 2	6.9	6.5	8.5	NS
Batai	7.2	6.8	6.0	NS
Chandipura	9.5	9.3	9.0	7.7

[a] All titers expressed as 50% end points of infectivity per 0.1 ml.
NS = Not susceptible.

The growth patterns of representative members of Groups A, B, Simbu and Bunyamwera in *A. albopictus* cells are shown in Fig. 76. Approximately 100 to 100.000-fold increases from the original inocula were observed and maximum titers were reached within 4 days with most of the viruses tested.

C. Cytopathic Effect

Cytopathic effect was observed with JE, WN, DEN-1, 2, 3, and 4 viruses only in the *A. albopictus* cell line (Paul et al., 1969). The main features of cytopathic effect induced by these viruses were (1) cytolysis of individual small cells (2) development of large syncytial masses (3) gradual increase in number of multinucleated giant cells (4) active phagocytosis of dead and dying cells and (5) ultimate recovery of the infected cultures (Fig. 77A–F). All the viruses which produced cytopathic effect in *A. albopictus* cell cultures belonged to group B of mosquito-borne arboviruses.

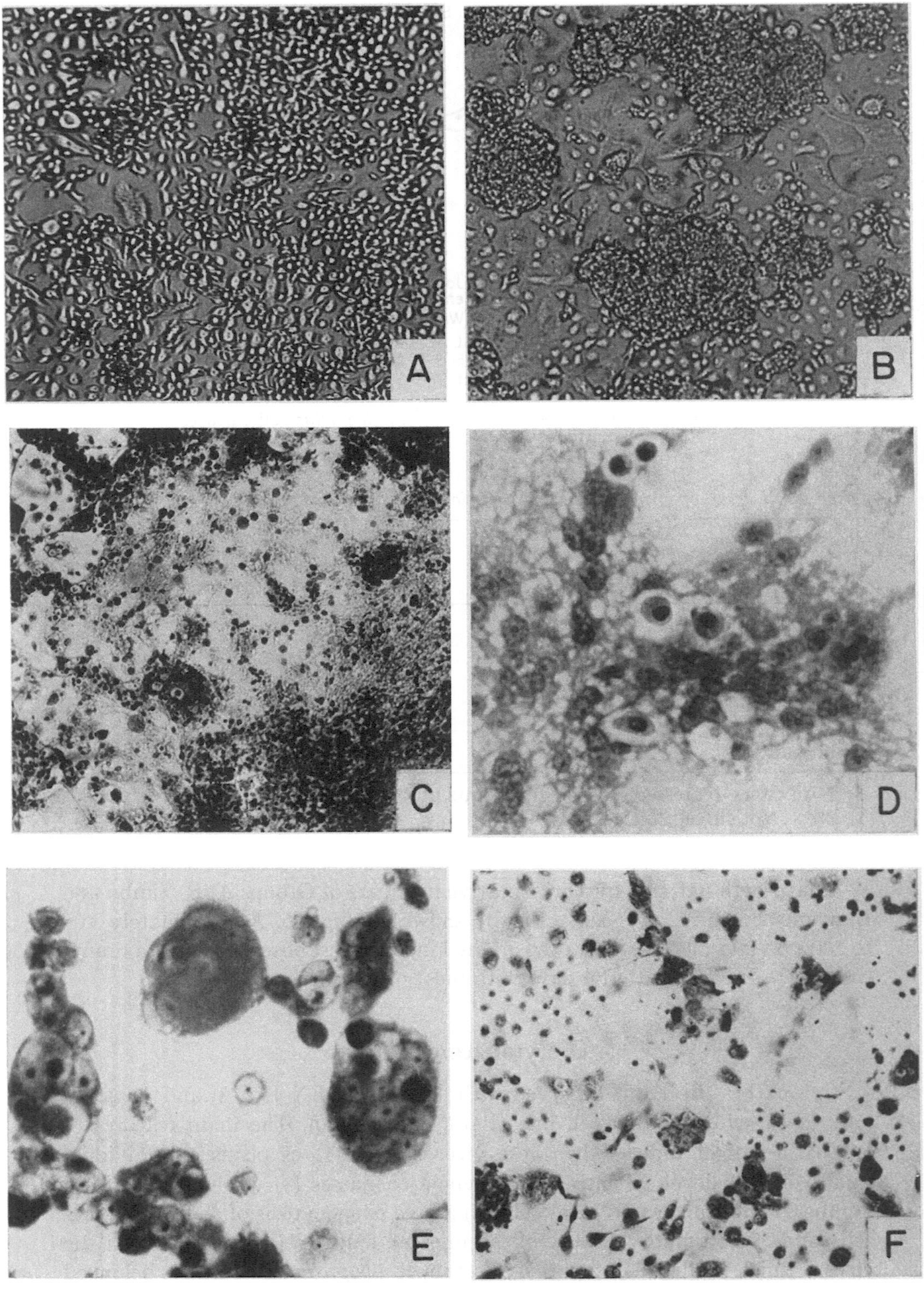
A
B
C
D
E
F

D. Primary Isolation of Arboviruses

Singh and Paul (1969), have demonstrated that dengue viruses of the four serotypes could be detected in human sera and mosquito suspensions on the basis of cytopathic effect produced in the *A. albopictus* cell line. The four serotypes of dengue virus multiply in the cell line to fairly high titers and the tissue culture fluid could directly be used for preliminary identification of viruses with complement fixation tests. It was also observed that preinoculation of the infected human sera in mosquitoes further facilitated the isolation and identification of dengue viruses, probably through enhancing the infective virus titer.

During the dengue isolation experiments in *A. albopictus* cell line, the human sera were diluted 1 : 10, before inoculation into the culture tubes, because undiluted sera had a gross anticellular effect, leading to rapid cell lysis in these cultures. A similar effect was noticed with sera from normal rabbits, monkeys and guinea pigs, but not with normal mouse serum. The anticellular effect produced with these sera was very rapid, leading to lysis of cell sheet within half an hour after inoculation into the cell cultures. However, dilution of serum resulted in a decrease in the anticellular activity that was completely eliminated by heating the sera at 56° C for 30 min.

E. Carrier Cultures

Carrier cultures of *A. albopictus* cell line infected with JE, WN and CHIK viruses have been established. A very interesting finding in the studies of carrier cultures was that virus harvested from CHIK carrier culture after a few serial passages lost its virulence for mice (Fig. 78). However, the virulence of the other two viruses, i.e., JE and WN for mice did not change.

Infant mice, inoculated by intracerebral or peripheral routes with the avirulent CHIK carrier culture virus, were immunized against intracerebral challenge with virulent CHIK virus. The avirulent strain of CHIK virus at 16th carrier culture passage was indistinguishable from the original mouse adapted strain in the neutralization test (Banerjee and Singh, 1969).

As CHIK virus belongs to group A of arboviruses, a carrier culture infected with Sindbis, another group A virus, was established to determine if the variation in mouse pathogenicity was common to all group A arboviruses. Although 26 subcultures have been made, the virus harvested from the first 20 subcultures has not shown any change in its mouse pathogenicity.

Fig. 77 A–F. *A. albopictus* cell culture infected with West Nile virus, showing various features of cytopathic effect. A Normal cell sheet (living culture), 100 ×. B Denudation of cell population, formation of multinucleated cells and syncytial cell masses, 4 days after inoculation (living culture), 100 ×. C A large syncytial mass merging into surrounding cell clumps (H and E stain), 100 ×. D A portion of the syncytial cell mass. Phagocytosis of small cells also seen (WG stain), 650 ×. E Multinucleated cells (WG stain), 450 ×. F Recovery of infected cell culture (H and E stain), 100 ×

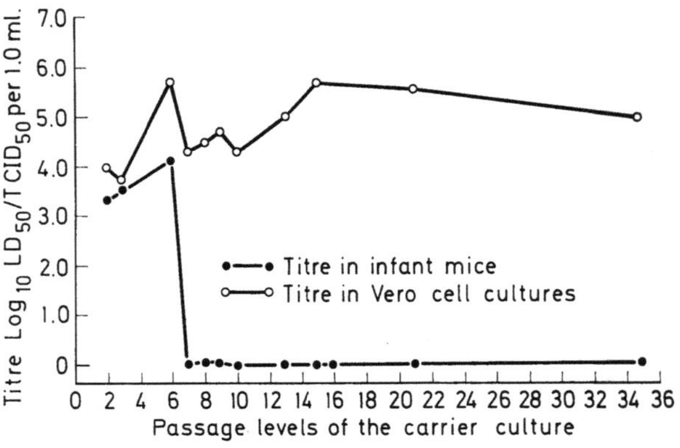

Fig. 78. Titers in mice and Vero cell cultures of the tissue culture fluid harvested from different passages of the Chikungunya carrier culture

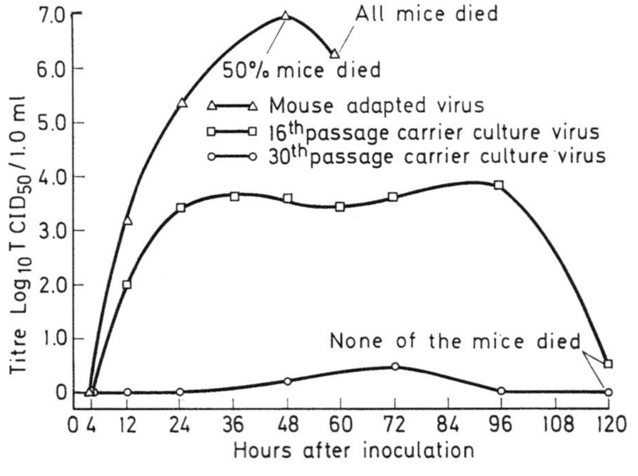

Fig. 79. Titers in Vero cell cultures of mouse brains harvested at various intervals from mice inoculated with mouse adapted and carrier culture strains of Chikungunya virus

Growth pattern of virulent and the avirulent *A. albopictus* adapted strains of CHIK virus in mouse brain was investigated (Fig. 79). Virus titer in mice infected with mouse-adapted virulent CHIK virus increased rapidly resulting in death of 50% of the mice at 48 hrs and 100% by 60 hrs. Mice inoculated with the *A. albopictus* adapted strain survived. Maximum virus-titers were 10^3 to 10^4 in mice inoculated with 16th *A. albopictus* passage virus but only a trace of virus was detected in mice infected with the 30th passage indicating further attenuation of the virus (BANERJEE and SINGH, 1969).

To learn whether the attenuated CHIK virus could infect mosquitoes, *Aedes aegypti* females were inoculated intrathoracically and after 10 and 15 days of incubation, they were fed on mice. The mice remained normal but on challenge with virulent CHIK virus were found to be immune. These findings showed that although CHIK virus had lost its virulence for mice, it could still be maintained in mosquitoes and in turn the mosquitoes could transmit the virus to mice without affecting the non-virulent properties of the virus.

It is evident from these and other studies on the growth of arboviruses in insect cell cultures, that arthropod tissue culture and particularly cell cultures from mosquitoes have become virtually at par with vertebrate tissue cultures, and that it has become possible to propagate arboviruses in mosquito cell cultures and to carry out detailed studies on the cell-virus relationship *in vitro*.

II. Multiplication of Chikungunya and O'nyong-nyong Viruses in SINGH'S *Aedes* Cell Lines

SONJA M. BUCKLEY

A. Introduction

We have been interested in insect cells as a tool to throw some light on virus-vector relationships. In studies conducted during 1968 and 1969 (BUCKLEY, 1969a and b), 32 strains of arboviruses were examined for their ability to infect *Aedes* cells. Twenty-seven viruses belonged to 9 different antigenic groups and 5 were ungrouped. The findings of SINGH and PAUL (1968a) that mosquito-borne viruses including chikungunya (CHIK) multiply either in both or at least in one of the cell lines were confirmed. However, we have found exceptions to the tentative generalization made by SINGH and PAUL that tick-borne arboviruses fail to infect the *Aedes* cells. In our hands, Colorado tick fever virus (CTF), an ungrouped tick-borne virus, multiplies in the *Aedes albopictus* cell line, but not in the *A. aegypti* cell line (BUCKLEY, 1969a). These results confirm the findings of YUNKER and CORY (1969). In addition to CTF, four viruses of the Kemerovo group, namely Kemerovo, Lipovnik, Tribec and Chenuda, non-group B tick-borne viruses, infected both *Aedes* cell lines (BUCKLEY, 1969b). Tick-borne viruses of group B such as Kyasanur Forest disease (SINGH and PAUL, 1968b), Langat (BUCKLEY, 1969a) and Kadam (W. N. MUGO, personal communication) do not multiply in the *Aedes* cells, whereas 5/18 non-group B tick-borne viruses infect either one or both of SINGH'S *Aedes* cell lines. These findings complement the results of REHACEK (1965b) who reported the multiplication of some group A and group B arboviruses in primary cell cultures of the tick *Hyalomma dromedarii* Koch.

The multiplication of CHIK and o'nyong-nyong viruses in *Aedes* cells is described in this report. O'nyong-nyong, a group A arbovirus, is antigenically closely related to CHIK virus (PORTERFIELD, 1961), but can be distinguished from CHIK virus by various serologic tests (WILLIAMS et al., 1962). Different vectors are involved in their transmission in nature, CHIK virus was isolated from *A. aegypti*, o'nyong-nyong virus was obtained from *Anopheles funestus* and *Anopheles gambiae* mosquitoes. It is due to the difference in vectors that we were interested in the multiplication of these viruses in the *A. aegypti* and *A. albopictus* cells.

We know of no studies comparing the behavior of these two viruses in mosquitoes.

B. Materials and Methods

In addition to the prototype CHIK and o'nyong-nyong strains at hand at the Yale Arbovirus Research Unit, three group A isolates from Ibadan, Nigeria, were included[1]. One is a topotype CHIK strain, Ib H 35. The other two isolates, provisionally designated Igbo-Ora virus, appear serologically to be o'nyong-nyong-like. The virus inoculum consisted of a 10% mouse brain suspension. With the exception of o'nyong-nyong, all the strains were isolated from man. Low and high mouse brain passages were employed with CHIK virus. The other viruses were passed only a few times in newborn mice since their original isolation.

Technical details as regards propagation of the *Aedes* cell lines and the maintenance of the cells following viral inoculation in this laboratory were previously described (Buckley, 1969a). Briefly, cells were grown in stationary tubes in the medium of Mitsuhashi and Maramorosch (1964), containing 20% inactivated fetal bovine serum. Prior to viral inoculation, a maintenance medium with 3% inactivated fetal bovine serum was substituted. Replicate cultures of *A. aegypti* cells were inoculated with 10^{-2} or 10^{-4} dilutions of infected mouse brain suspension, whereas cultures of the more sensitive *A. albopictus* cells were infected with a 10^{-5} dilution. Tubes containing maintenance medium without cells were inoculated for control purposes at the same time. Preparations were assayed at intervals for presence of virus by subinoculation into baby hamster kidney (BHK-21) cells (MacPherson and Stoker, 1962). Technical details pertaining to the propagation of arboviruses in this cell line have been described (Karabatsos and Buckley, 1967).

C. Results and Conclusion

The morphology of the *Aedes* cell lines was described by Singh (1967). No cytopathic effect (CPE) was observed under the experimental conditions used here. Figs. 80 and 81 illustrate results obtained with the prototype strains o'nyong-nyong and CHIK (high mouse brain passage) in both *Aedes* cell lines. Viral assays of combined cell and fluid phases were carried out over a period of one month. Figs. 82 and 83 list the data obtained with Igbo-Ora (IB10964 and IB12628) and CHIK virus (prototype, low mouse brain passage and topotype Ib H 35, low mouse brain passage) in both *Aedes* cell lines. Fluid phases only were assayed for the presence of virus over a period of up to 14 or 25 days.

In *A. albopictus* cells, o'nyong-nyong and Igbo-Ora viruses grow well and can not be distinguished from CHIK virus. O'nyong-nyong and Igbo-Ora viruses do not multiply in *A. aegypti* cells, however.

While o'nyong-nyong and Igbo-Ora viruses are clearly different from CHIK virus in *A. aegypti* cells, they can also be separated on the basis of their growth in newborn mice and Vero cells (Yasamura and Kawakita, 1963). As shown

[1] These viruses have not as yet been described in the literature; their name and laboratory designation are used here with the kind permission and cooperation of the Arbovirus Project, University of Ibadan, Ibadan, Nigeria.

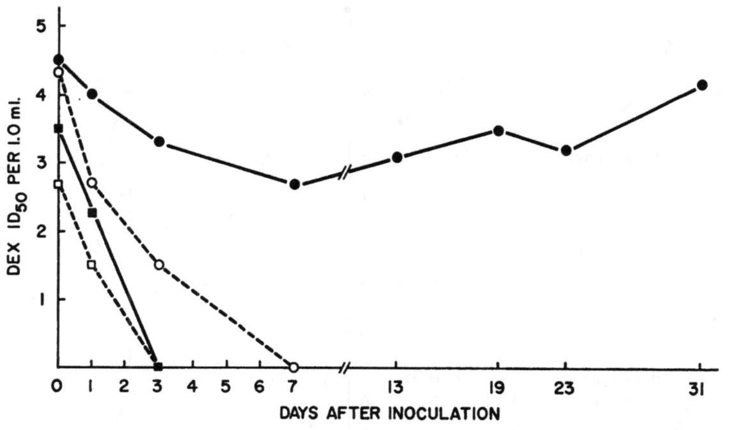

Fig. 80. Growth curves of chikungunya and o'nyong-nyong viruses in *Aedes aegypti* cells and medium without cells. •——• chikungunya virus in cells; o----o chikungunya virus in medium without cells; ▪——▪ o'nyong-nyong virus in cells; □----□ o'nyong-nyong virus in medium without cells

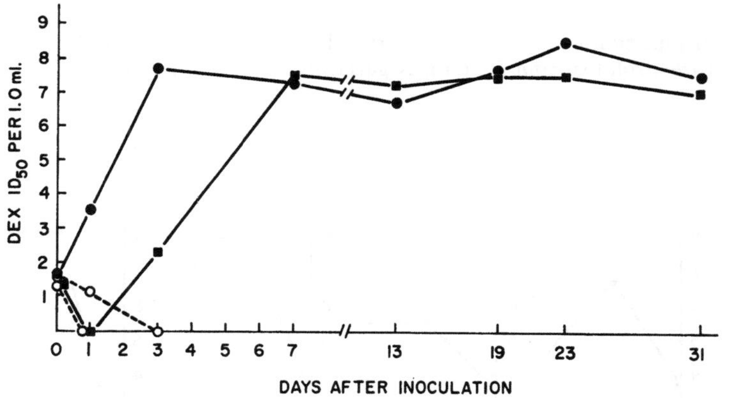

Fig. 81. Growth curves of chikungunya and o'nyong-nyong viruses in *Aedes albopictus* cells and medium without cells. •——• chikungunya virus in cells; o----o chikungunya virus in medium without cells; ▪——▪ o'nyong-nyong virus in cells; □----□ o'nyong-nyong virus in medium without cells

in Table 30, o'nyong-nyong and Igbo-Ora viruses killed mice less rapidly than CHIK virus. As shown in Table 31, plaques produced in Vero cells (SIMIZU et al., 1967) by o'nyong-nyong and Igbo-Ora viruses are smaller than those of CHIK virus. However, the viruses could not be distinguished on the basis of their cytopathic effect in BHK-21 cells.

In conclusion, o'nyong-nyong virus can be clearly separated from CHIK virus by the absence of multiplication of o'nyong-nyong virus in the *Aedes aegypti* cells.

In comparative studies in mice and vertebrate cell cultures, o'nyong-nyong virus produces symptoms more slowly in newborn mice and differs in plaque morphology in Vero cells from CHIK virus.

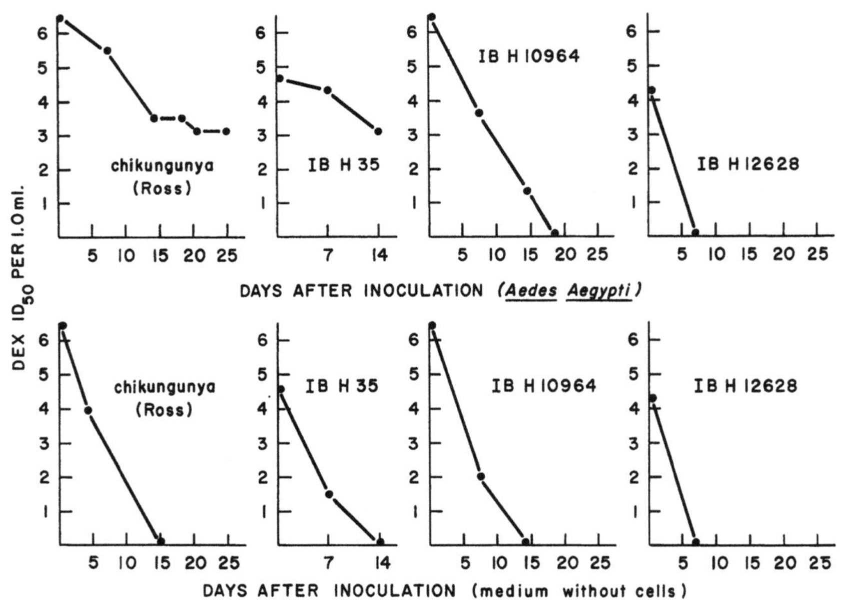

Fig. 82. Multiplication of chikungunya virus (Ross strain low mouse brain passage) and three isolates from Ibadan, Nigeria, in *Aedes aegypti* cells

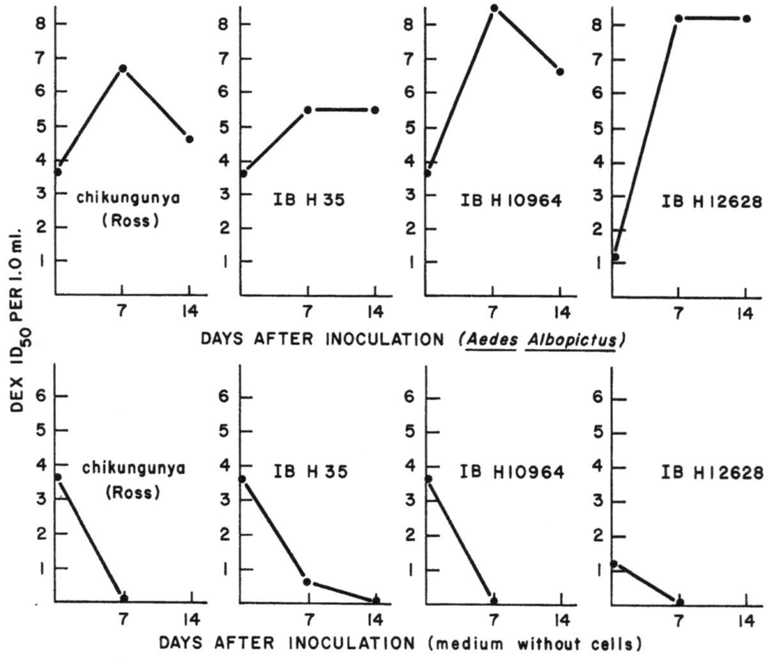

Fig. 83. Multiplication of chikungunya virus (Ross strain low mouse brain passage) and three isolates from Ibadan, Nigeria, in *Aedes albopictus* cells

Table 30. *Behavior pattern of viruses*[a] *in newborn mice*

Virus	Strain (No. of mouse passages)	IC LD_{50}[b] / IP LD_{50}	Average days survival after inoculation IC	IP
Chikungunya (1)	Ross (17)	2.1 dex[c]	2.4	4.6
Chikungunya (2)	Ross (175, 176)	1.6	2.0	4.4
Chikungunya	Ib H 35 (6)	0.4	2.4	5.0
O'nyong-nyong	MP 30 (11,12)	2.6	2.5	7.5
Igbo-Ora	IbH10964 (3, 4)	2.6	3.0	8.0
Igbo-Ora	IbH12628 (3, 4)	3.4	2.9	6.6

[a] Group A.
[b] IC = intracerebral; IP = intraperitoneal.
[c] Logarithms to the base of 10.

Table 31. *Cytopathic effect in BHK-21 cells and plaque morphology in Vero cells*

Virus (strain)	BHK-21 cells (CPE)			Vero cells (plaques)		
	days of first appearance	rating	TCD_{50} (dex/ml)	days of first appearance	maximum diameter (mm)	PFU[a] (dex/ml)
Chikungunya (1) (Ross)	1	4+	9.5	2	19	9.4
Chikungunya (2) (Ross)	1	4+	9.7	2	20	9.6
Chikungunya (IbH 35)	1	4+	10.2	2	17	10.2
O'nyong-nyong (MP 30)	1—2	4+	9.2	2	9	9.5
Igbo-Ora (IbH10964)	1—2	4+	9.2	2	9	8.3
Igbo-Ora (IbH12628)	1—2	4+	7.7	2	8	7.3

[a] Plaque forming units.

Two isolates provisionally designated Igbo-Ora virus behaved like the prototype strain o'nyong-nyong in insect cells, mice, and Vero cells. Serologically, the Igbo-Ora viruses appear to be o'nyong-nyong-like (J. DAVIS, personal communication).

Acknowledgment. This work was supported by funds from the Rockefeller Foundation, USPHS-NIH grant No. 1-R22-AI-08215-01A1, and the World Health Organization.

III. Growth Studies of California Encephalitis Virus in Two *Aedes* Mosquito Cell Line Cultures

ELINOR WHITNEY and RUDOLF DEIBEL

The mouse-adapted New York strain of California encephalitis (CE) (WHITNEY et al., 1969) virus grew readily on initial passage in cultures of SINGH'S (1967) *Aedes aegypti* and *A. albopictus* cell lines. The infection persisted over the observation period of 6 weeks. Subsequent experiments with another CE

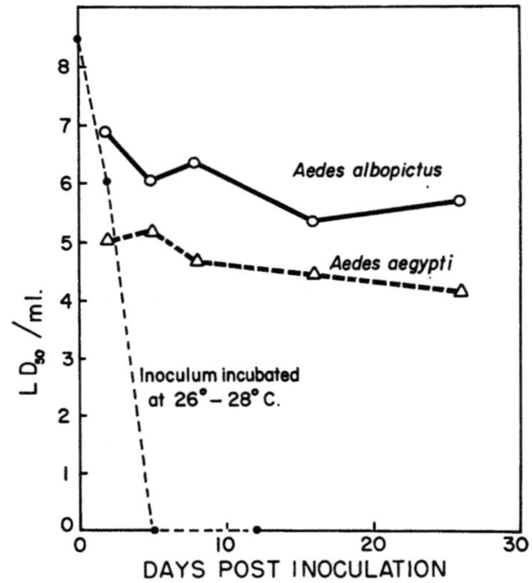

Fig. 84. Extracellular virus in mosquito cell cultures inoculated with N.Y.S. strain CE virus smb_5, incubated at 26—28° C

strain have shown that cells on passage still produce virus and are persistently infected. Cytopathic effects were not observed during the first and subsequent 8 passages after which the experiments were discontinued. Fig. 84 shows a comparison of the growth of CE virus on initial passage in two *Aedes* cell lines. A slight and gradual drop in titer of extracellular virus was demonstrated. The virus titer in *A. albopictus* cells was 5.9 LD_{50}/ml and generally 1 $\log_{10}$ higher than that observed in *A. aegypti* cells. In the absence of cells, no virus was detected by the fifth post-inoculation day in the original inoculum incubated at the same temperature.

The virus titers taken at the end of 7 serial weekly passages in both cell lines were with few exceptions between 10^4 to $10^{6.5}$ LD_{50}/ml.

Fig. 85 demonstrates the development of extracellular and cell-associated CE virus in *A. aegypti* cells. The inoculum was $10^{5.3}$ LD_{50} giving a multiplicity of 4×10^{-2} LD_{50}/cell. An eclipse phase of 4 hours' duration was noted followed by a steep increase in cell-associated virus. The curve suggests that a single growth cycle requires approximately 10 to 14 hrs. Extracellular virus was usually found in slightly higher titers. Maximum titers were in the range of 10^6 to 10^7 LD_{50}/ml. A second experiment with the higher input of virus of 1.5 LD_{50}/cell did not give higher titers after 3 to 7 days.

When *A. albopictus* cells were used, the results were similar, but the eclipse phase of the virus was 8 hrs, longer than that in the *A. aegypti* cells.

Analysis of growth curve experiments showed that the rate of virus adsorption was low: Between 0.1 to 0.3 % of the inoculum was adsorbed within 1 hr. In the first growth cycle 300 to 1 500 LD_{50} were produced per infected

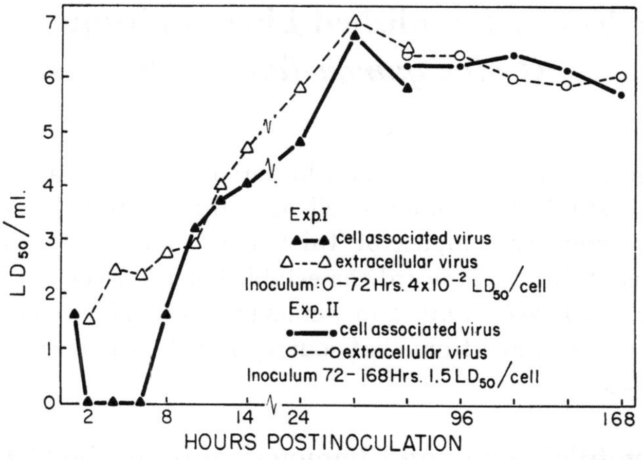

Fig. 85. Growth of N.Y.S. strain of CE virus in *Aedes aegypti* cells

cell. The virus was readily released from the cell and approximately 50 to 90 % of infectivity was found in the supernatant culture fluid. A relatively higher virus production was observed in the second growth cycle.

Giemsa-stained preparations failed to reveal any morphological changes in mosquito cells infected with CE virus. Immunofluorescence studies indicated that only a small fraction of the cells in a culture were infected. These cells showed small perinuclear granules, bright cytoplasmic bodies, or diffuse cytoplasmic bodies, or diffuse cytoplasmic staining.

Propagation of the CE virus in the mosquito cell lines was sufficiently great to allow these mosquito cell adapted strains to be used in comparative studies with strains adapted to mammalian cells.

Chapter 6

Growth of Arboviruses in Arthropod Cell Cultures: Comparative Studies

I. Preliminary Observations on Growth of Arboviruses in a Newly Established Line of Mosquito Cell (*Culex quinquefasciatus* Say)

S. H. Hsu

During the past two years several hundred attempts have been made by the author to establish continuous cell lines from various mosquitoes of the *Culex tritaeniorhynchus* complex of Taiwan, but all resulted in failures. Attempts were next directed toward the establishment of cell lines from ovarian tissue of *Culex quinquefasciatus* and were successful. This report describes the development of a cell line from that source, and the susceptibility of the cells to 9 arboviruses.

A. Establishment and Characterization of the Cell Line

Laboratory-reared 3–5 day old female adult mosquitoes, a Taiwan strain of *Culex quinquefasciatus* Say, were sterilized by immersion for 2 min in 0.25% solution of mercuric chloride in 70% ethyl alcohol and washed twice with sterile distilled water. Ovarian tissue was removed aseptically under a dissecting microscope and placed in 12×75 mm test tubes containing 0.5 ml of culture medium. The tubes were sealed with paraffin and incubated at 28° C in a slanted position.

The medium employed was prepared as follows: Solution A (mg per 100 ml): KCl 80, NaCl 450, $CaCl_2$ 35, $MgSO_4 \cdot 7H_2O$ 40, KH_2PO_4 25, $NaHCO_3$ 100, lactalbumin hydrolysate 2000, D-glucose 160, sucrose 600, TC-Yeastolate 200, Bactopeptone 500, l-malic acid 60, α-ketoglutaric acid 40, succinic acid 6, fumaric acid 6, were dissolved in demineralized distilled water to a final volume of 100 ml. The pH of the mixture was adjusted to 6.6 with KOH and sterilized by passage through a Millipore filter unit with a pore size of 0.22 μ. Medium No. 199 (solution B) and fetal bovine serum (solution C) were obtained from Grand Island Biological Co. Prior to use, solution B was diluted to 1:10 without the addition of $NaHCO_3$ and the fetal bovine serum was inactivated at 56° C for 30 min. The ratio of the volume of the various components of the completed culture medium, A, B, and C, was 7:2:1 and for convenience has been designated as 721 medium. The final pH was adjusted to 6.8.

Twenty-four hours after the cultures were initiated, the ovarian tissue showed active contraction. After 12 days, cells began to proliferate from the explant, spread out, and attach to the glass (Fig. 86). Multiplication became

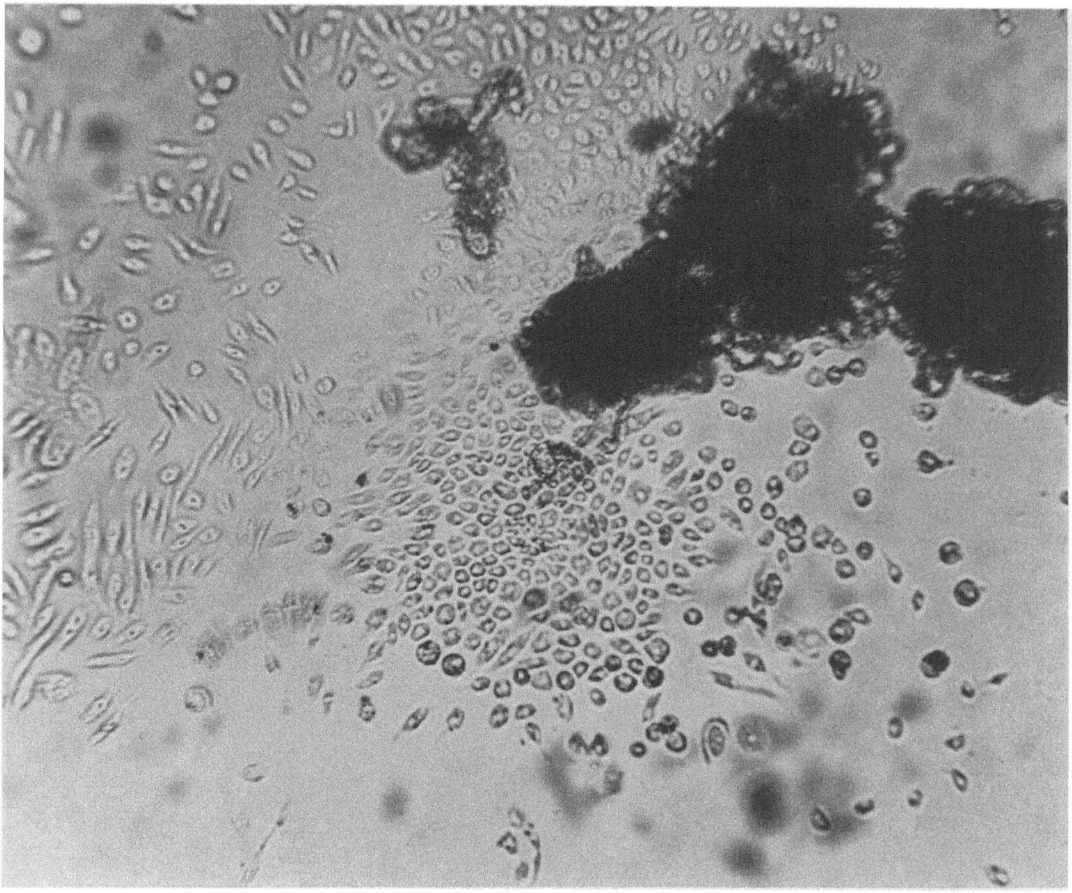

Fig. 86. Cell outgrowth from ovarian tissue of *C. quinquefasciatus*, 21 days after culture in test tube. ×375

more rapid after 14 days and at 21 days the first subcultures were made. The subcultured cells continued to multiply and small colonies developed which eventually enlarged and merged with neighboring colonies to form a complete monolayer (Fig. 87). Serial subcultures were performed at 14–21 day intervals during the first 3 passages, but after the third passage it became necessary to transfer the cells every 3 or 4 days. From January 1969 until February 1970, the cells were carried through 116 passages.

Morphologically, the cells are epithelial-like and form monolayers containing two distinct cell types. The more predominant type has clear cytoplasm, is spindle-shaped and measures approximately 21.5 × 12.9 μ. The other is round, measures approximately 12.1 μ in diameter and the cytoplasm is granular. The remaining cells, seen in Fig. 87, are regarded as morphological variations of either spindle-shaped or round cells.

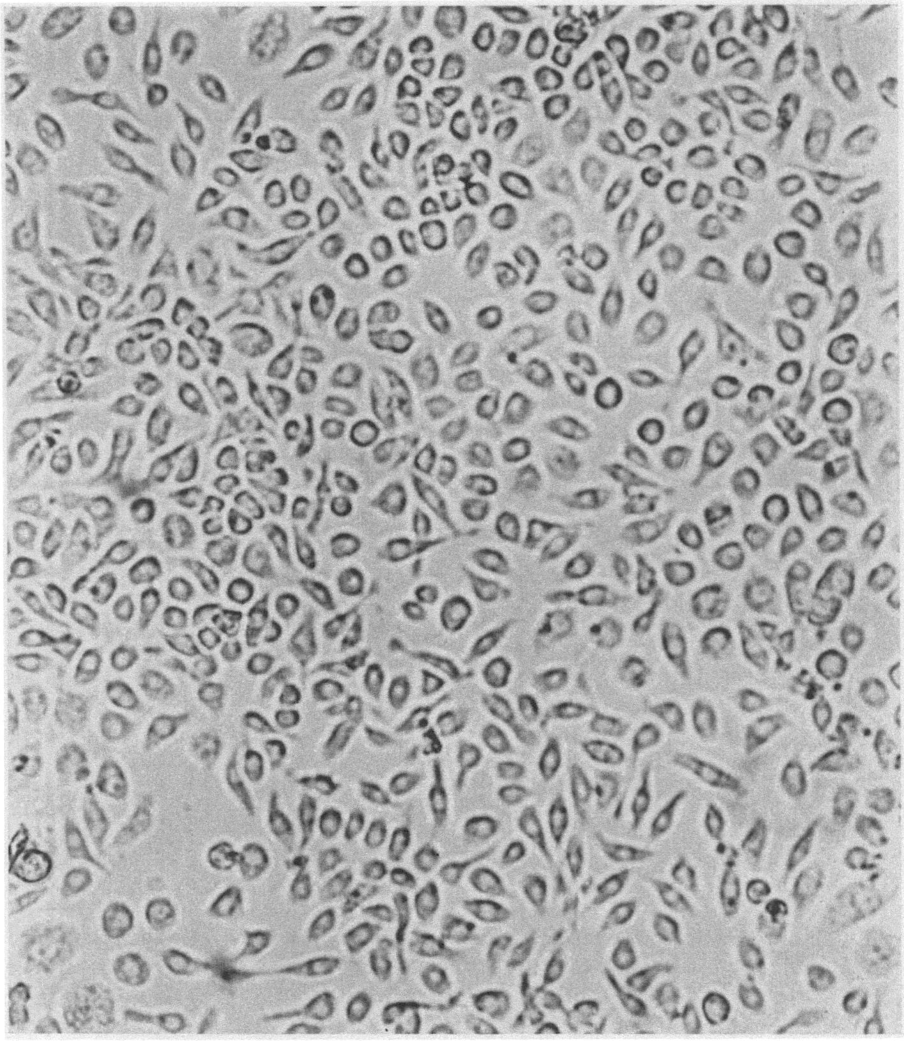

Fig. 87. Confluent sheet of *C. quinquefasciatus* cells in 44th passage. ×480

The chromosome complement of the cells was studied 4 and 14 months after they became established in culture, and most of the cells examined were diploid ($2n=6$). This is similar to the number reported by Breland (1961) who studied brain cells of *C. quinquefasciatus in vivo*.

B. Susceptibility to Infection with Arboviruses

Virus Stocks. The viruses[1] employed in the experiments were EEE, SAG, and SIN of group A, MVE, JBE, WN, and DEN-2 and 4 of group B, and BUN. All of

[1] Abbreviations for arboviruses will be those used in Chapter 4, Table 23, p. 114 or in a more complete list cited in that Chapter: McLean et al. (1969).

the viruses were obtained from the collection of Naval Medical Research Unit No. 2, and all have undergone numerous intracerebral passages in suckling Swiss albino mice.

Table 32. *Growth of arboviruses in Culex quinquefasciatus cells*

Virus (group and strain)	Age of mice used in titrations	No. trials	Virus inoculation	Highest titer in culture	Virus increase
A Group					
EEE	Adult	4	5.2[a]	Neg.	—
SAG	Suckling	2	4.6	6.5	1.9
SIN	Suckling	2	6.3	7.0	0.7
B Group					
JBE	Adult	4	2.9	7.8	4.9
MVE	Adult	2	3.0	4.5	1.5
WN	Adult	1	<3.0	6.6	>3.6
DEN-2	Suckling	2	4.4	4.8	2.4
DEN-4	Suckling	2	4.5	6.0	1.5
BUN Group					
BUN	Adult	2	4.4	6.5	2.1

[a] Figures indicate $\log_{10}$ of mouse IC LD_{50} per 0.03 ml tested in adult mouse and per 0.01 in suckling mouse.

Virus Inoculation. Seed viruses were prepared from infected mouse brain as 20% suspensions in phosphate buffered saline containing 0.5% bovine albumin (PBS/BA), and were stored at −70° C until used. After removal of the medium, cell cultures in flasks were inoculated with 0.4 ml of a 10^{-1} dilution of virus suspension in medium 721. A similar virus suspension was inoculated into control flasks containing 3.6 ml of culture medium but without cells. The controls received no further treatment except for incubation at 37° C. After the cultures were reincubated at 37° C with shaking at 10-min intervals for an adsorption period of 1 hr, the virus suspension was removed, the cells were washed twice with culture medium to remove unadsorbed virus, and 4.0 ml of fresh culture medium were added. The cells from one of the cultures were titrated immediately in mice, and the results obtained were designated in the growth curves as the value at time 0.

Virus Assay. The cultures were titrated for virus content after 24 and 48 hrs incubation, and every second day thereafter for a test period of 8 days. At the selected times, sterile glass beads were placed in the flask which was shaken vigorously to disrupt the cells. Suspensions were removed, centrifuged, and serial 10-fold dilution of the supernatant made in PBS/BA. Aliquots of the harvests were inoculated intracerebrally in 2–3 day old mice (0.01 ml), or 21-day old mice (0.03 ml), depending upon which virus was used (Table 32). The control cultures, except for the addition of glass beads, were treated similarly. The mice were observed daily and the 50% mortality end point ($MICLD_{50}$) was estimated by the method of Reed and Muench (1938).

Results. This preliminary study has shown that all of the viruses tested, except EEE and SIN, were able to propagate in *C. quinquefasciatus* cell cultures in various degrees. The growth curves for the 9 viruses are shown in Figs. 88–96.

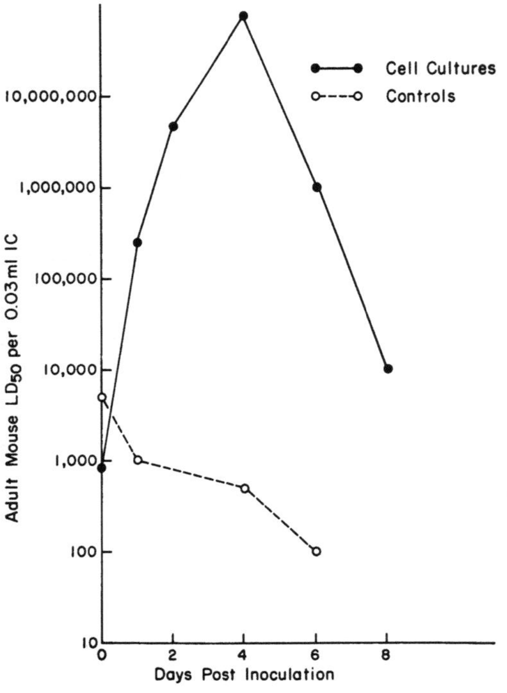

Figs. 88—96. Growth of viruses in *C. quinquefasciatus* cell cultures

Fig. 88. JBE virus

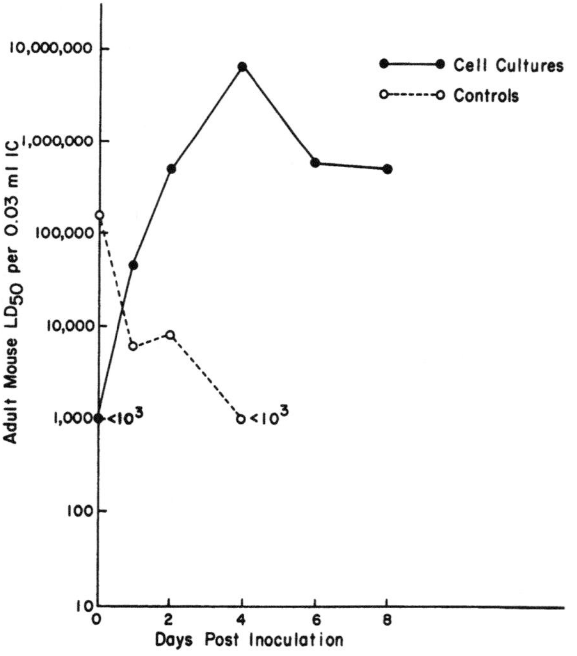

Fig. 89. West Nile virus

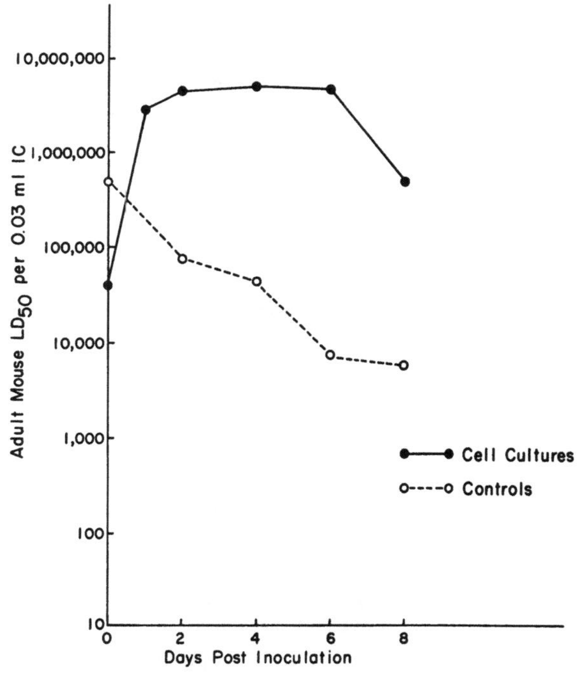

Fig. 90. Bunyamwera virus

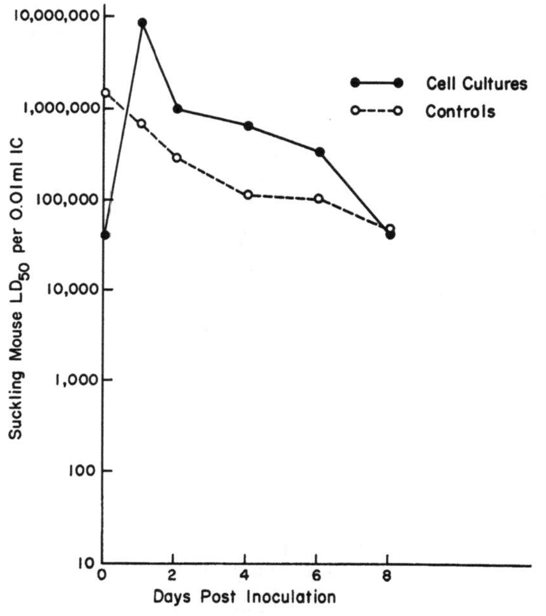

Fig. 91. Dengue-II virus

JBE (Fig. 88) propagated rapidly and reached a maximum titer of $10^{7.8}$ $MICLD_{50}$ on the fourth day, a level 100,000 times greater than that of the inoculum. After the fourth day, however, the titers decreased rapidly.

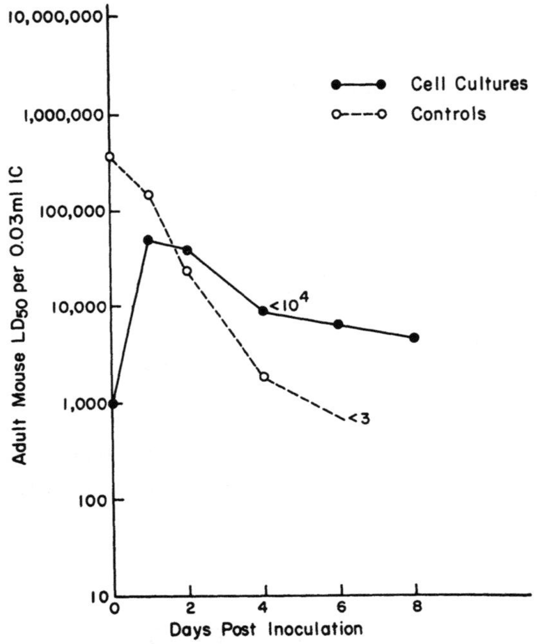

Fig. 92. Murray Valley Encephalitis virus

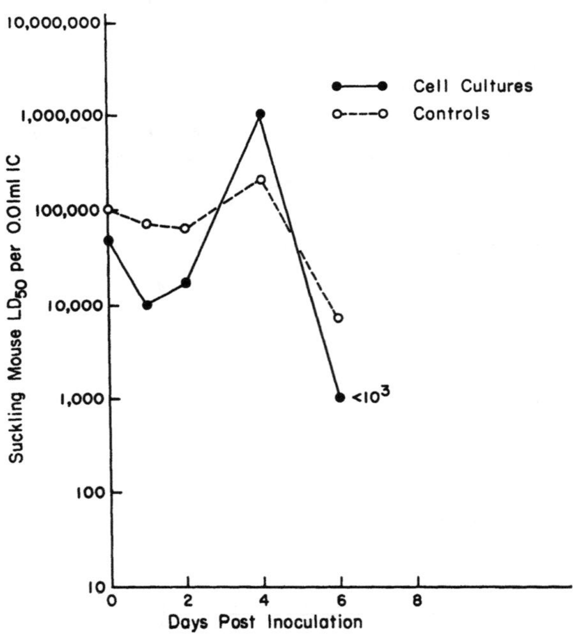

Fig. 93. Dengue-IV virus

The propagation of WN (Fig. 89) also reached the maximum at four days exhibiting a 4000-fold increase greater than the initial titer. The titer decreased by the sixth day and remained elevated until the eighth day. BUN

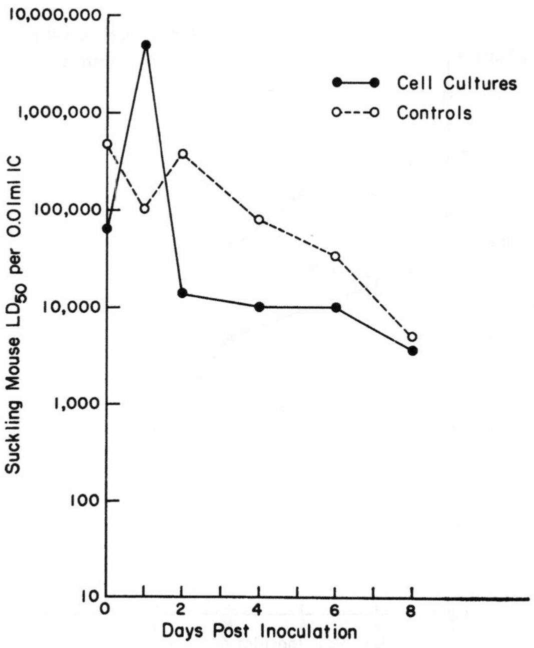

Fig. 94. Sagiyama virus

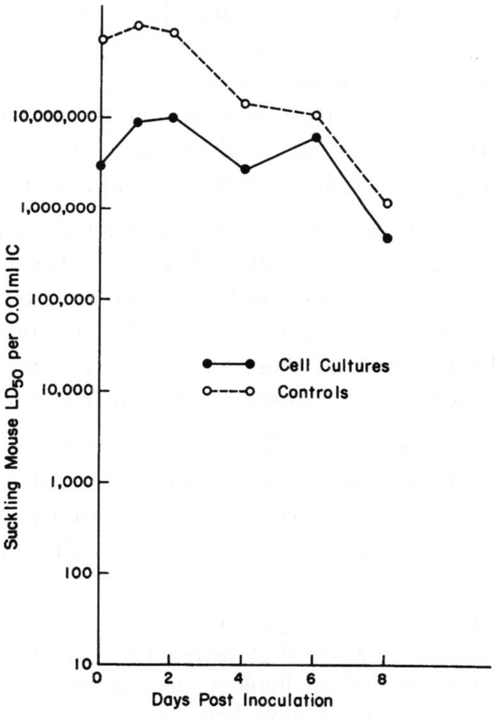

Fig. 95. Sindbis virus

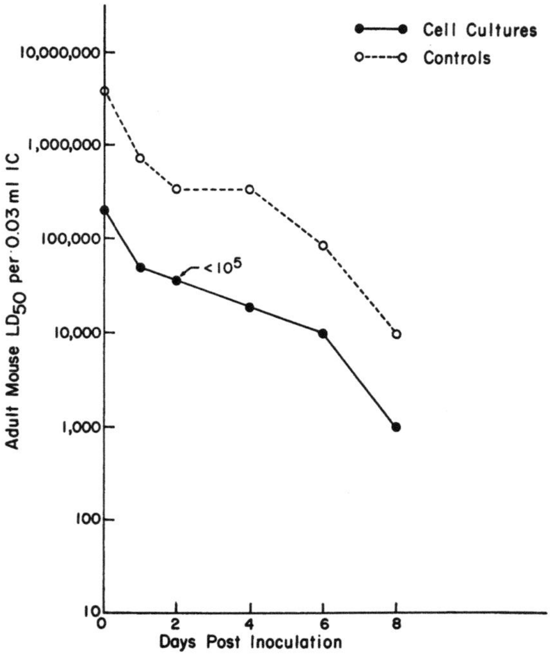

Fig. 96. Eastern Equine Encephalitis virus

(Fig. 90) multiplied rapidly during the first day, reached a maximum by the fourth and sixth days, and decreased on the eighth day. DEN-2 (Fig. 91) and MVE (Fig. 92) demonstrated maximum replication after 24 hrs but decreased thereafter. DEN-4 virus (Fig. 93) on the other hand showed an initial decrease in titer on the first and second day, followed by a 32-fold rise by the fourth day, and a sharp decrease again on the sixth day.

The three group A viruses, SAG, SIN, and EEE, did not grow well in the presence of *C. quinquefasciatus* cells. SAG (Fig. 94) multiplied for 2 days, then decreased rapidly and finally the titers decreased to a level below that of the initial inoculum. SIN (Fig. 95) and EEE (Fig. 96) completely failed to multiply and the survival of these viruses in the presence of mosquito cells was almost identical with viruses in the control medium without cells.

In the control cultures only DEN-4 and SIN viruses exhibited moderate rises in titer above the initial inoculum. The titers of all other viruses in the control cultures, for the most part, decreased rapidly. None of the viruses appeared to produce cytopathic effects upon the cells during the period of study.

Acknowledgment. The author is indebted to Captain R. H. Watten, MC, USN, Commanding Officer of Naval Medical Research Unit No. 2, and Dr. J. H. Cross, Head of the Department of Medical Ecology, for encouragement and helpful suggestions. This work would also not have been possible without the able technical assistance of Mr. W. H. Mao, Miss M. S. Huang, Miss S. Y. Li, and Mr. J. R. Wang.

II. Attempts to Grow Tacaribe and Junin Viruses in GRACE'S Continuous Line of Moth Cells

WILLIAM D. HANN and ROBERTSON B. CLARKE

SUITOR'S (1966b) success in propagating JBE in GRACE'S moth (*Antheraea eucalypti*) cell line suggested that it might be desirable to determine if this line could support the growth of viruses of the TCR group.

Cells were grown by SUITOR in GRACE'S insect cell culture medium (GMA) (GRACE, 1962) with 10 mg/ml bovine plasma albumin and 10% *Antheraea*

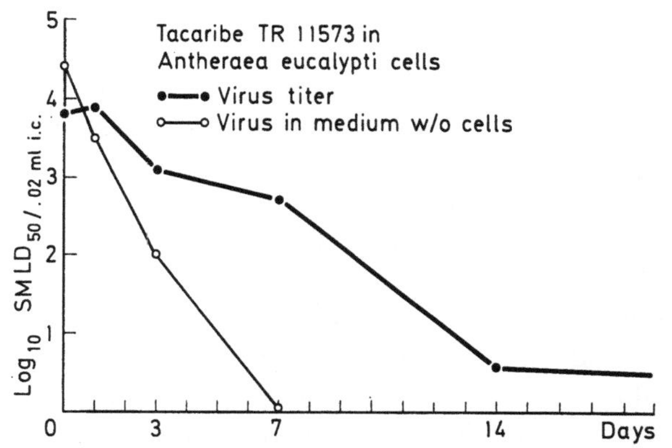

Fig. 97. Survival of TCR in GRACE'S moth (*Antheraea eucalypti*) cell line. The open circles show the virus in medium without cells. Virus was detected at 21 days but not at 28 days

pernyi hemolymph. Virus strains available as early passages in laboratory animals were selected. Our strain of TCR (TR 11573) had 20 mouse passages and one hamster passage. JUN strain RC had 12 mouse passages and JUN strain MC2 had only 3 mouse passages.

Ten percent viral mouse brain suspension was diluted to 10^{-2} in GMA for inoculation of the cell cultures which were incubated at 28° C. Samples were taken periodically up to 6 weeks and stored at —70° C. Virus was titrated by intracerebral inoculation in one-day old suckling mice and by plaquing in Vero cell (YASAMURA and KAWAKITA, 1963) monolayers.

TCR survived for approximately one week in the medium alone, but minimal amounts of virus persisted for 21 days when associated with cells. These results, obtained by mouse, titration are depicted in Fig. 97. Virus could not be detected from 21–42 days, when the experiment was terminated. When the harvests were titrated in Vero cells comparable results were obtained.

JUN, RC strain, gave similar results (Fig. 98) but the MC2 strain (Fig. 99) did not show significantly increased survival time over virus contained in the medium alone.

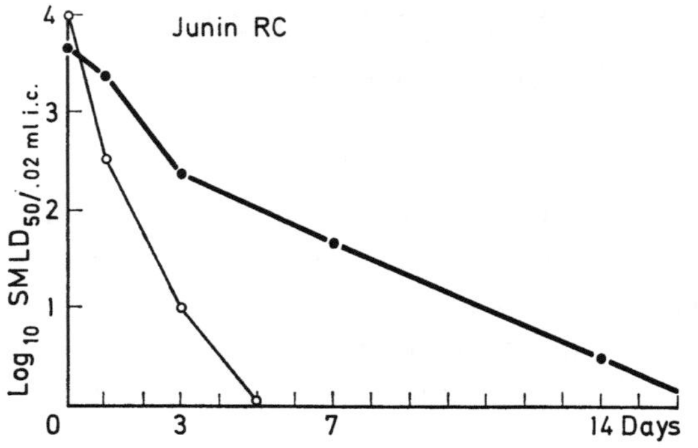

Fig. 98. Survival of JUN virus in Grace's moth cells

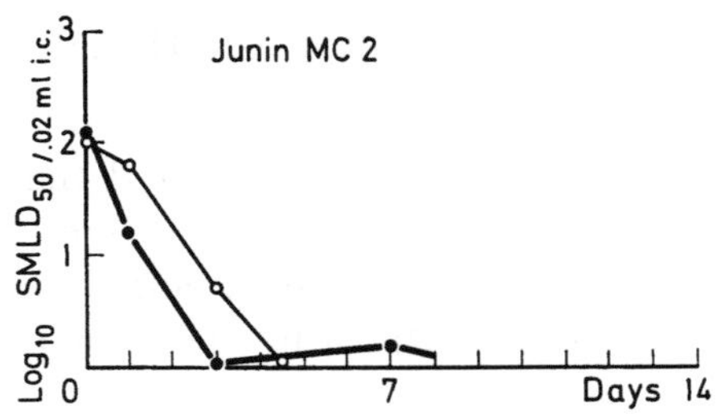

Fig. 99. Survival of JUN virus in Grace's moth cells

Parodi and others (1966) have reported that JUN is inactivated after 3 days at 25° C and this agrees with other unpublished studies. Although we were unsuccessful in propagating any of the TCR group in moth cells, we feel that the increased survival of the virus in the presence of the cells suggests a possible use for the cells for unrefrigerated transport of virus, similar that to suggested by Woke and Rosenberger (1957).

Others (Converse and Nagle, 1967; Yunker and Cory, 1968a) have reported the ability of these cells to support the growth of various arboviruses after the cells were adapted to hemolymph-free medium. The adapted cells are totally different cells; thus the role of hemolymph might be of importance in virus survival in nature in insect hosts not propagating that virus.

Acknowledgment. E. C. Suitor, Jr. provided the cells and collaborated in these studies.

III. A Comparative Study of the Viral Susceptibility of Monolayer and Suspended Mosquito Cell Lines

B. H. Sweet and H. D. Unthank

Our studies were first performed with suspended cell cultures, utilizing Grace's line of *A. aegypti* (1966) cloned and adapted to grow in a hemolymph-free medium (Suitor et al., 1966a; F. J. Paul, unpublished work) and *C. inornata* and *A. vexans* lines developed in our laboratories. In Table 33 are shown the viruses used and mouse passage levels employed. Replicate cultures containing 35000 cell/ml in Grace's medium (1962) with 10% fetal calf serum were inoculated at time 0. The number of tissue culture infective doses ($TCID_{50}$) used was determined by observation of the cytopathogenic end point of a dilution series inoculated into roller tube cultures of a stable line of porcine kidney cells (PS-Y15). Tubes containing no cells, but inoculated with the same dose of virus, were kept as controls and handled in a similar

Table 33. *Viruses used in suspended mosquito cell cultures*

Virus and strain designation	Mouse passage level	$TCID_{50}$/0.1 ml in PS-Y15 cells	Virus/cell ratio at time zero
EEE, Mass	463	9.0	3000:1
CE, BFS-283	20	6.5	10:1
CV, Original	10	6.5	10:1
SLE, Hubbard	103	6.5	10:1
JBE, Nakayama	44	7.0	30:1

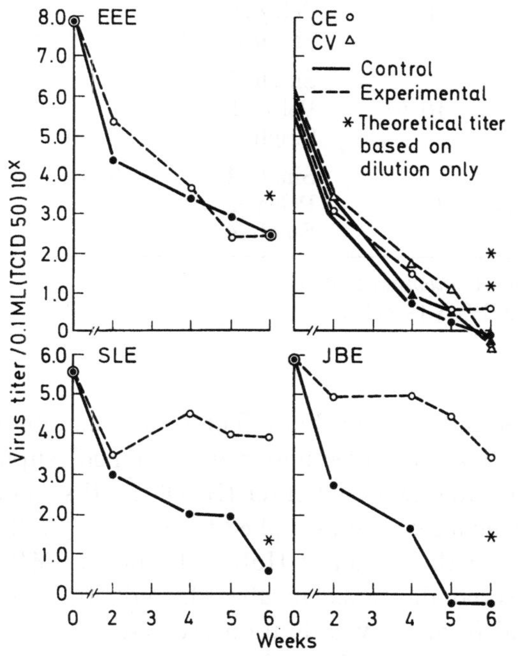

Fig. 100. Propagation of various arboviruses in *C. inornata* suspended cell cultures at 28° C

manner. Inoculated cultures and control tubes were maintained for 6 weeks at 28° C. At weekly intervals one-half the medium was withdrawn for viral assay and replaced with fresh medium.

Results obtained with *C. inornata* cells are given in Fig. 100. Similar results were obtained with *A. vexans* and *A. aegypti* cell lines. Evidence of low level viral multiplication was obtained only with SLE and JBE. We found no changes in viable cell counts over the period of these experiments. If cytopathic effect (CPE) occurred, it was not obvious. Somewhat similar studies with selected dengue viruses have thus far been negative.

With the availability of the monolayer lines developed by Singh (1967) and Peleg (1966), we carried out studies with both unadapted and mouse adapted strains of several arboviruses. Table 34 shows results of studies with

Table 34. *Evidence of multiplication of SLE viruses in monolayer cultures of A. aegypti and A. albopictus cell lines*

Virus strain and designation	Titer in PS-Y15	Results of titration in indicated cell line		Titers of inocula and harvests	
		cell line	CPE, and highest dilution positive	virus input	virus yield
SLE, mosquito pool, NCDC TH438C	$10^{-1.0}$	Singh *A. a.*	0	$10^{3.0}$[a]	$\geq 10^{6.5}$[a]
		Peleg *A. a.*	0	$10^{3.0}$[a]	$\geq 10^{6.5}$[a]
		Singh *A. albo*	+, $10^{-4.0}$	...	...
SLE, fox isolate, mouse passage 1	$10^{-7.5}$	Singh *A. a.*	...	...	...
		Peleg *A. a.*	...	...	...
		Singh *A. albo*	+, $10^{-8.5}$		
SLE, Hubbard, mouse passage 124	$10^{-6.5}$	Singh *A. a.*	0	$10^{4.5}$[b]	$< 10^{4.0}$
		Peleg *A. a.*	0	$10^{4.5}$[b]	$10^{4.0}$
		Singh *A. albo*	0	$10^{3.5}$	$\geq 10^{6.5}$

[a] Determined in *A. albo*pictus cells.
[b] Determined in PS-Y15 cells.

SLE and PELEG's and SINGH's lines. Three different preparations were employed—a known positive infectious mosquito pool supplied to us by the National Communicable Disease Center (NCDC), a first mouse-passage (MP-1) of SLE isolated from a fox brain supplied to us by the California State Department of Health, and the standard Hubbard strain in MP-124. Cultures were inoculated with serial dilutions, incubated, and observed for 10 days. The tissue culture fluid from a given dilution was harvested and titrated in either *A. albopictus* or PS-Y15 cells. The infected mosquito pool and MP-1 produced cytopathology with excellent yields of virus in SINGH's *A. albopictus* cells only. No CPE was observed in either *A. aegypti* line. However, by comparing titrations of viral inoculum and harvests (Table 34) there was definite evidence of multiplication of "wild virus" in both *A. aegypti* cell lines. Mouse passaged (Hubbard) virus produced no CPE and showed a significant level of multiplication only in SINGH's line of *A. albopictus*.

The role of mouse passage in determining the occurrence of CPE with some viruses was further demonstrated with studies of dengue (DEN) viruses. Utilizing DEN-1 strains in the form of several acute phase human sera and mouse adapted virus at various passage levels, the extent of viral growth in *A. albopictus* cells was studied (Table 35). Titrations were performed by plaquing in PS-Y15 cells or by inoculation of mice, and/or in *A. albopictus* cell lines, depending upon the virus used. Marked cytopathology and excellent titers were found in *A. albopictus* cells with unadapted virus and virus that had been passaged 5 times in mice. Further, a virus subline which exhibits

Table 35. *Growth of DEN-1 strains in* SINGH'S *line of A. albopictus cells*

Description of DEN-1 strains	Plaque titer; log no. plaques/ 0.1 ml in PS-Y15	Mouse titer; i.c.	$TCID_{50}$/0.1 ml in *A. albopictus* cells
Human acute phase sera	Neg.	4.1—5.1 [a]	≥5.5—6.5
Mouse pass. 5	3.0	5.4 [a]	5.5
Mouse pass. 24	3.0	5.6 [b]	Neg.
Mouse pass. 124	4.5	6.6 [b]	Neg.

[a] Log ID_{50}/0.1 ml in suckling mice.
[b] Log ID_{50}/0.1 ml in weanling mice.

even greater virulence for this cell line has been established from some of these starting seeds. Despite numerous attempts and manipulations with viral inputs, media, and cell/virus ratios, we were not able to demonstrate any marked CPE that could be passed when we used virus that had undergone 24 or more passages in mice. Prototype mouse adapted DEN-3 and 4 viruses have also failed to yield CPE in our *A. albopictus* cells. On the other hand, we have demonstrated that mouse passaged DEN-1 and DEN-4 strains do multiply in *A. albopictus* cells as determined by plaquing of harvests in PS-Y15. We have not been able to show this for mouse passaged DEN-3.

In contrast to the results with DEN-1, 3 and 4, all type 2 strains we have tested, regardless of stage of adaptation to mice or vertebrate tissue culture, and including acute phase sera, multiplied and produced marked CPE in *A. albopictus* cells. Mouse and tissue culture titers were similar (Table 36). Type 2 mouse-adapted DEN virus also multiplied in PELEG'S and SINGH'S *A. aegypti* lines without concomitant CPE.

The remaining viruses used in the initial studies with suspended cell lines were also tested for their ability to replicate in the monolayer lines. Table 37

Table 36. *Cytopathogenic effects of DEN-2 strains in* SINGH'S *line of A. albopictus cells*

Designation of dengue strain	Plaque titer; log no. plaques/ 0.1 ml in PS-Y15	Mouse titer; i.c.	$TCID_{50}$/0.1 ml in *A. albopictus* cells
Human acute phase sera	Neg.	≤3.0 [a]	3.5—≥5.5
NG "C" MP5	...	4.0 [a]	4.0
NG "C" MP26	6.0	6.6 [b]	6.5—7.5
TH—36 MP24	5.0	6.4 [b]	>5.5
TR-1751	6.0	6.6 [b]	6.0

[a] Log ID_{50}/0.1 ml in suckling mice.
[b] Log ID_{50}/0.1 ml in weanling mice.

Table 37. *Multiplication of several arboviruses in Aedes cell lines in monolayer cultures*

Virus and designation	Mouse passage no.	Viral input (log.)	Virus yield in indicated cell lines[a]		
			SINGH *A. a.*	SINGH *A. albo*	PELEG *A. a.*
WEE	MP-1	4.0	<4.0	≥6.5	6.5
WEE	MP-28	4.0	4.5	≥6.5	5.0
EEE	MP-46	5.0	<5.0	6.0	<5.0
CE (BSF261)	MP-20	4.5	5.0	≥6.5	4.5
CV	MP-10	4.5	4.0	6.5	5.5
JBE	MP-24	4.0	4.0	≥9.5 + CPE	4.5

[a] Log $TCID_{50}$/0.1 ml in PS-Y15 cells.

summarizes the findings. With the one exception noted, no CPE was observed over a 10 day period following inoculation in these cell lines. Therefore, harvests were titered in PS-Y15 cells to determine if multiplication had occurred. With WEE, multiplication to a significant degree occurred with both early and late passage virus in *A. albopictus* cells and in PELEG's *A. aegypti* The former line appeared to be the more sensitive. Mouse passage 46 of EEE showed a low level of multiplication (10-fold) in *A. albopictus* cells only. The prototype mouse adapted CV multiplied in *A. albopictus* cells and to a lesser degree in PELEG's line of *A. aegypti*. Mouse passaged JBE was markedly cytopathogenic for *A. albopictus* cells and titers were routinely $10^{9.5}$/0.1 ml. In this respect this virus behaves differently than SLE and mouse-adapted types 1, 3, and 4 dengue. It behaves, instead, like dengue-2. This virus failed to multiply significantly in the *A. aegypti* lines.

Acknowledgment. This work was supported in part by U.S.P.H.S. Grant No. 4 R-22 IA 08208-02 and the Louisiana State Science Foundation.

Chapter 7

Growth of Viruses in Arthropod Cell Cultures: Applications

I. Attenuation of Semliki Forest (SF) Virus in Continuously Cultured *Aedes aegypti* Mosquito Cells (Peleg) as a Step in Production of Vaccines

J. Peleg

The fact that some of the most widely employed vertebrate cells for the production of human vaccines might be contaminated with various viruses, including the highly pathogenic B virus and the SV_{40} hamster tumorogenic viruses (Hilleman, 1969), led us to investigate the possibility of utilizing cultured mosquito cells for human vaccine production.

The mosquito cells employed in these studies were derived from *Aedes aegypti* embryos. The culture medium, and the techniques employed in continuous cell culture, were reported in detail elsewhere (Peleg, 1969b). These cells were found to be free of detectable contaminating agents. This finding was not unexpected since (a) mammalian viruses are not transmitted transovarially in mosquitoes to their embryos (Chamberlain and Sudia 1961), and (b) the cells were cultured in a medium free of invertebrate hemolymph, a potential source of contamination. Furthermore, these cells have not acquired malignant properties even after 130 passages (tested by inoculating newborn hamsters and mice with cells), as was reported to happen with continuously cultured vertebrate cells (Hayflick and Moorhead, 1961).

The arboviruses which we attempted to attenuate were Semliki Forest (SF) in group A, and West Nile (WN) and Japanese B encephalitis (JBE) in group B.

The passage history of these viruses was unknown, but all of them have undergone at least 15 mouse brain passages in our laboratory. Virus passage was initiated by inoculating tube cultures of mosquito cells with 0.1 ml of mouse brain virus suspensions containing about 10^5 mouse LD_{50}. Pre- and post-infected cultures were incubated at 28° C. Serial passage of virus was carried out at intervals of 3–5 days, by inoculating fresh mosquito cell cultures with 0.1 ml of culture fluid from previous passages. To detect changes in virus virulence, fluids harvested after each 5 passages were titrated in suckling mice by the subcutaneous (SC) and in adult mice by the intracerebral (IC) routes of infection using 0.03 ml as an inoculum. The LD_{50} of the virus was calculated by the method of Reed and Muench (1938). Loss of virulence for adult mice following IC inoculation was taken as an indication of virus attenuation.

JBE and WN have undergone 47 and 62 mosquito cell passages respectively, but both remained virulent for adult mice. On the other hand, SF lost virulence for adult mice inoculated IC after 20 serial passages, but remained virulent for suckling mice by all routes of inoculation. The gradual loss of virulence of the SF virus for adult mice is shown in Fig. 101. Virus,

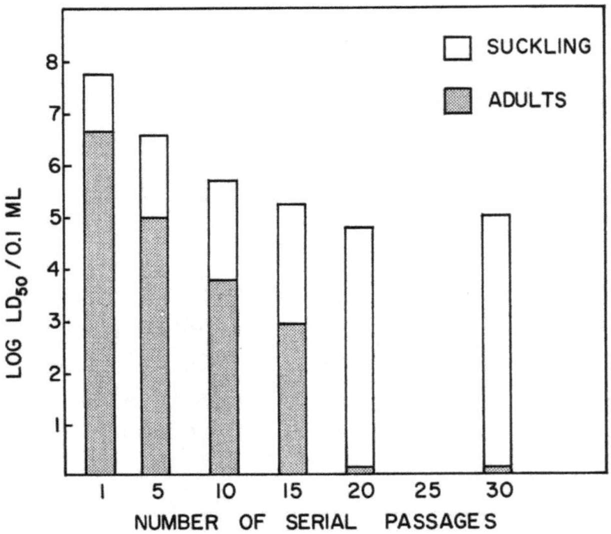

Fig. 101. The effect of serial passage of Semliki Forest virus in *Aedes aegypti* cells upon virulence for suckling and adult mice

although avirulent, was present in the brains of inoculated adult mice. This was demonstrated by inoculating suckling mice with brain suspensions prepared from these mice, and by their resistance to challenge with virulent SF virus.

The properties of the attenuated SF virus and virulent SF viruses are compared in Table 38. Both viruses were virulent for suckling mice, but the

Table 38. *Comparison of properties of parent*

Virus	In vertebrates (virulence)				
	mice		hamsters		chick embryos[a]
	suckling	adults	suckling	adults	
Parent	+	+	+	+	+
Attenuated	+	−	+	−	−

[a] Inoculated into the chorioallantoic membrane.

attenuated virus became avirulent for adult hamsters and adult mice. The attenuated virus was avirulent for 11-day-old chick embryos inoculated by the chorioallantoic route, whereas embryos of the same age were highly susceptible to infection with parent virus. Both viruses could produce cytopathic effects (CPE) and plaques in chick fibroblast cell cultures, but the plaques produced by the parent were about 7 mm in diameter whereas the attenuated virus produced plaques measuring about 3 mm or less (Fig. 102a and b). *Aedes aegypti* mosquitoes supported the proliferation of both viruses to the same extent. Virus transmission by the mosquitoes infected with the attenuated virus was lifelong, whereas mosquitoes infected by the virulent parent virus ceased to transit it at more advanced stages of infection, due to salivary gland damage. The attenuated virus was inactivated at 56° C after 60 min, whereas the parent virus was inactivated after 90 min.

The immunizing capacity of the attenuated virus was tested by: (1) challenging vaccinated animals with the virulent SF virus, and (2) testing the sera of vaccinated animals for neutralizing antibodies.

Vaccination was performed with the virus progenies of the 22d and 23rd serial mosquito cell passages, using adult mice and adult hamsters. Animals were inoculated either IC with a single dose of the attenuated virus, or intraperitoneally (IP) with one or two doses of the virus, administered 7 days apart. The volumes inoculated IC into mice and hamsters were 0.03 and 0.05 ml, respectively, and 0.1 ml IP. Fourteen days later, most of the animals were challenged by the IC route with 5000–10,000 LD_{50} of the parent virus. The remaining mice were bled, and their sera tested for neutralizing antibody.

Intracerebral inoculation of mice with undiluted attenuated virus, or of hamsters with a dilution of 1:10 protected them against the challenge virus (Table 39). As virus inoculum was diluted, its efficiency as a protective agent decreased, and mice inoculated with a dilution of 10^{-3} were susceptible to the challenge virus to the same extent as unvaccinated control mice.

Mice inoculated IP with 2 doses of undiluted virus, or hamsters receiving a single dose of virus diluted 1:10 were also protected against the challenge virus (Table 40). Dilution of the virus inoculum resulted in decreased potency in protecting animals. The inoculation of a second dose of the attenuated virus did not greatly increase protection against the challenge virus over that observed after a single vaccination.

and attenuated SF virus

In monolayers (chick embryo fibroblasts)	In mosquitoes (*A. aegypti*)		*In vitro*
plaque size	proliferation	transmission	Time required for inactivation at 56° C
7 mm	+	±	90 min.
3 mm	+	+	60 min.

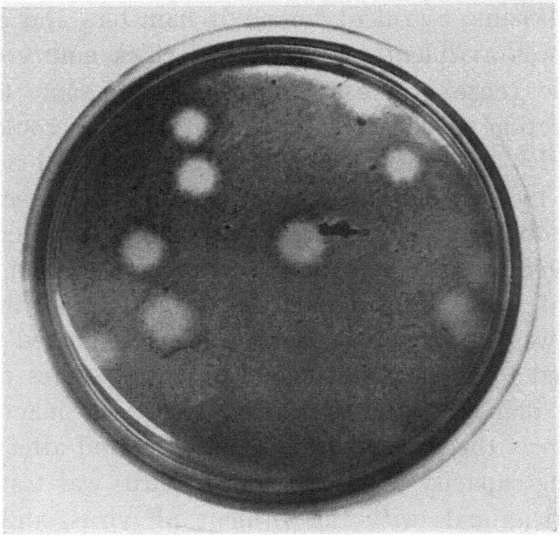

a

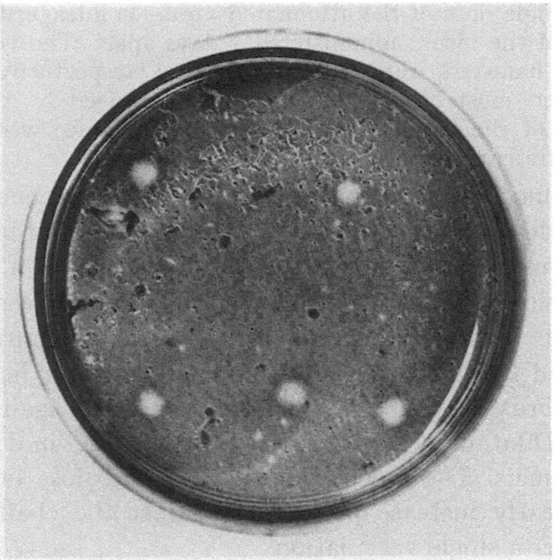

b

Fig. 102a and b. Plaques produced by Semliki Forest virus on chick embryo cell monolayers. a Parent strain. b Attenuated strain

The immune response to vaccination by attenuated virus was more pronounced when the sera of vaccinated animals were tested for neutralizing antibodies.

Prior to test, sera were heated to 56° C for 30 min. In neutralization tests, constant serum versus varying doses of the parent SF virus was employed. The virus-serum mixtures were incubated at 37° C for 45 min, then titrated either in

Table 39. *Challenge of intracerebrally vaccinated animals*[a]

Animals vaccinated	Dilution of attenuated virus	Survival after challenge (dead/total inoculated)		Per cent protection
		unvaccinated	vaccinated	
Mice	10^{-0}	60/60	0/342	100
	10^{-1}	60/60	98/323	70
	10^{-2}	30/30	113/140	19
	10^{-3}	30/30	155/155	0
Hamsters	10^{-1}	9/9	0/24	100

[a] The animals were challenged IC with 5000—10,000 mouse LD_{50} of virulent SF virus 14 days after inoculation with the attenuated virus. The results of three separate experiments are combined.

Table 40. *Challenge of intraperitoneally vaccinated animals*[a]

Animals vaccinated	Vaccination		Survival after challenge (dead/total inoculated)		Per cent protection
	no. of inoculations	dilution of virus	unvaccinated	vaccinated	
Mice	1	10^{-0}	12/12	16/108	85
		10^{-1}	30/30	71/212	66
		10^{-2}	16/16	82/107	23
	2[b]	10^{-0}	16/16	0/240	100
		10^{-1}	12/12	30/120	75
		10^{-2}	10/10	52/80	35
Hamsters	1	10^{-1}	9/9	0/24	100

[a] The animals were challenged IC with 5000–10,000 mouse LD_{50} of virulent SF virus 14 days after the last inoculation with the attenuated virus. The results of two separate experiments are combined.

[b] Administered 7 days apart.

suckling mice by SC, or in 18-day-old mice by IP routes. In control experiments, antiserum was replaced by normal serum from the same animal species. The neutralization index of each antiserum was calculated by subtracting the LD_{50} obtained in its presence from that obtained in control experiments.

Regardless of the route of inoculation, the number of doses administered. animals vaccinated with undiluted virus or with a dilution of 1:10, produced antibody sufficient to neutralize between 10^5 to 10^6 mouse LD_{50} of the virulent SF virus (Table 41). The immune state elicited by the attenuated virus lasts for prolonged periods. Hamsters bled 40 days after vaccination, yielded similar high levels of antibodies. Immunity was induced only by the living virus, heat inactivation abolished its immunizing capacity.

Propagation of attenuated SF virus in intact *A. aegypti* mosquitoes was next investigated to learn whether the virus is capable of proliferating in

Table 41. *Levels of neutralizing antibodies in the sera of vaccinated animals*[a]

Animals vaccinated	Immunization			SF virus titer[b] in the presence of		Serum neutralizing index[c]
	route of inoculation	dilution of virus	no. of inoculations	immune serum	normal serum	
Mice	IC	10^{-0}	1	4.0	9.3	5.3
	IP	10^{-1}	2	3.0	9.0	6.0
Hamsters	IP	10^{-1}	1	4.0	9.5	5.5
	IP	10^{-1}	2	3.0	9.5	6.5

[a] The animals were inoculated with the attenuated SF virus and bled 14 days later. Their sera were tested in neutralization experiments, using suckling mice.

[b] Log LD_{50}/0.1 ml.

[c] Titer in the presence of immune serum subtracted from that obtained in the presence of normal serum.

mosquitoes, and if so, whether it retains its attenuated properties, or it reverts to virulence.

To answer these questions, *A. aegypti* mosquitoes were allowed to feed on viremic suckling mice inoculated 48 hrs before with the attenuated SF virus. Subsequently, at various intervals suckling as well as adult mice were exposed to bites of these mosquitoes. Then the mosquitoes were triturated, and the suspensions were titrated by the SC route in suckling mice and IC in adult mice.

Suckling mice, exposed to the bites of infected mosquitoes died regularly as a result of virus transmission, whereas all the adult mice, employed in transmission experiments, remained alive (Table 42). Suspensions prepared from infected mosquitoes killed suckling mice in high dilutions. On the other hand all the adult mice, including the ones inoculated with undiluted suspensions, survived, and were resistant to IC challenge of about 500 LD_{50} of

Table 42. *Properties of attenuated SF virus propagated in A. aegypti mosquitoes*[a]

Days after infection	Mice exposed to infected mosquitoes		Virus content of mosquitoes[b]
	suckling	adult	
0	not tested	not tested	4.8
7	died	survived, resistant to challenge C	6.7
14	died		6.3
21	died		6.6
26	died		6.4

[a] Mosquitoes were allowed to feed on viremic suckling mice and at intervals allowed to feed on normal mice.

[b] Determined by SC inoculation of suckling mice. Log LD_{50}/0.1 ml.

[c] IC with 500 LD_{50} of parent virulent virus.

virulent virus. The experiments provide evidence that association of attenuated virus with *A. aegypti* mosquitoes did not cause reversion to virulence.

In conclusion, the preliminary results presented here indicate that in certain cases mosquito cells can be utilized for the attenuation of arboviruses and that the attenuated viruses may serve as good immunogens. The safety of mosquito cells for vaccine production, tested by the methods at present available, may be greater than vertebrate cells, but whether they contain some hitherto unknown substances or contaminants which might adversely affect human beings, remains to be investigated.

Acknowledgment. This investigation was supported by Research Grant 06-325-01 from the U.S. Public Health Service, Bureau State Services.

II. Cultivation of Oncogenic Viruses in Mosquito Cells *in Vitro*

J. Rehacek, T. Dolan, K. Thompson, R. G. Fischer, Z. Rehacek, and H. Johnson

During the last few years studies have been conducted on the role of arthropods as vectors of oncogenic viruses. In our laboratory insects were studied as vectors of murine leukemia virus, murine sarcoma virus and a reticuloendotheliosis virus causing avian leukemia. Preliminary results indicated that certain insects in some instances can act as mechanical vectors of these viruses. To complement these studies experiments were performed *in vitro* to determine if cultured mosquito cells can serve as host cells for the replication of oncogenic viruses.

A. Materials and Methods

Viruses. Friend leukemia virus having an unknown number of passages in BALB/c mice, murine sarcoma virus (Moloney-RP) in its 121st passage in suckling mice and reticuloendotheliosis virus, strain T, in its 35th passage in White Rock chicks, were used. Plasma or tumor extract from infected animals were used as the cell culture inoculum.

Mosquito tissue cultures. Aedes albopictus cell line derived by Singh (1967) was obtained in the 61st passage from Dr. Yunker. An additional 14 passages were made during these experiments. The medium used was HLH, consisting of lactalbumin hydrolysate in Hanks' salts (Grand Island Biological Co.), 0.1% bovine albumin (Fraction V), and 10% fetal calf serum with antibiotics. All cultures were incubated at 28° C.

Infection of tissue cultures and virus assay. Mosquito tissue cultures which had not formed complete monolayers (approximately 6×10^5 cells/ml) were infected with varying concentrations of viruses. The cells were not washed after inoculation.

Samples of culture media and cells were assayed for virus daily during the first week, then at 3–7 day intervals until the end of the second month following inoculation.

Friend virus infectivity titers were determined in 3–4 week-old BALB/c mice by intraperitoneal (IP) inoculation. The response to Friend virus infection was determined by spleen focus formation, splenomegaly and death of mice (Chirigos et al., 1967). The mice were sacrificed 3 weeks following infection, dissected spleens were weighed and immersed in fixative for about 2 min to expose foci.

The T strain of reticuloendotheliosis virus titer was determined in a similar manner. The samples of 0.1 ml were injected IP into test chickens. Three weeks later, the surviving animals were sacrificed, autopsied and examined histologically for specific changes in the spleen and liver.

Stocks of murine sarcoma virus contain leukemia virus and a mixture of competent and defective sarcoma virus (O'CONNOR and FISCHINGER, 1968). Two virus particles are necessary for focus formation, the murine sarcoma genome and a leukemia virus (HARTLEY and ROWE, 1966). The sarcoma genome is defective and requires the murine leukemia virus to supply the protein coat to become infectious. To assay for the replication of the defective sarcoma virus in mosquito cells, samples of culture medium or cells were assayed in secondary NIH Swiss mouse embryo cell cultures with added Rauscher leukemia as a helper virus to recover defective virus (FISCHINGER and O'CONNOR, 1968). Competent murine sarcoma virus does not require helper leukemia virus for focus formation. The competent virus can be considered as two packets of nucleic acid (the sarcoma and the leukemia genomes) surrounded by a common membrane. The virus is detected without helper virus by direct assay in Swiss mouse embryo cells.

In two experiments, mosquito cells infected for 9 or 25 days with mouse sarcoma virus were co-cultivated with normal secondary mouse embryo cells. The experiments were performed in a manner similar to that used for "rescue" of sarcoma virus from infected hamster cells (HUEBNER et al., 1966). Mosquito cells were removed from the tissue culture flasks by scraping with a rubber policeman and primary mouse cells were detached from vessels by the use of trypsin. Approximately 5.0×10^5 cells of each type were planted into plastic dishes (Falcon Plastics). The medium was modified McCoy's 5A with 15% fetal calf serum (Grand Island Biological Co.). As a control 1.0×10^6 mosquito cells were planted into the dishes without mouse cells. At the time of the planting, all dishes were divided into two groups, with one group receiving an additional 0.1 ml of a 1:5 dilution of Rauscher leukemia virus. Cultures were incubated at 28° C for 24 hrs and placed in a 35° C incubator. After 7 days, fluids were removed and frozen at −70° C, the media replaced and fluids again harvested after one month.

B. Results

Friend virus was found to persist in mosquito cell culture up to 5 days following inoculation (Table 43). The period of persistence correlated with the amount of virus inoculated. When cultivating the virus in growth medium without cells similar results were observed. Virus titers gradually decreased in the mosquito cells as well as in the control medium. In no case did the virus yield from tissue cultures exceed the virus inoculum.

The same results were obtained in 6 experiments with murine sarcoma virus. Titers of murine sarcoma in *A. albopictus* cells rapidly decreased and virus was not generally detected after 24 hrs of inoculation. Thermal inactivation of the virus occurred in 24 hrs when placed in cell-free growth medium. Co-infection of samples with Rauscher leukemia virus in the assay system failed to reveal replication of the defective virus.

In the experiments using combined mouse and mosquito cultures, it was noted that the mosquito cells grew very well for about 14 days with the secondary mouse embryo cells. At this time the control cultures of mosquito cells alone

Table 43. *Experiments on cultivation of Friend virus (murine leukemia) in Aedes albopictus cells in vitro*

Experiment No.	Samples tested	Days after infection												
		0[a]	1	2	3	4	5	6	7	8	9	10	11	12—70 (in 7-day intervals)
1	Mixture of cells and medium	2.5[b]		1.2		—		—			—			—
2	Mixture of cells and medium	1.9		$\leq$ 0.3		—			—				—	—
3	Mixture of cells and medium	3.1				$\geq$ 1.1							—	—
4	Mixture of cells and medium	3.4		0.9		0.9			—				—	—
5	Mixture of cells and medium	2.2		$\leq$ 0.3		—					—		—	—
6	Mixture of cells and medium	2.3		1.2		—		—						—
7	Mixture of cells and medium	4.3	2.1	$\leq$ 0.3	$\leq$ 0.5	$\leq$ 0.3	$\leq$ 0.4	—	—		—			—
8	Mixture of cells and medium	2.8	1.3	—	—	—	—	—	—					—
9	Mixture of cells and medium	2.9	2.2	$\leq$ 0.5	—	$\leq$ 0.3	—	—	—	—			—	—
	Cell-free medium (control)	2.9	1.9	1.3	$\leq$ 0.3	—	—	—	—	—			—	—
10	Mixture of cells and medium	3.2	1.7	0.9	$\leq$ 0.3	—	—	—	—			—		—
	Cell-free medium (control)	3.2	1.7	1.1	$\leq$ 0.3	—	—	—	—			—		—
11	Mixture of cells and medium	3.1	1.3	—	—	—	—	—	—			—		—
	Cell-free medium (control)	3.1	2.5	$\leq$ 0.5	$\leq$ 0.5	—	—	—	—			—		—

[a] Samples withdrawn immediately after inoculation.
[b] Log mouse LD_{50}/1 ml. — Virus not detected.

degenerated rapidly as did the mosquito cells co-cultivated with mouse cells. The mouse cells then formed a monolayer and did not show any cytopathological changes. All fluids harvested from the various cultures were found to be negative when assayed for sarcoma virus.

Similar results were obtained with reticuloendotheliosis virus, T strain, in four experiments. The samples were assayed for 60 days after the inoculation of mosquito cultures. All samples were negative for virus.

Acknowledgment. This study was conducted under Contract PH43-66-8 within the Special Virus Cancer Program of the National Cancer Institute, NIH, USPHS.

III. RNA of Uninfected and Sindbis Virus-Infected *Aedes albopictus* Cells

VICTOR STOLLAR, THOMAS M. STEVENS and THOMAS SHENK

Little is known of the sequence of biochemical events of arbovirus replication in insect cells. Therefore, we have begun to examine the properties of ribonucleic acid (RNA) extracted from uninfected and Sindbis virus infected *Aedes albopictus* (SINGH) cells. STEVENS (1970) has described the growth of Sindbis virus in this cell line. Although no obvious ill effects were produced on the cells, infectious titers as high as 10^8 plaque forming units (pfu) per ml were obtained.

A. Properties of RNA Extracted from *A. albopictus* Cells

RNA was extracted from monolayers of mosquito cells by the hot phenol method of SCHERRER and DARNELL (1962) and fractionated by sucrose gradient centrifugation. Vertebrate cells contain three species of RNA separable by sucrose gradient centrifugation: the 28S and 18S ribosomal RNA and 4S transfer RNA. On the other hand, *A. albopictus* RNA extracted with hot phenol shows only two RNA species (Fig. 103).

The sedimentation coefficients of the two *A. albopictus* RNA species were estimated by comparison in sucrose gradients with ribosomal and trans er RNA obtained from a hamster cell line, BHK-21 (Fig. 104). Mosquito cells were labeled with ^{3}H-uridine for 24 hrs and the hamster cells with ^{14}C-uridine for the same period. The cells were mixed and the RNA extracted with hot phenol. The hamster cell RNA contained the expected 28S, 18S and 4S species. The mosquito cell RNA contained 18S and 4S species. The 28S ribosomal RNA was not observed.

Similar observations were described for RNA extracted from silk moth pupae, *Hyalophora cecropia* (APPLEBAUM et al., 1966) and from *Drosophila virilis* (GREENBERG, 1969). In both cases "28S RNA" could be demonstrated with gentler methods of extraction. Accordingly, RNA was extracted from *A. albopictus* cells with sodium dodecyl sulfate (SDS) at room temperature (BECKER and JOKLIK, 1964). The RNA product of this extraction was frac-

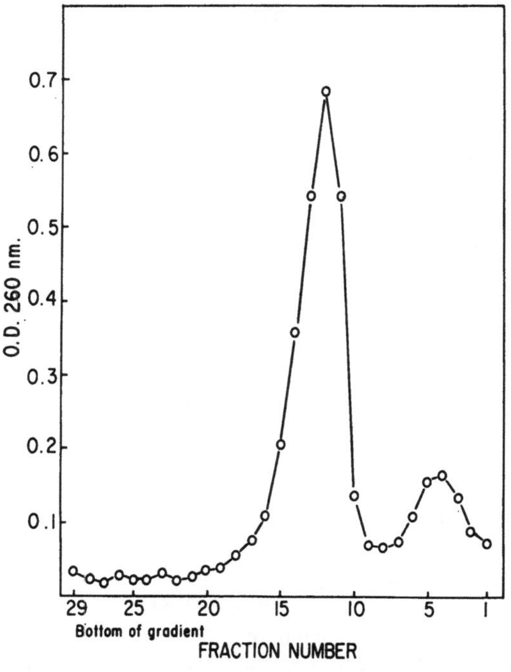

Fig. 103. Sucrose gradient centrifugation of *A. albopictus* cell RNA extracted by the hot phenol method. Centrifugation was in the Spinco SW25-1 rotor at 23,000 rpm for 16 hrs at 4° C. The sucrose gradient was 5–20% w/w

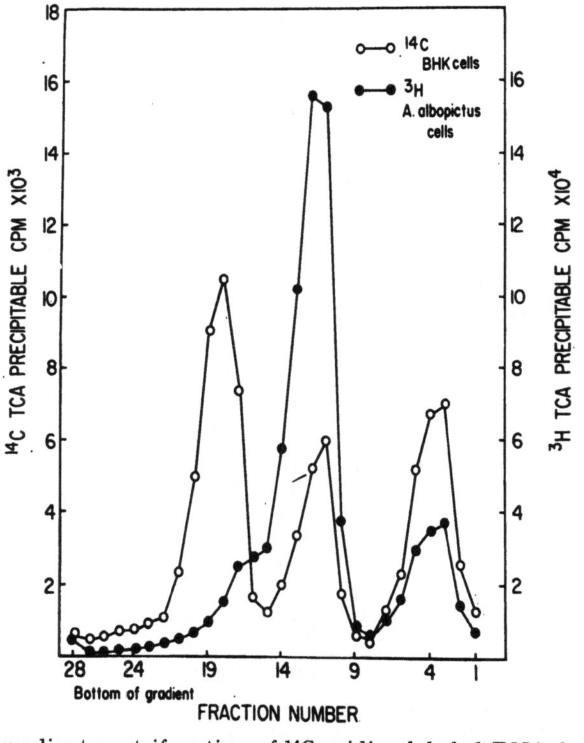

Fig. 104. Sucrose gradient centrifugation of ^{14}C-uridine labeled RNA from BHK-21 cells and ^{3}H-uridine labeled RNA from *A. albopictus* cells. Cells were labeled for 24 hrs, mixed and extracted together using the hot phenol method. The conditions of centrifugation were as described in the legend to Fig. 103

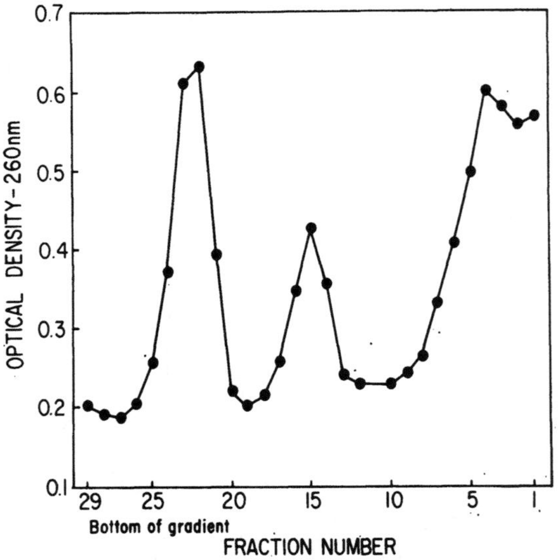

Fig. 105. Sucrose gradient centrifugation of SDS-extracted RNA from the cytoplasm of *A. albopictus* cells. Cytoplasmic RNA was extracted by the method of Becker and Joklik (1964) using only SDS. Centrifugation was in the SW 25-1 rotor at 19,000 rpm for 18 hrs at 22° C. The sucrose gradient was 15–30% w/w and contained 0.5% SDS

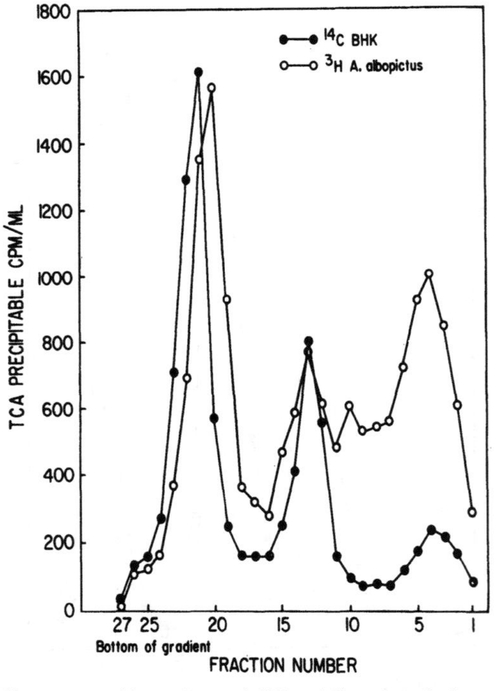

Fig. 106. Sucrose gradient centrifugation of ^{14}C-uridine labeled cytoplasmic RNA from BHK-21 cells, and ^{3}H-uridine labeled cytoplasmic RNA from *A. albopictus* cells. Cells were labeled for 24 hrs, and extracted separately using only SDS. RNA from the two cell types were mixed and examined in a single sucrose gradient. The conditions of centrifugation were as described in the legend to Fig. 105

tionated on sucrose gradients containing SDS. With this extraction procedure, *A. albopictus* cells were found to contain the expected three major types of cellular RNA, including the 28S species (Fig. 105).

To confirm the sedimentation values observed for the RNA species isolated from the mosquito cells, an experiment was performed with a mixture of RNA extracted with SDS derived from ^{3}H-uridine labeled *A. albopictus* cells and ^{14}C-uridine labeled BHK-21 cells (Fig. 106). The 18S RNA from the

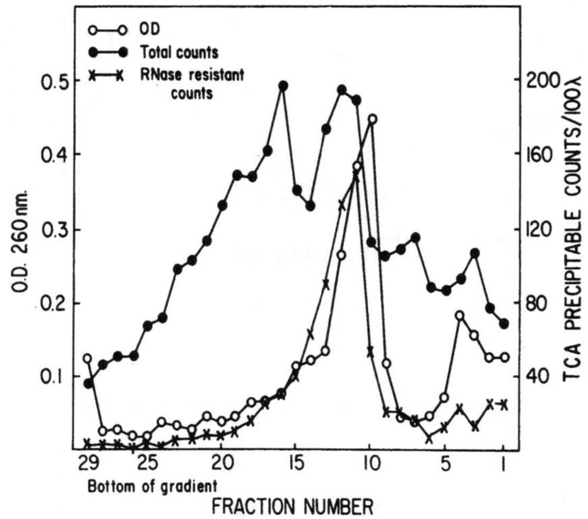

a

Fig. 107a–c. Sucrose gradient centrifugation of RNA from Sindbis virus-infected *A. albopictus* cells. Monolayers of *A. albopictus* cells were infected with Sindbis virus, and 24 hrs later treated with actinomycin D. After 1 hr exposure to actinomycin, cells were labeled with ^{3}H-uridine for 10 min (a), 40 min (b) and 420 min (c). RNA was extracted by the hot phenol method. Centrifugation was in the SW 25-1 rotor at 22,000 rpm for 16 hrs at 4° C. The sucrose gradient was 5–20% w/w

mosquito cells co-sedimented with the hamster cell 18S RNA, whereas the 28S mosquito cell RNA sedimented slightly more slowly than the BHK-21 28S RNA. These experiments show that *A. albopictus* cells do contain a 28S RNA but it is much less stable than that extracted from mammalian cells. This is the first report of an unstable 28S RNA in established mosquito cell lines.

B. Properties of Sindbis Virus-specific RNA Extracted from Infected *A. albopictus* Cells

Viral RNA synthesis in *A. albopictus* cells infected with Sindbis was measured in the presence of actinomycin D. It was demonstrated in a preliminary experiment that actinomycin D inhibited well over 90% of *A. albopictus* cell RNA synthesis.

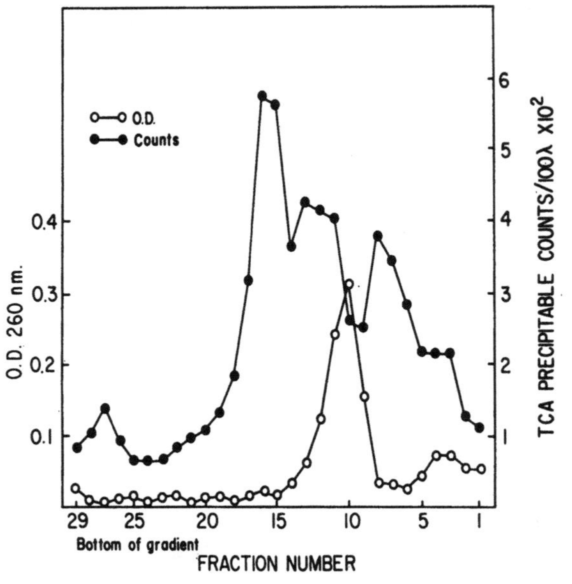

Fig. 107b

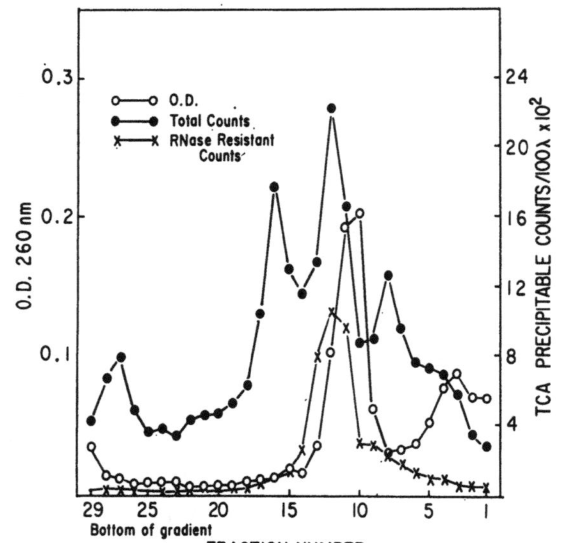

Fig. 107c

To study the sequence of viral RNA synthesis, *A. albopictus* cells were infected with Sindbis virus for 24 hrs, treated with actinomycin D for 1 hr, and then labeled with ^{3}H-uridine for varying periods of time. RNA was extracted with hot phenol and examined on sucrose gradients. RNA from cells labeled for 10 min contained two small peaks (4S and 12–14S) encountered also in uninfected actinomycin treated cells. Two larger RNA species with sedimentation coefficients of approximately 22S and 32S were found only in the infected cells (Fig. 107a). The 32S peak showed considerable trailing on the heavy side. When fractions of this gradient were treated with ribonuclease, a ribonuclease-resistant peak was found directly under the 22S peak. This peak is presumably the double-stranded replicative form of viral RNA (FRANKLIN, 1967).

The sucrose gradient pattern of RNA from cells labeled for 40 min is shown in Fig. 107b. Three major virus-specific RNA peaks were observed, the 22S, 32S and 42S RNA. In this gradient there was no trailing on the heavy side of the 32S RNA. The 42S RNA probably represented newly synthesized mature viral RNA.

The sucrose gradient pattern of RNA from cells labeled for 7 hrs (Fig. 107c) is very similar to that observed at 40 min except that the amount of each species was increased. The 22S RNA peak was the most prominent and retained its ribonuclease resistance. In another experiment, not shown, it was demonstrated that all the RNA peaks were degradable by alkali.

In summary, these experiments show that *A. albopictus* cells contain an inherently unstable 28S RNA, which was destroyed by hot phenol extraction, but could be extracted by SDS. Sindbis virus infection of *A. albopictus* cells, treated with actinomycin D, resulted in the appearance of three virus-specific RNA species with sedimentation constants of 22S, 32S, and 42S. The 42S RNA probably represents mature viral RNA and the 22S RNA the replicative form of the viral RNA. The nature of the 32S RNA is not clear at this time.

Acknowledgment. This work was supported by research grant AI-05920 from the National Institutes of Health, U.S. Public Health Service.

Chapter 8

Homoptera Cell Culture and Its Application to the Study of Plant Pathogens

H. HIRUMI

I. Introduction

Following the first attempt to grow testicular follicle cells of a moth (GOLDSCHMIDT, 1915), a large number of experiments on the *in vitro* cultivation of insect cells and tissues were carried out. Although the earlier attempts met with unexpected difficulties, extensive progress has been made during the past decade as described in a number of reviews (GOODCHILD, 1954; HARTZELL, 1958; DAY and GRACE, 1959; MARTIGNONI, 1960, 1963; JONES, 1962; MARAMOROSCH, 1962; SCHNEIDER, 1967; VAGO, 1967; GRACE, 1968, 1969; VAUGHN, 1968; MITSUHASHI, 1969a; BLACK, 1969; HIRUMI and MARAMOROSCH, in press) and in previous chapters of this volume. The present chapter will be focused on recent developments in Homoptera cell cultures and their application to the study of plant pathogens.

A large number of plant viruses are transmitted by Homoptera. Insects, such as leafhoppers, planthoppers and aphids, are known to be vectors of economically important and widely distributed pathogens, which in some instances ean multiply in vector cells as well as in plant cells. To understand the vector-virus interactions, extensive and exhaustive studies have been carried out previously (BLACK, 1959; MARAMOROSCH, 1963). Electron microscopic studies (SHIKATA and MARAMOROSCH, 1969a; MARAMOROSCH et al., 1969) and serological tests (SINHA, 1969) have revealed the localizations of virions and the fate of plant pathogenic viruses in insect vectors. Nevertheless, information on the mechanism of virus penetration into insect cells and of virus multiplication is still fragmentary.

Recently, mycoplasma-like microorganisms have been described as the suspected causative agents of mulberry dwarf, potato witches' broom, aster yellows and paulownia witches' broom (DOI et al., 1967). Their findings led to the search for similar microorganisms in other diseased plants of the yellows type, which were previously considered as being caused by viruses. Since then, mycoplasma-like bodies have been reported as associated with more than thirty plant diseases, including many which are transmitted by homopterous vectors (WHITCOMB and DAVIS, 1970; MARAMOROSCH et al., 1970). Similar microorganisms have also been found in various organs of yellow-infected leafhopper vectors, suggesting the possible multiplication of the microorganisms in the insects (NASU et al., 1967; GIANNOTTI et al., 1968a, b; GRANADOS et al., 1968a; MARAMOROSCH et al., 1968; HIRUMI and MARAMOROSCH, 1969; SHIKATA and MARAMOROSCH, 1969b; BRČÁK and KRÁLÍK, 1969). At present, there can be little doubt that they belong to the Mycoplasmatales and that many of them can multiply in both insect vector and host plant cells as do Homoptera-borne plant pathogenic viruses.

During the past decade, *in vitro* cultivation of Homoptera cells, particularly of leafhopper vectors, has been attempted in several laboratories. However, greater difficulties were encountered in the cell culture of paurometabolous insects, than of holometabolous insects, such as the Lepidoptera.

The first breakthrough in the development of leafhopper cell culture came in 1964, when the advantage of embryonic tissues in the blastokinetic stage of *Macrosteles fascifrons*, a vector of aster yellows agents, for culture was discovered (HIRUMI and MARAMOROSCH, 1964a, b). Since then, the embryonic cell culture technique has become a useful method for homopterous insect cell cultures. To date, successful cell growths have been obtained from 13 different species of leafhoppers (Table 44). More than ten types of cells have been described in these cultures.

The second major breakthrough was achieved by CHIU and BLACK in 1967, when they first established leafhopper cell lines from 4 different species (Table 44). Some cell lines of *Agallia constricta* (Van Duzee), a vector of wound tumor virus, became stable and are now being used in many laboratories. Recently, another leafhopper cell line was established from *Colladonus montanus* (Van Duzee), a vector of western X-disease agents, by D. D. JENSEN (personal communication).

The leafhopper cell culture system can now be adequately applied to the study of virus-vector or mycoplasma-vector interactions. Initial studies along this line have been carried out recently, using leafhopper cultures with the aster yellows agent (HIRUMI and MARAMOROSCH, 1963b; MARAMOROSCH et al., 1965), rice dwarf virus (MITSUHASHI, 1965b; MITSUHASHI and NASU, 1967), wound tumor virus (SINHA, 1965; CHIU et al., 1966; CHIU and BLACK, 1967, 1969; GÁMEZ and CHIU, 1968; HIRUMI and MARAMOROSCH, 1968), wound tumor virus and reovirus (STREISSLE et al., 1967) and *Chilo* iridescent virus (an insect virus) (MITSUHASHI, 1967b). These efforts have been directed at studying virus penetrations into insect cells, to elucidate the initial stages of virus multiplications, to clarify host specificities and to obtain methods for accurate quantitative analysis of Homoptera-borne plant-pathogenic viruses. Currently, studies on the multiplication of plant-pathogenic mycoplasmas, newly implicated plant pathogens of yellows type diseases, are being intensified using leafhopper cell cultures.

Cell cultures of aphids and of planthoppers have been studied less extensively in comparison to those of leafhoppers and are now in early stages of their development. To date, preliminary attempts with the pea aphid, *Acyrthosiphum pisum* (Harris), a vector of pea enation mosaic virus (TOKUMITSU and MARAMOROSCH, 1966; HIRUMI, unpublished data), the sowthistle aphid,

Table 44. *Successful in vitro cultivation of cells, tissues and organs of Homoptera*

Vector	Plant pathogen[a]	Culture[b]	References
Leafhopper			
Aceratagallia sanguinolenta	PYDV	Cell line (E)	Chiu and Black (1967)
Agallia constricta	WTV, PYDV	Survival (N, A)	Hirumi and Maramorosch (1963a)
		Primary (E, N, A)	Mitsuhashi and Maramorosch (1964)
		Cell line (E)	Chiu and Black (1967)
Agallia quadripunctata	WTV, PYDV	Cell line (E)	Chiu and Black (1967)
Agalliopsis novella	WTV	Cell line (E)	Chiu and Black (1967)
Cicadella viridis		Primary (A)	Vago and Flandre (1963)
Colladonus montanus	WXM[c]	Cell line (E)	Jensen (personal communication)
Dalbulus maidis	CSM[c]	Primary (E, N, A)	Mitsuhashi and Maramorosch (1964)
Inazuma dorsalis	RDV	Primary (E)	Yamada et al. (1969)
Macrosteles fascifrons	AYM[c]	Survival (N, A)	Hirumi and Maramorosch (1963a)
		Primary (E)	Hirumi and Maramorosch (1964a)
Macrosteles sexnotatus	EAYM[c];	Primary (A)	Vago and Flandre (1963)
Nephotettix apicalis	RDV, RYDM[c]	Primary (E)	Mitsuhashi (1969a)
Nephotettix cincticeps	RDV, RYDM[c]	Primary (E)	Mitsuhashi (1965a)
Philaenus spumarius	LDV	Primary (A)	Vago and Flandre (1963)
Aphid			
Acyrthosiphum pisum	PEMV	Survival (E, N, A)	Tokumitsu and Maramorosch (1966)
		Primary (E)	Hirumi (unpublished data)
Hypermyzus lactucae	SYVV	Primary (E)	Peters and Black (Black, 1969)
Planthopper			
Laodelphax striatellus	RSV, RBSV	Primary (E)	Mitsuhashi (1969b)

[a] Important plant pathogens (for further information, see Heinze, 1959; Nielson, 1968; Ishihara, 1969).

[b] Cultures and materials used in *in vitro* cultivations. Cell line, an established line; Primary, active cell growth of cells, attached to culture flasks; Survival, maintained *in vitro* for several weeks. Developmental stages of insect materials are indicated as follows: A, adult; E, embryo; N, nymph.

Hypermyzus lactucae (L.), a vector of sowthistle yellow vein virus (PETERS and BLACK, 1970), and the smaller brown planthopper, *Laodelphax striatellus* Fallen, a vector of rice stripe virus, rice black streaked virus and northern mosaic virus of barley, wheat and oat (MITSUHASHI, 1969b) have been carried out (Table 44).

II. *In Vitro* Cultivation

A. Early Work

The first attempts using tissue culture techniques to study *in vitro* multiplications of the aster yellows agent in its leafhopper tissues were carried out by MARAMOROSCH (1956). Since neither adequate culture medium nor specific techniques were available at that time, he simply used body fragments of yellows-infected nymphs of *M. fascifrons*.

After a minimum incubation period of 29 days, the insects injected with the supernatant obtained from the yellows-infected cultures became infective. Since it was known that the aster yellows pathogen could not maintain its infectivity *in vitro* for more than a few hours at room temperature, the experiment suggested that the body fragments survived *in vitro* for at least 10 days and sustained the survival of the aster yellows agent. Although no cell growth of the leafhopper was obtained, the experiment led to further studies of the *in vitro* cultivations of leafhopper cells, tissues and organs.

Following the first experiment, several attempts to obtain *in vitro* growth of cells and tissues of *M. fascifrons* and *Dalbulus maidis* Del. & W., a vector of corn stunt agents, were carried out by GRACE (1959). Although these attempts met with little success and resulted in no cell growth, the gut tissues derived from these leafhoppers survived in a culture medium for several days. He concluded that the medium used in these experiments was not adequately defined for leafhopper cell growth.

In 1963, HIRUMI and MARAMOROSCH (1963a) succeeded in the cultivation of ovaries, testes, gut, brain, salivary glands and Malpighian tubules obtained from surface sterilized nymphs and adults of *M. fascifrons* and *A. constricta*.

Survival of the organs for periods of over 3 months was obtained. Active cell proliferation and mitotic divisions were observed upon staining with acetic Orcein after periods ranging from 5 days to 8 weeks, although these cells did not adhere to the glass surface and remained in suspension in the medium.

[c] Previously considered as a virus but recently implicated as a mycoplasma agent. AYM, aster yellows (mycoplasma); CSM, corn stunt (mycoplasma); EAYM, European aster yellows (mycoplasma); LDV, lucerne dwarf virus; PEMV, pea enation mosaic virus; PYDV, potato yellow dwarf virus; SYVV, sowthistle yellow vein virus; RBSV, rice black streaked virus; RDV, rice dwarf virus; RSV, rice stripe virus; RYDM, rice yellow dwarf (mycoplasma); WTV, wound tumor virus; WXM, western X disease (mycoplasma).

VAGO and FLANDRE (1963) also attempted to grow leafhopper cells *in vitro*. By means of a color indicator in a capillary, they established that the pH of the hemolymph of cicadellids, *M. sexnotatus* Fallen, a vector of European aster yellows agents, and *Philaenus spumarium* L., a vector of lucerne dwarf virus, was 6.4–6.6. Proliferation of fibroblast-like cells from the hypodermis and ovaries, and of epithelial cells from the digestive tracts of *Cicadella viridis*, *M. sexnotatus* and *P. spumarius* were observed. Their experiments provided cell growths in the semi-plasma clot cultures. However, this technique had to be discontinued because of the inconvenience of obtaining prolonged cultures, subculturing, and studying of the plant-pathogenic viruses *in vitro*.

B. Primary Culture of Leafhopper

Growth potentiality. One explanation for the great difficulties encountered in the first insect tissue culture experiments might have been the lack of undifferentiated cells, because most tissues of insects differentiate in early stages of development. Since the use of tissues and organs obtained from nymphs and/or adults met with little success, HIRUMI and MARAMOROSCH (1964a) employed the embryonic tissues of *M. fascifrons* for *in vitro* cultivations and succeeded in obtaining very active cell growth. Following this experiment, they examined the growth potential of different embryonic stages (HIRUMI and MARAMOROSCH, 1964b).

To obtain a large number of embryos at the same developmental stage, male and female adults, approximately 200 insects per cage, were caged on young rye plants, *Secale cereale* L., in 10-cm pots. The insects were transferred to fresh plants every day. The plants with deposited eggs were maintained at 25° C with a 16-hr photoperiod provided by 6600 lu/m² of fluorescent light. Under these conditions, first-instar nymphs of *M. fascifrons* usually hatched 11 days after oviposition.

Embryos on the 1st, 3rd, 5th, 7th, 8th, 10th and 11th day after oviposition were tested. Active cell growth was obtained from 7- and 8-day-old embryos, while no cell growth was observed from the early and late stage embryo development. These stages could be determined by the location and size of the eye under a dissecting microscope. Small, light pinkish eyes appeared in the posterior region of the egg on the 6th day after oviposition, and then moved to the anterior region during the next several days, increasing in size and density of color (Fig. 108A–E). The stage that provided the most suitable source for the cell culture was identified as the stage of blastokinetic movement, that is between the 6th and 9th day after oviposition (Fig. 108A–C). After the proper stage had been established, the adult insects were confined on rye plants for a week. Then, during the following 7 days, the eggs in the blastokinetic movement stage were collected and used for *in vitro* cultivation. However, the developmental periods varied among various species and differed with the environmental conditions.

Later, it was demonstrated by MITSUHASHI and MARAMOROSCH (1964) that leafhopper cells also could grow *in vitro* from tissues of other developmental stages. These experiments, and the earlier work by VAGO and FLANDRE (1963), revealed that leafhopper tissues from other than the blastokinetic stage, including nymphal and imaginal tissues, also have some growth potentialities *in vitro*. Still, embryonic tissues in the blastokinetic stage are widely recognized as providing the most useful materials for the cell culture of many leafhopper species (Table 44).

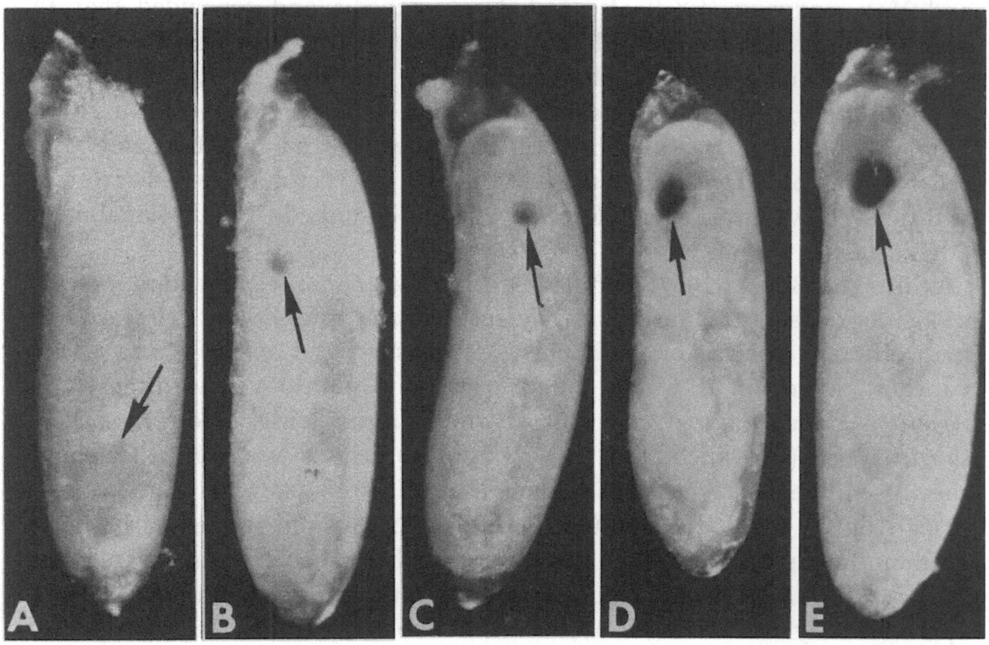

Figs. 108 A–E. Eggs of *Macrosteles fascifrons*, showing different developmental stages of the embryos (70 ×). Arrows indicate the location of eyes. A, 7-day-old egg, at an early stage of blastokinesis; B, 8-day-old egg; C, 9-day-old egg, at a late stage of blastokinesis; D, 10-day-old egg; E, 11-day-old egg, just before hatching. Embryos during blastokinetic stages (A–C) provide the most adequate material for *in vitro* cultivation

Eggs are collected into a 12 ml glass centrifuge tube containing EARLE's BSS. After an adequate number of eggs, usually about 150–200, settle on the bottom of the tube, the EARLE's BSS is gently removed. The eggs are resuspended in 5 ml of 70% ethanol for 60 sec, and then rinsed three times with EARLE's BSS.

Surface sterilization of leafhopper nymphs and adults is relatively difficult; therefore, excised tissues are immediately disinfected after dissection of the insects. VAGO and FLANDRE (1963) use a series of five antibiotic baths, consisting of bacitracin (5 IU/ml), colimycin (25 µg/ml), penicillin (200 IU/ml), streptomycin (0.05 mg/ml), chloramphenicol (5–10 µg/ml), and mycostatin (optional) on excised cicadellid tissues. After surface sterilization in 0.1% Hyamine for 3 min, leafhopper nymphs and adults are dissected in drops of 0.1% Hyamine solution (HIRUMI and MARAMOROSCH, 1963 a, b). The excised tissues are rapidly washed in fresh 0.1% Hyamine solution, and then rinsed several times with culture medium.

Culture media. It was difficult to devise a suitable culture medium for leafhopper cells because of the inadequate analysis of the hemolymph of the small insects. All leafhopper culture media used for previous work were developed through trial and error by modifying established culture media. As a result of the initial work of testing nine different culture media, modified TC 199 (Morgan et al., 1950), with modified Medium B. M. 22, originally designed for *Bombyx mori* L. by Vago and Chastang (1958), was used for the primary culture of *M. fascifrons*, and provided active cell growth *in vitro* (Hirumi and Maramorosch, 1964a, b, c, 1968; Hirumi, 1965). Mitsuhashi and Maramorosch (1964) also tested seven media and concluded that the basic medium No. 1 provided most adequate cell growth of *M. fascifrons*, *A. constricta* and *D. maidis*. This medium has been used for the cell cultures of mosquitoes, *Aedes albopictus* (Skuse) and *A. aegypti* (L.) (Singh, 1967). Later, Mitsuhashi (1965a) revised the basic medium and used it for embryonic cell cultures of *Nephotettix cincticeps* Motsch., a vector of rice dwarf virus and rice yellows dwarf agent. Recently, he has described two additional media (Mitsuhashi, 1969a).

Chiu and Black (1967) modified Schneider's *Drosophila* medium (Schneider, 1964). Using this medium, they succeeded in establishing cell lines from four different species of leafhoppers, *A. constricta*, *A. quadripunctata* (Provancher), a vector of wound tumor virus and potato yellow dwarf virus, *Agalliopsis novella* (Say), a vector of wound tumor virus, and *Aceratagallia sanguinolenta* (Provancher), a vector of potato yellow dwarf virus. Originally they used the basic medium designed by Mitsuhashi and Maramorosch (1964). Later, when the modified Schneider's medium proved to be better for maintaining cell lines, they carried out further subcultures in this medium. Chiu and Black (1969) reduced the concentrations of inorganic salts to approximately half strength as shown in Table 45. In this laboratory, leafhopper cell cultures, either of the AC20 cell line, obtained through the courtesy of Dr. L. M. Black, or primary cultivations of *M. fascifrons*, *A. constricta* and *D. maidis*, are carried out with this medium. Details of other leafhopper culture media have been described elsewhere (Hirumi and Maramorosch, in press).

The addition of homologous hemolymph to culture media for invertebrate cells was considered an important factor for successful cultivations. Although several insect extracts and/or heterologous arthropod sera had been added to the culture media (Hirumi and Maramorosch, 1963a; Sinha, 1965; Chiu and Black, 1967), their replacement by fetal bovine serum provided the necessary growth factors (Hirumi and Maramorosch, 1964a, b, c, 1968; Hirumi, 1965; Mitsuhashi and Maramorosch, 1964; Chiu and Black, 1967). Thus, neither homologous nor heterologous arthropod serum seems to be essential for leafhopper cell culture. It was also found that newborn calf serum could be substituted for fetal bovine serum, but that normal calf, chicken or horse sera were unsatisfactory (Mitsuhashi and Maramorosch, 1964).

"Sitting-drop" culture. The sitting-drop culture technique, used by Hirumi and Maramorosch (1964a, b) for the embryonic tissue culture of *M. fascifrons*, became a standard method for leafhopper cell culture, and has since been

Table 45. *Chiu and Black's leafhopper cell culture medium (pH 7.0)*[a]

Component	Amount (mg/liter)
$CaCl_2 \cdot 2H_2O$	400.0
KCl	800.0
KH_2PO_4	300.0
$MgSO_4 \cdot 7H_2O$	1853.0
NaCl	1053.0
$NaHCO_3$	353.0
Dextrose	4000.0
Lactalbumin hydrolyzate	6500.0
Yeastolate	5000.0
Fungizone	2.5
Neomycin	50.0
Penicillin G	100,000.0 (IU)
Streptomycin	100.0
Fetal bovine serum	200.0 (ml)

[a] Based on Chiu and Black (1969).

used for many other species, in some cases with minor modifications (Hirumi and Maramorocsh, 1964c, 1968; Mitsuhashi and Maramorosch, 1964; Hirumi, 1965; Mitsuhashi, 1965a, b, 1966a, 1969a; Tokumitsu and Maramorosch, 1966, 1967, 1968; Chiu et al., 1966; Chiu and Black, 1967; Mitsuhashi and Nasu, 1967; Streissle et al., 1967; Yamada et al., 1969).

Trypsin treatment has been widely used for primary cultivation from the leafhopper. It improved initial growth of cells, but trypsinization did not seem to be essential for cell proliferation; such cells were also grown without the trypsin treatment (Mitsuhashi and Maramorosch, 1964; Chiu et al., 1966; Chiu and Black, 1967; Tokumitsu and Maramorosch, 1967). A possible explanation is that the mild treatment may dissolve the basal laminae which tightly cover the outer surface of most organs, allowing cells to migrate not only from the cut surfaces of the tissues but also from the uncut surfaces. The basal laminae of the untreated tissues may inhibit cell migration from the explants, resulting in the delay of cell growth. Approximately 0.5 ml of culture medium containing resuspended tissue fragments derived from about 10 embryos for each flask, were placed on the center of the culture flasks.

Culture vessels. In comparison with vertebrate materials, a considerably smaller amount of tissue is generally involved in primary cultures of small insects. Thus, most cultures of leafhoppers have been carried out with "sitting-drop" techniques using small culture vessels. In order to facilitate the handling of a small volume of the materials, a special vial was designed by Vago and Flandre (1963). The V-H type flask has been modified from this vial to obtain better observations under high magnifications of phase contrast (Hirumi, 1963; Vago, 1967).

Small culture vessels, consisting of a microslide ring and two cover glasses sealed on the top and bottom of the ring, have been used for leafhopper cell cultures (MITSUHASHI and MARAMOROSCH, 1964; MITSUHASHI, 1965a; TOKUMITSU and MARAMOROSCH, 1967; HIRUMI and MARAMOROSCH, 1968). Such chambers were used with the sitting-drop culture technique.

To obtain a large cell population, 15 drops of the culture medium containing the tissue fragments were placed on the bottom of a plastic T-type flask, 30 ml capacity, in the same manner as in sitting-drop cultures. Seven days after the initial cultivation, 5 ml of the culture medium was added. With this technique, primary cultures of *M. fascifrons* cells continued to grow for more than 150 days (HIRUMI and MARAMOROSCH, 1968). A combination of sitting-drop culture and coverslip culture techniques was used by CHIU, REDDY and BLACK (1966) in Sykes-Moore culture chambers, or, for larger numbers of cells, CHIU and BLACK (1967) used 30–60 mm sealable Petri dishes.

Each type of culture vessel, described above, has advantages and disadvantages. The V-H flask is quite convenient for continuous observation under high magnifications using phase contrast and for changing the culture medium, but it is difficult to clean and is expensive. The microslide ring chamber also permits observation at high magnification and is economical, but is inconvenient for changing the medium. Condensation, which prevents phase contrast observation, appears on the inner surface of the top cover glass of this chamber. The sealable Petri dish and T-type flask can handle a large number of cells and are inexpensive, but the thickness of their bottom walls prevents observation at high magnification. Thus, culture vessels have to be selected on the basis of the objective of each experiment.

C. Establishment of Leafhopper Cell Lines

As mentioned earlier, a major breakthrough in Homoptera cell cultures was the establishment of leafhopper cell lines by CHIU and BLACK (1967).

They explanted 20 or more embryonic tissue fragments in 60 mm sealable Petri dishes with a small amount of medium of MITSUHASHI and MARAMOROSCH (1964). These primary cultures were maintained at room temperature (about 24° C), and the medium was changed at weekly intervals. After more than 100 days the cultures were treated with 0.05% trypsin in RINALDINI's salt solution for 10 min. Trypsinization was stopped by adding an equal volume of growth medium and the cells dispersed by gentle pipetting. When most of the living cells had become attached to the culture dish, the medium was replaced with fresh medium, and the degenerated cells were removed. Subculturing by the same procedure was repeated four times at 7–8 day intervals in the same Petri dish. Afterwards, subcultures were carried out regularly by a procedure which involved trypsinization for 6–8 min in 0.05% trypsin, centrifugation for 2 min at $200 \times g$ and resuspending the cells in the fresh medium. With a seeding density of 4.8×10^6 cells in 4 ml of the medium, confluent monolayers were formed in several days at 27° C in a plastic T-type flask with a useful flat surface of 25 cm^2, or 3 ml of the resuspended medium in a 50 mm Petri dish.

Early attempts to subculture the cells with the medium of MITSUHASHI and MARAMOROSCH failed. Successful subculturing was obtained with CHIU and BLACK's medium as modified from SCHNEIDER's medium. The authors pointed out that successful subculturing could also be achieved when cell growth became extensive about 3–4 weeks after starting the primary culture and an established cell line could be obtained in another month.

Using these techniques, nine cell lines from *A. constricta* and two cell lines from *A. quadripunctata* were established. At the time when the results were reported, *A. constricta* cell line 2 (AC2) had been successfully subcultured more than 55 times and continuously maintained in several laboratories. The two cell lines of *A. quadripunctata* had passed the ninth and twentieth generations. Attempts at establishing cell lines of *A. sanguinolenta* and *A. novella* also have been reported by CHIU and BLACK (1967). A cell line of *C. montanus* was recently obtained by D. D. JENSEN using CHIU and BLACK'S technique, and carried through more than 20 passages (personal communication).

In order to establish leafhopper cell lines, it seems necessary to obtain a large number of cells in the primary cultures. Whenever the first subculture was made with a small number of cells, using the same techniques and medium described above, the subcultured cells failed to grow continuously. However, as a large amount of cells was collected from 5–10 primary cultures, without repeated trypsinization in 50 mm sealable Petri dishes, and used for the subcultures, these cells continued to grow actively.

D. Growth of Leafhopper Cells

Primary Cultures. In primary cultures of leafhopper cells, only minor differences of cell growth have been described among different species.

During the initial few hours of cultivation, a number of dipolar single cells became attached to the surface of the culture flasks. Usually they multiplied up to 10-fold during the next 2 days, but seldom survived for long. Within 24 hrs of cultivation, a type of fibroblast-like cell began to appear from tissue explants, and continued to grow actively, forming networks (Fig. 109). Many mitoses were observed in the outgrowth. The cells multiplied along the line of dipolar cytoplasmic processes from the center of the outgrowth to the periphery (HIRUMI and MARAMOROSCH, 1964a, b, c, 1968; HIRUMI, 1965; MITSUHASHI and MARAMOROSCH, 1964; MITSUHASHI, 1965a; CHIU et al., 1966; CHIU and BLACK, 1967). Two other types of fibroblast-like cells were also distinguished. One of them was observed in old cultures. Their most characteristic feature was their large size. The cells often developed irregular cytoplasmic pseudopodia at both ends, and mitotic divisions were fairly frequent (MITSUHASHI and MARAMOROSCH, 1964). Another type of cell appeared after 6–8 days of cultivation, forming networks with irregular pseudopodia and loosely spread nuclei. Although the outgrowths of this type were not extensive, the cells continued to divide mitotically for a long period (HIRUMI and MARAMOROSCH, 1964c; HIRUMI, 1965).

During the initial 48 hrs of cultivation, epithelial cells became attached to the surface of the culture flask and began to grow. They multiplied rapidly, forming cell sheets (Figs. 110–112). These cell sheets expanded gradually and eventually engulfed the original tissue fragments. Numerous mitotic divisions

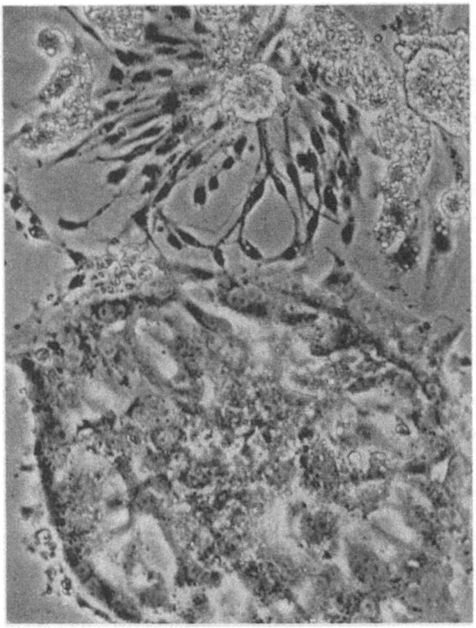

Fig. 109

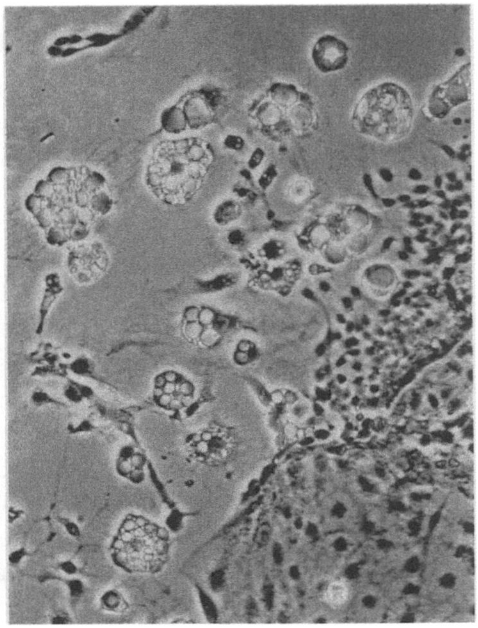

Fig. 110

Figs. 109–112. Cells of *M. fascifrons*, growing *in vitro* in primary cultures

Fig. 109. Fibroblast-like cells and epithelial cells on the 2nd day of cultivation (200 ×)

Fig. 110. One type of large phagocyte-like cells in the 14-day-old culture. Two types of epithelial cells are also seen (175 ×)

were observed. These epithelial cells varied in size, and changed from one size to another during cultivation, especially at the periphery of the outgrowth. Two different types were distinguished (Figs. 111 and 112) (Hirumi and Maramorosch, 1964c; Hirumi, 1965). Mitsuhashi and Maramorosch (1964) later classified them into five types, largely on the basis of their size.

Both the fibroblast-like and the epithelial cells occasionally grew out of the same tissue fragments, but often only one of these two types appeared. Generally, cells derived from small tissue explants began to degenerate on the 15th day; however, cells derived from large tissue fragments, which usually showed contractile movements, continued to grow for a long period.

After 24 hrs of cultivation, two distinct types of large phagocyte-like cells appeared in the culture (Fig. 110). No mitotic division in the cells was observed. Both types of cells increased in number by migration from the original tissue fragments (Hirumi and Maramorosch, 1964c; Hirumi, 1965).

Two types of wandering cells, multiplying by mitosis, were described in the culture of *M. fascifrons* (Mitsuhashi and Maramorosch, 1964). In the culture of *N. cincticeps*, three different types of wandering cells were seen; however, no mitoses could be observed in these cells. Cells similar to wandering

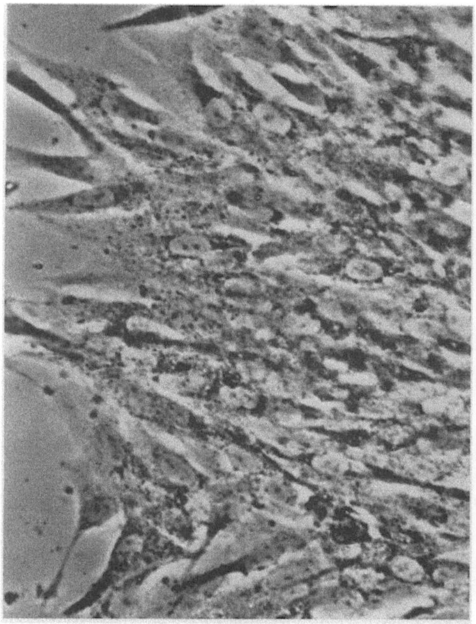

Fig. 111

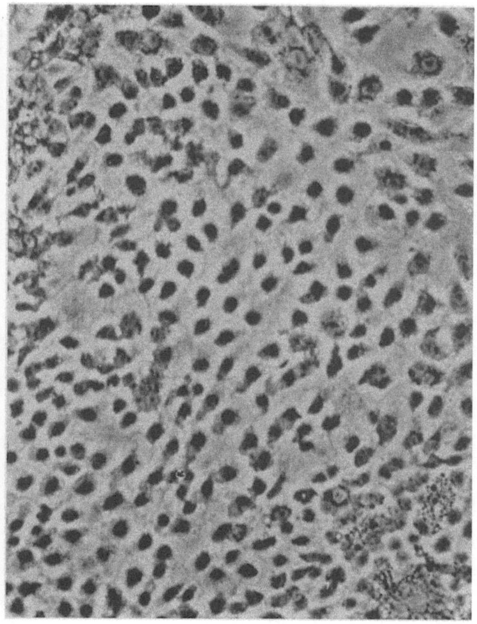

Fig. 112

Fig. 111. Large epithelial cells dominate the primary culture, on the 8th day of cultivation (235 ×)

Fig. 112. Other types of dominating epithelial cells in the primary culture, on the 13th day of cultivation (225 ×)

cells which were obtained in the culture of *M. fascifrons* did not appear in the culture of *N. cincticeps* (MITSUHASHI, 1965a).

In the nymphal and imaginal tissue cultures, various types of cells began to proliferate from the tissue fragments and grew in the same manner as the embryonic cells. Some of the organs kept their functions for more than two months *in vitro*. Leg muscles continued contractive movements, while cells proliferated from their cut openings; alimentary tracts and oviducts continued peristaltic movements for a long period. The stomach was found to attach to the surface of the culture flasks and proved to be the most easily cultured organ. No cell growth was obtained from brains, subesophageal ganglia, thoracic and abdominal ganglia, salivary glands, muscles, fat bodies, testes, Malpighian tubules, mycetomes or epidermis (MITSUHASHI and MARAMOROSCH, 1964).

Main cell types. A type of fibroblast-like cell and two types of epithelial cells usually became predominant in the primary cultures. The growth of other types of cells was not extensive and did not become dominant (HIRUMI and MARAMOROSCH, 1964a, b, c, 1968; MITSUHASHI and MARAMOROSCH, 1964; HIRUMI, 1965; MITSUHASHI, 1965a).

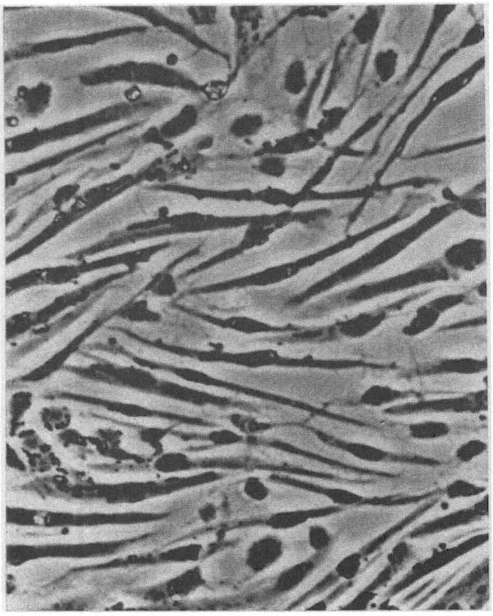

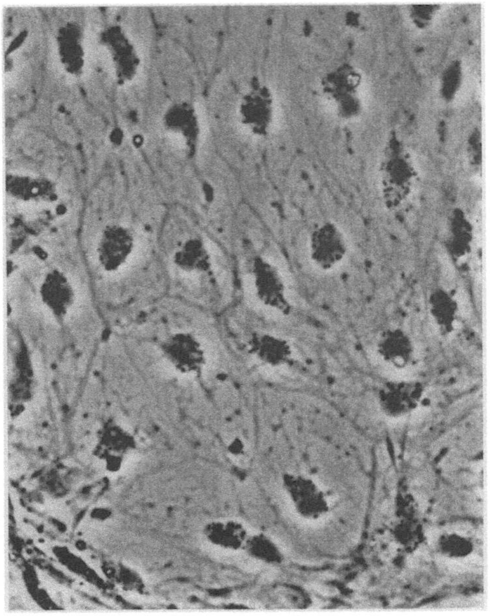

Fig. 113 Fig. 114

Figs. 113–116. Cells of *Agallia constricta*, growing *in vitro*

Fig. 113. Networks, consisting of fibroblast-like cells, on the 59th day of primary cultivation (400×)

Fig. 114. Portion of the compact cell sheet, consisting of small type epithelial cells on the 59th day of primary cultivation (400×)

The fibroblast-like cells, derived from original explants during the 24 hrs after initial cultivation, grew actively for more than 150 days, forming networks. Cells of this type had long, thin, dipolar pseudopodia approximately 40 μ in length and 8 μ in width at the widest part. The nuclei were small and compact, about 7 μ in diameter (Fig. 113). These cells appeared in the cultures prior to other types, and were the most common type during the initial period of cultivation.

A type of epithelial cell, which began to grow between 24 and 48 hrs after initial cultivation, usually dominated the primary cultures. This cell type consisted of large cells and formed cell sheets (Figs. 111 and 115). The cells of the outgrowths had well-spread cytoplasm, in which many cytoplasmic granules were observed. These granules were apparent as early as the 4th day and gradually increased in number from the center to the periphery of the outgrowth. The average size of the cells was somewhere between 75 μ and 120 μ and their nuclei were approximately 30 μ in diameter. The cells continued to grow with very active mitoses for a long period.

The other type of epithelial cells, which also became dominant, was smaller than the first type and free from large cytoplasmic granules or oil droplets (Figs. 112 and 114). This type began to grow slightly later than the first type;

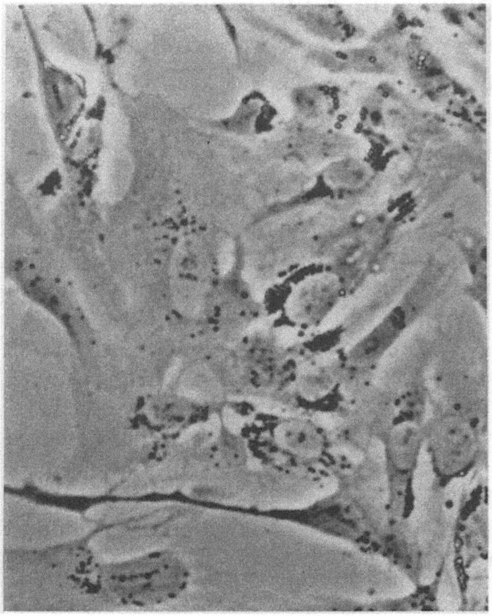

Fig. 115

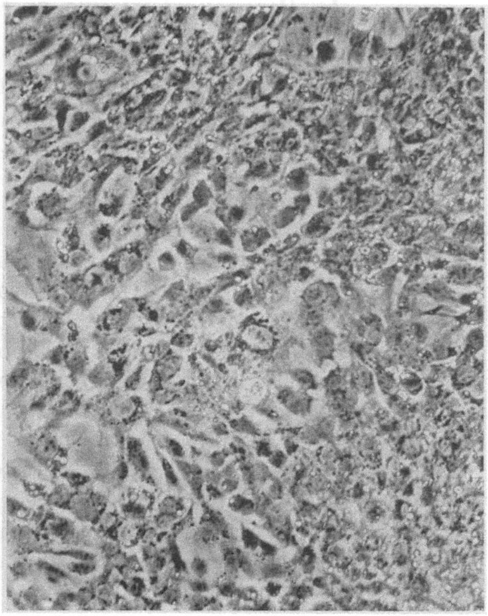

Fig. 116

Fig. 115. Portion of an outgrowth consisting of large type epithelial cells on the 10th day of primary cultivation (425 ×)

Fig. 116. AC-20 cell line, established by CHIU and BLACK (1967). Cell sheets of two types of epithelial cells on the 14th day after subculturing (175 ×)

however, it multiplied very rapidly. Many cells of this type were polygonal in shape, forming fairly compact cell sheets. These cells also continued to grow for more than 150 days.

Cell line. All nine cell lines of *A. constricta* consisted mainly of epithelial cell types. However, some differences concerning cell morphology and growth habit were noticed among the cell lines. One cell line was comprised of more slender cells than others. Although a steady growth rate was reached, none of the nine cell lines attained uniformity in cell morphology (CHIU and BLACK, 1967). The morphology of a subline, which was originally obtained from AC20 established by CHIU and BLACK, was examined in this laboratory. This cell line was composed of two types of epithelial cells. The morphological resemblance between these cells and the two types usually dominating in the primary cultures was striking (Fig. 116). The first type of epithelial cells, represented by the large cells, was more dominant than the second type, which was smaller.

In the early stages of the establishment of the *M. fascifrons* cell line, the subcultured cell population was composed of almost all types of cells which were observed in the primary cultures. However, two types of epithelial cells which were similar to those of the subline of AC20 became more dominant than other types. Thus the origin of the AC20 cell line presumably was the epithelial cells which dominated the primary culture.

E. Primary Culture of Aphids and Planthoppers

Aphids. Cell culture of Aphididae has not been studied extensively. Preliminary attempts have been carried out with cells of the pea aphid, *A. pisum*, using the sitting-drop culture techniques in microslide ring chambers (TOKUMITSU and MARAMOROSCH, 1966; HIRUMI, unpublished data).

Since the chorions of embryos, obtained by Caesarean sections, were very delicate, surface sterilization of the embryos had to be avoided. At the suggestion of C. VAGO (personal communication), aseptic rearing of *A. pisum* similar to that described for leafhoppers (MITSUHASHI and MARAMOROSCH, 1963) was employed. Parthenogenetic apterous females, reared in mass cultures in an insectary, were sterilized by immersion in 70% ethanol for approximately 5 sec. After insects recovered, they were transferred to pea plants, *Pisum sativum* L., and grown aseptically in test tubes on agar media. After a delay of a few days, the surface sterilized females delivered nymphs. In this manner aseptic aphid colonies were obtained.

Nymphs from aseptic colonies, as well as ovaries, embryos and alimentary tracts excised from surface sterilized females, were cut into small tissue fragments, then placed in a culture medium modified from GRACE's Lepidoptera medium (GRACE, 1962). In some cases, tissue fragments were treated in 0.25% trypsin solution for 15 min at room temperature.

After 48 or more hrs of initial cultivation, two types of fibroblasts began to migrate from the tissue fragments and scattered on the surface of the culture flasks. One type of cell migrated mainly from the small trypsinized explants. Another, smaller type of fibroblast migrated from tissue fragments of the alimentary tract. Multinucleate cells were observed in the cultures for about 20 days. Of all tested stages, young nymphs provided the best material for *in vitro* cultivation. However, no cell divisions were observed (TOKUMITSU and MARAMOROSCH, 1966).

Recently active growth of pea aphid cells was obtained from embryonic tissues in a culture medium consisting of 10 parts of SCHNEIDER's medium, 10 parts of modified VAGO and CHASTANG's *Bombyx mori* medium 22, and 3 parts of fetal bovine serum with antibiotics. Although no prolonged cultivation has been achieved and studies of morphological aspects of the cultured cells have not been completed, four different types of fibroblast-like cells, an epithelial type, and large multinucleate cells were observed. Shortly after migration from the original tissue explants, large multinucleate cells were spindle-shaped and uniform in size. Within a few hours after migration, they attached to the surface of the culture flasks, and began to show irregular cytoplasmic protrusions (Fig. 117). Sometimes these cytoplasmic processes were conspicuously broad. One of the fibroblast-like cells was comprised of more or less spindle-shaped cells which formed colonies (Fig. 118). Mitotic divisions were observed in these cells (HIRUMI, unpublished data).

BLACK (1969) reported that primary cultures were obtained by PETERS and BLACK (1970) from sowthistle aphids *Hypermyzus lactucae* (L.). Ovarian tissues and embryos were removed from apterous viviparous insects and cut into small fragments. The fragments were treated with pronase, which was removed by washing, and the tissue fragments were explanted in a suitable

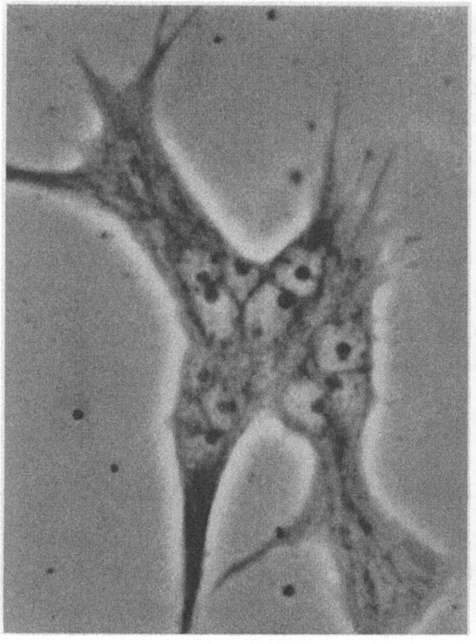

Fig. 117

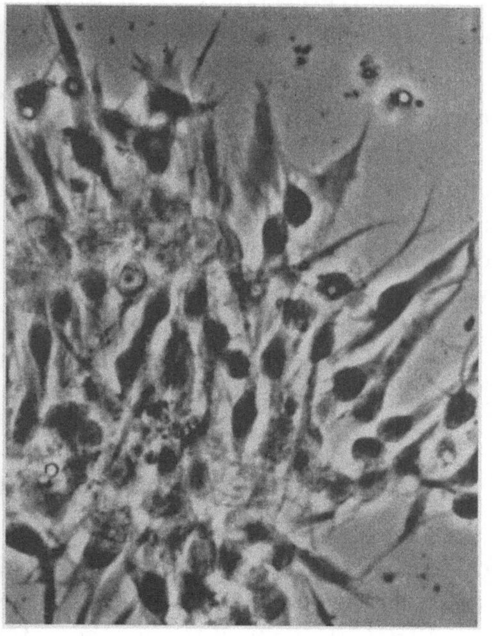

Fig. 118

Figs. 117 and 118. Primary cultures of the pea aphid, *Acyrthosiphum pisum*

Fig. 117. A large multinucleate cell with broad cytoplasmic processes on the 3rd day of cultivation (800 ×)

Fig. 118. Fibroblast-like cells on the 7th day of cultivation (800 ×)

culture medium on coverslips. Cells with extending pseudopodia were successfully maintained for up to 10 days *in vitro*. The primary cultures were used in assaying sowthistle yellow vein virus, as described later in this chapter. At present (March 1970) no further information is available concerning PETERS and BLACK'S techniques of *in vitro* cultivation of the aphid cells.

Planthopper. Some planthoppers are also known to be vectors of certain plant pathogens. Although it is important to establish the *in vitro* cultivation of cells of these insects in order to study the virus-vector interaction, no attempt to grow the vector cells had been made until recently.

MITSUHASHI (1969b) reported the first preliminary *in vitro* cultivation of the smaller brown planthopper, *L. striatellus*, a vector of rice stripe virus, rice black streaked virus and northern mosaic virus of barley, wheat and oat.

As no cell growth from the tissues of *L. striatellus* was observed with the techniques and medium used for leafhopper cell culture (MITSUHASHI and MARAMOROSCH, 1964), MITSUHASHI examined each process of the culture techniques and the composition of the medium. It was found that tissues derived from the embryos which had not developed beyond completion of blastokinesis provided the most suitable source for *in vitro* cultivation. Trypsinization (0.1% for about 5 min at 25° C) of tissue fragments enhanced the adherence of tissue fragments of the culture flasks.

Almost no cell migration was observed from whole embryos. Cell migration from tissue explants occurred in adequate media when embryos were cut into small tissue fragments. Various culture media were tested. The most promising result was obtained with GRACE'S medium (GRACE, 1962), supplemented with 20% fetal bovine serum and 1% *L. striatellus* egg extract.

Within 24 hrs of the initial cultivation, fibroblast-like cells and a few epithelial-like cells appeared. No extensive cell growths such as were obtained in the leafhopper cell cultures were seen. Hollow spherical vesicles, consisting of epithelial-like cells, were often observed. Mitoses were observed in the cells which formed the vesicles, but not in the cells that migrated. These cells and explants were cultivated up to about 40 days when they started to deteriorate. Although the experiment was only a preliminary attempt to grow planthopper cells *in vitro*, it should lead to further studies of planthopper cell cultures.

III. Application to the Study of Plant Pathogens

This section will be limited to the studies of plant pathogens in Homoptera cell cultures. The host specificity and pathogenicity of plant pathogens as well as their physical and chemical nature have been thoroughly reviewed by BLACK (1959, 1962), MARAMOROSCH (1960, 1963), MARAMOROSCH et al. (1970), WHITCOMB and DAVIS (1970) and others.

A. Aster Yellows

Aster yellows is a widely distributed plant disease and its etiologic agent is transmitted by several leafhoppers. Since KUNKEL (1926) demonstrated the virus-like interactions between the causative agent, host plants and its leafhopper vectors, virus etiology of this disease had been generally accepted (BLACK, 1959; MARAMOROSCH, 1963). Initial studies of this pathogen in tissue cultures of leafhopper vectors had been carried out assuming that a virus caused the disease.

Following the work by MARAMOROSCH (1956) described earlier, HIRUMI and MARAMOROSCH (1963b) attempted to identify the organs and tissues that support the multiplication and are the storage sites for the aster yellows pathogen in *M. fascifrons*. The fate of the ingested pathogen in the vector was examined by recovering the agent from freshly dissected organs of yellows-infected leafhoppers, as well as from leafhopper vector organs maintained *in vitro*. After a 5-day feeding period and 14 days of incubation, salivary glands, gut, Malpighian tubules, ovaries and testes were removed. The same organs, removed after 5 days of acquisition, were maintained for 14 days in the culture medium, consisting of 85% of VAGO and CHASTANG'S *Bombyx mori* medium 22 and 15% of newborn calf serum. Retention of infectivity in these organs was tested by the insect injection technique (HIRUMI and MARAMOROSCH, 1963b). The aster yellows agent was recovered from fresh gut removed after 5 days, but not from gut removed after 19 days of total incubation period; while salivary glands removed after 19 days provided large amounts of the pathogen. The aster yellows agent was also recovered from Malpighian tubules maintained *in vitro* for 14 days, but no recovery was obtained from other organs. These experiments provided certain information concerning the fate of the aster yellows agent in the leafhopper body. Moreover, the recovery of the agent from the separated organs *in vitro* suggested the feasibility of maintaining the pathogen in tissue cultures of its leafhopper vector.

As *in vitro* culture of leafhopper cells became available, the growing of the aster yellows agent in cell cultures of *M. fascifrons* (MARAMOROSCH et al., 1965) was attempted. Although epithelial cells from *A. constricta* embryos, inoculated with reovirus, retained the virus for 3 weeks, no infective agent of aster yellows was recovered from homogenized cells of *M. fascifrons*. The nature of aster yellows "virus" remained obscure until the discovery of the mycoplasmal etiology of yellows type diseases (DOI et al., 1967).

During the past three years, major efforts have been made in many laboratories in the search for mycoplasma-like microorganisms by means of electron microscopy in diseased plants as well as in infective insect vectors. As mentioned earlier, mycoplasma-like microorganisms have been found in more than 30 plant diseases of the yellows type. The agents of over 25 of these diseases are known to be transmitted by leafhopper vectors (WHITCOMB and DAVIS, 1970; MARAMOROSCH et al., 1970). The search for a microorganism in insect vectors revealed the presence of mycoplasma-like bodies in the salivary glands, intestine, fat body and nervous system (GIANNOTTI et al., 1968a, b; GRANADOS et al., 1968a; MARAMOROSCH et al., 1968; HIRUMI and MARAMOROSCH, 1969; SHIKATA and MARAMOROSCH, 1969b; SHIKATA et al., 1969; BRČÁK and KRÁLÍK, 1969). The finding of four general types of pleomorphic bodies in the cytoplasm of salivary glands of aster yellows-infected *M. fascifrons* suggested that the mycoplasma-like bodies complete their life cycle in salivary gland cells (HIRUMI and MARAMOROSCH, 1969). This has encouraged the study of interactions between the aster yellows pathogen and its vectors in leafhopper cell cultures. Currently, studies on the multiplication of the aster yellows mycoplasma agent are being attempted in the primary culture of the vector, *M. fascifrons*, and in a cell line of the non-vector, *A. constricta*. If the cell line of *M. fascifrons* becomes available, it will provide a very useful tool for studying the aster yellows pathogen in a cell culture system.

B. Rice Dwarf

Rice dwarf virus is one of several economically important viruses transmitted by several species of the deltocephaline leafhoppers. It is known that the virus can pass transovarially to the progeny (FUKUSHI, 1933; NASU, 1965, 1969). The host relationships (FUKUSHI, 1969) and morphology (FUKUSHI et al., 1962; FUKUSHI and SHIKATA, 1963; KIMURA and SHIKATA, 1968) of this virus as well as its chemical and physical nature (SUZUKI, 1969) have been studied extensively. Although FUKUSHI and SHIKATA (1963) demonstrated the localization of the virus in the leafhopper vector, the mechanism of virus multiplication remains unknown.

Mitsuhashi (1965b) studied the multiplication of rice dwarf virus (RDV) in a tissue culture of *N. cincticeps*. Viruliferous, as well as virus-free, embryonic tissues were cultivated *in vitro*. The cell growth obtained from viruliferous tissues was very similar to that of non-viruliferous tissues; however, distinct cytoplasmic granulations were observed in the epithelial cells derived from viruliferous tissues. Electron microscopical examination revealed the presence of virions in almost all of the epithelial cells in one of the viruliferous cultures after 27 days of cultivation. Virus-free cultures were also inoculated with the virus material prepared from RDV-infected females. To obtain the inoculum, fat bodies were removed aseptically from 5 viruliferous insects, and then homogenized in 0.5 ml of the culture medium. After centrifugation for 5 min at 3000 rpm the clear part between the lipid layer and sediment was used as inoculum. Cultured cells were inoculated for 24 hrs at 25° C. Following inoculation, marked granulations were observed in the epithelial cells. Many small virus-like particles, 30 mμ in diameter, were observed by electron microscopy on the 9th day after inoculation. The small particles were considered to be a small form of rice dwarf virus. A small number of particles 70 mμ in diameter, which is the regular size of this virus, were also found. No infectivity tests with the virus from the inoculated culture were reported.

Later, Mitsuhashi and Nasu (1967) reported further evidence of virus multiplication in vector cells. The virus-containing inoculum was prepared from a 57-day-old culture of viruliferous embryonic tissues. The cultured cells were homogenized in the culture medium, and then centrifuged for 30 min at 4000 rpm. A 37-day-old non-viruliferous embryonic culture was infected with the supernatant for 24 hrs at 25° C. Granulations of the cells were not observed after inoculation. The authors suggested that the cytopathic effects which occurred in the previous experiments were not due to virus multiplication, but to some other substances in the fat body homogenate. Rice dwarf virus particles, 70 mμ in diameter, appeared in viral materials examined by electron microscopy on the 28th day after inoculation. Linearly arrayed virions within a sheath-like structure and crystalline aggregates also were observed. It was concluded that rice dwarf virus multiplied in the cultured cells; however, no recovery and infectivity tests with the virus were attempted. These experiments revealed that the vector cells could be artificially infected *in vitro* with rice dwarf virus.

C. Wound Tumor

The wound tumor virus has been extensively studied in several laboratories since its discovery by L. M. Black in 1941. The early work on the wound tumor virus has been thoroughly reviewed by Black (1959, 1962) and Maramorosch (1963, 1970). The virus causes tumors in several host plants and can be transmitted by certain leafhopper species of the subfamily Agallinae. This virus is also known to be transmitted transovarially to the progeny.

Systemic invasion of the leafhopper vectors by the virus has been studied by electron microscopy (SHIKATA et al., 1964; SHIKATA and MARAMOROSCH, 1965, 1966, 1967; HIRUMI et al., 1967; GRANADOS et al., 1967, 1968b; MARAMOROSCH et al., 1969) and by serological tests (SINHA, 1965, 1969). These studies provided notable information concerning the fate of the wound tumor virus in the vector body. However, the mechanisms of the virus multiplication in the insect cells still remain obscure. Cell cultures of the leafhopper vectors can be utilized to study these aspects at the cellular level. Since the cell lines of its leafhopper vectors have been established, as described previously, it is possible to make synchronous inoculations of vector cells to assay vector-virus interactions. Consequently, the studies of wound tumor virus, as well as potato yellow virus, described later, in cultured cells are most promising approaches.

CHIU et al. (1966) successfully demonstrated multiplication of wound tumor virus (WTV) in embryonic tissue cultures of a leafhopper vector, *A. constricta*, by fluorescent antibody staining techniques. Sitting-drop cultures of trypsinized tissue fragments were made on 25-mm coverslips with basic medium (MITSUHASHI and MARAMOROSCH, 1964) supplemented with 20% fetal bovine serum in Sykes-Moore culture chambers. Viruliferous inocula were prepared from WTV-infected adult leafhoppers. The insects, 30–60 in number, were homogenized in 2 ml of the growth medium with 5% of fetal bovine serum. Following clarification at $3100 \times g$ for 5 min, the supernatant was passed through Millipore filters of pore diameters of 0.65 and 0.30 μ. After 10 days of cultivation, the primary cultures were inoculated by replacing the growth medium with inoculum for 2 or 3 hrs of adsorption. The inoculated cultures were subsequently incubated at room temperature, and the medium changed at 2- or 3-day intervals. Utilizing fluorescent antibody staining techniques (NAGARAJ et al., 1961; SINHA and REDDY, 1964), they observed viral antigen reactions in a few cells in WTV-inoculated cultures on the 3rd day and the number of stained cells and the intensity of cytoplasmic staining increased later.

The infectivity of WTV in culture medium and cells was also tested at various intervals after inoculation by the injection of extracts into non-viruliferous insects. On the basis of the concentration of soluble antigen of the virus in the insects 21 days after injection, CHIU et al. (1966) concluded that the virus concentration in the cells increased at least 100-fold during 8 days of incubation. To obtain further evidence of WTV multiplications, the authors carried out a serial passage of WTV in cultures. In this experiment, the original inoculum was prepared from root tumors on *Mellilotus officinalis* (L.) and had a relative virus concentration of $10^{6.5}$. Six serial passages were made with a 1:32 ($10^{-1.5}$) dilution of virus at each passage. Had the virus not multiplied, the titer would have been reduced from $10^{6.5}$ to $10^{-2.5}$ at the end of the passages. Infectivity by the injection into non-viruliferous insects indicated that the relative concentration of virus from the cell samples at each passage was from $10^{5.4}$ to $10.^{6.1}$. Thus, it was concluded that WTV had multiplied during the serial passages.

Later, when the cell lines of leafhopper vectors had been established as described earlier, CHIU and BLACK (1967) demonstrated that the AC cell lines, also, were susceptible to inoculation with WTV as revealed by fluorescent antibody staining techniques. They pointed out that the infection could be detected as early 12 hrs at 30° C after inoculation and that there was no significant difference in the growth rate of healthy and infected cells.

GÁMEZ and CHIU (1968) established the minimum concentration and dose of WTV particles giving an infection in AC cell cultures. Cell monolayers 2 or 3 days old, subcultured from the cell line, were inoculated for 3 or 4 days at room temperature with viruliferous inocula obtained from root tumors of *M. officinalis*. The concentration of virions in the inoculum was counted by the spray droplet technique for negatively stained virus (GÁMEZ and BLACK, 1967, 1968) prior to assaying the minimum infective dose. The minimum concentration that produced positive infection was found to be between $10^{6.1}$ and $10^{5.4}$ virions/ml, and the minimum infective dose per coverslip in 0.02 ml used as inoculum was $10^{4.44}$ to $10^{4.68}$. This value is 10^2 times greater than the minimum infective dose calculated by insect injection techniques (GÁMEZ and BLACK, 1968). The authors considered that the difference in minimum infective doses obtained by the two techniques may be due to the difference in volume of solution that can be applied to the assay unit by the two methods.

Recently CHIU and BLACK (1969) attempted to develop a standardized technique for the assay of WTV in leafhopper cell cultures. After the confluent monolayers of AC cells were prepared on 15-mm coverslips in the manner described previously (CHIU and BLACK, 1967), they were inoculated with the virus materials obtained from root tumors. The cells were harvested at appropriate times, and WTV infection was detected by fluorescent antibody staining. Effects of length of incubation period were examined at two dilutions (1:2000 and 1:6000). After 12 hrs of incubation at 30° C, weakly fluorescent spots were noted in infected cells. These spots increased in intensity, number and size during 18–21 hrs. Although virus antigens were not observed within 27 hrs, they were detected in the samples harvested after 30 hrs of incubation.

Effects of length of virus adsorption on the number of infected cells were also investigated. CHIU and BLACK (1969) found that the maximum adsorption period was about 2 hrs at 30° C and that a further prolonged period did not increase virus adsorption. The adsorption was completed in 3 hrs at 24° C, while there was little adsorption at 9° C even when the period was prolonged to 6 hrs. On the strength of these findings, a standard assay technique was established as follows: (1) monolayers were prepared on coverslips 18–24 hrs prior to virus inoculations, (2) virus adsorption took place for 2 hrs at 30° C, (3) subsequent incubation followed for 25 hrs at 30° C. When an adsorption was carried out at 24° C, the adsorption period was prolonged to 3 hrs and was followed by an incubation period of 27 hrs at 30° C.

Utilizing this standardized technique, CHIU and BLACK (1969) demonstrated a linear relationship between the number of infected cells and relative virus concentrations, indicating that each infection is initiated by a single virus particle. However, the cell infective unit value differed from that obtained by another assay for WTV, that is by the virus particle counting technique (GÁMEZ and CHIU, 1968). This difference was considered due to overestimation of the cell infective unit obtained by fluorescent antibody cell counting. CHIU and BLACK (1969) expected that the sensitivity of the assay system could be further improved by cloning for the more susceptible cell types. They also reported that AS cells established from *A. sanguinolenta*, a non-vector of WTV but closely related to *A. constricta*, were susceptible to WTV *in vitro*. AS cells were less sensitive than AC cells for WTV and gave WTV titer 1 or 2 log units lower.

HIRUMI and MARAMOROSCH (1968) inoculated for 2 hrs the primary cultures of the leafhopper, *M. fascifrons*, which is not a vector of WTV and which is not closely related to *A. constricta*, with WTV inoculum prepared from WTV-infected *A. constricta* adults. A small number of WTV particles, presumably representing the original inoculum, were found in the periphery of the cytoplasm 18 hrs after virus inoculation. Following 7 days of an incubation period, the WTV viroplasmic matrices, containing scattered virus particles were observed in the inner region of the cytoplasm by electron microscopy. However, these particles were devoid of capsids. The authors also examined the epithelial cells of intestines excised from *M. fascifrons* initially confined to the WTV-infected plants. No virus particles were found in the cells. These findings indicated that WTV could multiply to a certain extent, but not extensively, in the non-vector cells.

D. Potato Yellow Dwarf

Potato yellow dwarf virus (PYDV) is rather unique among the leafhopper-borne viruses in its host relationships. Two varieties, the New York (NY-PYDV) and the New Jersey (NJ-PYDV) strains, are known to be transmitted specifically by two related leafhoppers, *A. sanguinolenta* and *A. constricta* respectively (BLACK, 1944). The viruses are also passed transovarially (BLACK, 1953). Details concerning the morphology (BLACK et al., 1965; MACLEOD et al., 1966; MACLEOD, 1968) and chemical nature (AHMED et al., 1964; BRAKKE, 1956) of the PYDV have been reported previously. The characteristic host specificities of these viruses can be an advantage in the assaying of virus-vector relationships.

Using a cell line established from *A. sanguinolenta*, the studies of the virus multiplication were carried out very recently by CHIU et al. (1970). A cell line (AS1) was used in the experiments after passing through about 30 subcultures. Using coverslip cultures of AS1 and AC, described earlier, cell lines were made by placing 0.04 ml of a cell suspension, 1.2×10^6/ml on 15-mm coverslips. A growth medium (CHIU and BLACK, 1967) containing 17.5–20% fetal bovine serum was used for the experiments. The virus inoculum was prepared from PYDV-infected *Nicotiana rustica* L. Although attempts to infect the AS1 cell cultures with purified PYDV prepared by filtration through Millipore filters failed, successful inoculations were achieved with purified virus materials prepared by a density gradient centrifugation procedure (BRAKKE, 1951). Monolayer cultures of AS1 and AC21 cell lines were inoculated with NY-PYDV and NJ-PYDV respectively.

A viruliferous AS1 cell line, infected with NY-PYDV, was established and used as the virus source for further experiments. Since the efficiency of inoculation with the virus source was very low, CHIU et al. (1970) attempted to enhance virus infectivity for the cultured cells by incorporating diethyl-

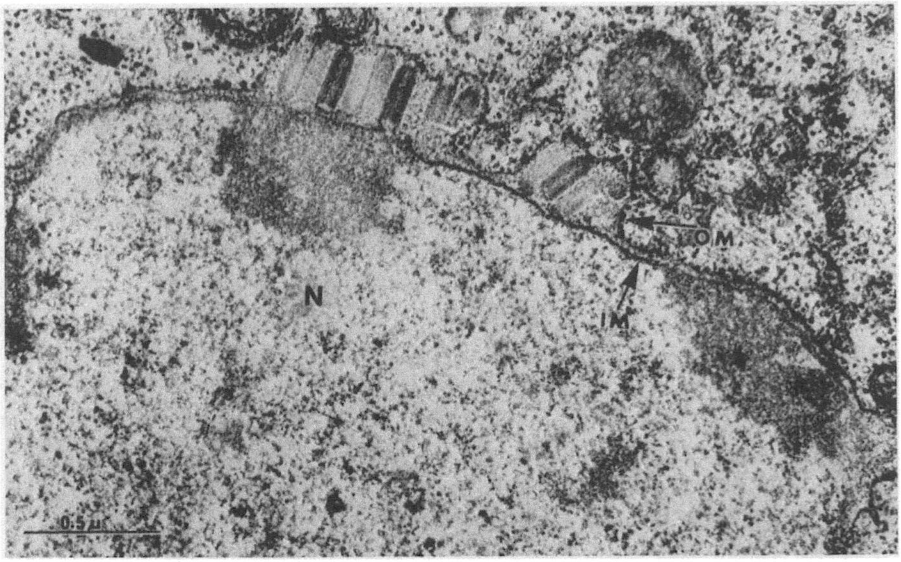

Fig. 119. Electron micrograph of an *Aceratagallia sanguinolenta* cultured cell infected by NY-PYDV, showing bacilliform virions at the periphery of the nucleus. *N* nucleus; *OM* outer nuclear membrane; *IM* inner nuclear membrane. (From CHIU, LIU, MACLEOD, and BLACK, 1970; courtesy of Dr. R. MACLEOD)

aminoethyl-dextran (DEAE-dextran) at various concentrations. Utilizing the fluorescent antibody staining technique (CHIU and BLACK, 1969), they found that the treatment with DEAE-dextran at the concentration 25 μg/ml greatly increased the efficiency of inoculation. Following the initial tests, the assays of NY-PYDV and NJ-PYDV (prepared directly from infected *N. rustica* plants incorporating 0.25 μg/ml of DEAE-dextran) in cell monolayers were made by fluorescent cell counting techniques. Linear relationships were demonstrated between the virus concentrations in the inocula and the number of infective cells in inoculated cultures in both instances of AS1 cells infected with NY-PYDV and of AC21 cells with NJ-PYDV. Their findings clearly revealed that the viruses multiplied in their vector cells, suggesting that the infections were initiated by single virus particles. It was also demonstrated that only the nuclei were stained with fluorescent antiserum in early stages of infection, while later some staining was observed in the cytoplasm.

Further evidence of the virus multiplication in the PYDV-infected cells was made by electron microscopic observation (Chiu et al., 1970). Bacilliform particles were observed in the perinuclear space of infected cells (Fig. 119). The morphology of these particles, having the average dimensions of 290 × 75 mμ, resembled that of virions observed in PYDV-infected plant cells MacLeod et al., 1966), although the sizes of virions were somewhat shorter. Chiu et al. (1970) also observed a budding process, presumably a developmental process, from the inner nuclear membrane. Cross-susceptibility of non-vector cells to both virus strains was examined, using AS1 cells infected with NJ-PYDV and AC20 cells with NY-PYDV. Fluorescent antibody staining made virus infections evident. These findings indicate that vector specificity may not be explained solely on the basis of the presence or absence of virus multiplication.

E. Sowthistle Yellow Vein

Sowthistle yellow vein virus (SYVV), transmitted by sowthistle aphids, *Hypermyzus lactucae*, was studied in primary cultures of vector cells by Peters and Black (1970). They inoculated the primary cultures with SYVV purified from SYVV-infected plants for 2 to 3 hrs at room temperature. Utilizing fluorescent antibody staining techniques, infected cells were first detected about 37 hrs after inoculation. Later, the number of infected cells increased as shown by more intense staining. This report represents the first successful study of an aphid-borne plant virus in its vector cells *in vitro* (Black, 1969).

F. Animal and Insect Viruses in Leafhopper Cell Cultures

Reovirus. Possible relationships between wound tumor virus and reovirus, a widely distributed animal virus, have been studied utilizing leafhopper cell cultures. Maramorosch et al. (1965) inoculated embryonic tissue cultures of *A. constricta*, a vector of WTV, with reovirus, and demonstrated virus retention in the epithelial cells for at least 3 weeks. Following the preliminary experiment, Streissle et al. (1967) carried out serial infections of reovirus in the primary cultures of *A. constricta*. They inoculated a 2-month-old cell culture with $10^6 TCIU_{50}$ of reovirus type three. After one week of inoculation, the medium containing reovirus was diluted 1:10 and placed in a virus-free culture. Serial inoculations were made until the quantity of virus became diluted to 10^{-6} of the original concentration. Although no detectable inactivation of reovirus occurred in control experiments with the virus incubated in cell-free culture medium, reovirus was recovered from cell cultures up to the fourth passage but not afterwards. During each passage, the amount of recovered virus did not increase. The authors concluded that the virus retention in cell cultures for over 3 weeks was not the result of virus multiplication, but was due to the stabilizing effect of the culture medium. They also pointed out that leafhopper cell culture techniques used for the experiments were not yet adequately developed at that time; therefore, the lack of virus multiplication may have been due to poor physiological conditions.

Chilo iridescent virus. *Chilo* iridescent virus (CIV) was discovered in larvae of the rise stem borer, *Chilo suppressalis* Walker (Lepidoptera: Pyralidae), by FUKAYA and NASU (1966). MITSUHASHI (1966b) demonstrated that the site of virus multiplication can be easily detected by Bragg reflection in infected tissues. He also demonstrated virus multiplication in a hemocyte cell line obtained from diapausing larvae of the host insect (MITSUHASHI, 1966c),

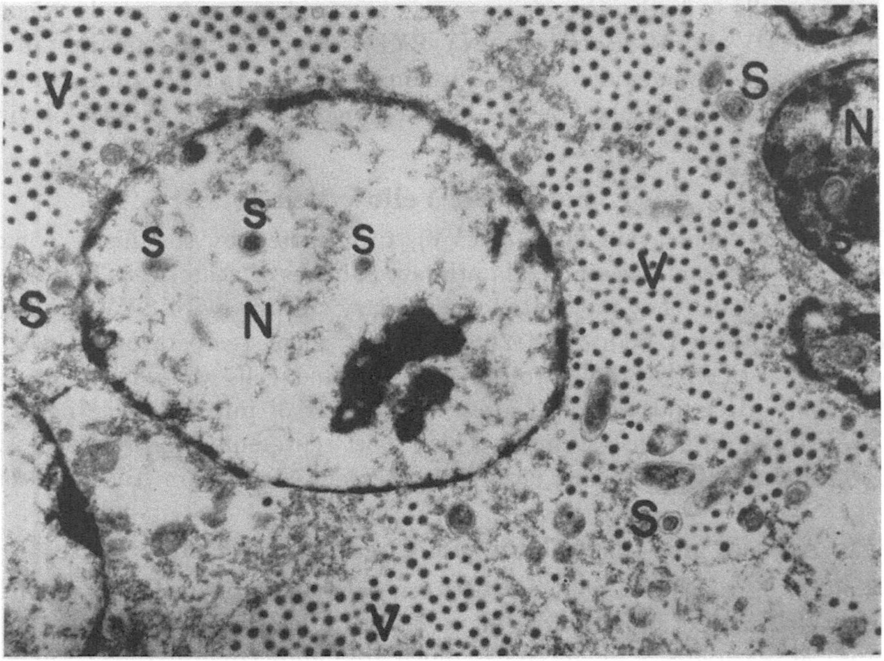

Fig. 120. An ultrathin section of a CIV-infected cell cultivated from *Nephotettix cincticeps* embryos: *N* nucleus; *S* symbiotic bacteria; *V* CIV particles (8760×). (From MITSUHASHI, 1967b; courtesy of Academic Press, Inc.)

and succeeded in the establishment of a cell line infected with the virus (MITSUHASHI, 1967a).

Since no insect viruses are known to multiply in homopterous insect cells, MITSUHASHI (1967b) attempted to infect an embryonic tissue culture of *N. cincticeps*, a vector of rice dwarf virus, with CIV which was prepared from CIV-infected larvae of *C. suppressalis*. Granules of various sizes and vacuoles were seen in the cytoplasm of the infected cells 24 hrs after inoculation, and iridescence appeared 7 days after the inoculation. Cytopathic effects were observed on the 10th day. After 14 days following inoculation, electron microscopic examination revealed the presence of numerous CIV particles in the cytoplasm of the cultured cells (Fig. 120). Although no quantitative assay of virus multiplication was reported, this experiment clearly demonstrated that the homopterous insect cells are susceptible to the insect virus which infects the Lepidoptera host in nature. The susceptibility of non-vector cells to some other plant pathogenic viruses has already been mentioned. These findings will be important in future studies of the virus-host range at the cellular level.

Acknowledgments. Work at the Boyce Thompson Institute was sponsored, in part, by U.S. Public Health Service Research Grants AI-04290 and AI-07687 from the Institute of Allergy and Infectious Diseases and CA-7453 from the National Cancer Institute, and by National Science Foundation Grants GB-1199, GB-5444 and GB-11861.

Chapter 9

Lepidoptera Cell Culture and Its Application to the Study of Plant Viruses and Animal Parasites

I. Cultivation of Gypsy Moth Hemocytes

H. M. MAZZONE

Hemocytes offer the investigator a readily accessible source of important cell culture material. Since the hemolymph pervades all areas of the invertebrate host, its cells provide a significant system for analysis with respect to pathogen infection. A number of studies have indicated the value of hemocyte cell cultures in health and disease states (GLASER, 1917; ARVY and GABE, 1946; MILLARIA, 1946; HORIKAWA and KURODA, 1959; MARTIGNONI and SCALLION, 1961; MITSUHASHI, 1967a). Our interest in use of viruses for the biological control of forest pest insects such as the gypsy moth (*Lymantria dispar, Porthetria dispar*) induced the initiation of *in vitro* hemocyte cultures of this insect.

A. Preparation and Observation of Primary Cultures

Cultures of hemocytes were prepared in Leighton tubes containing an inserted rectangular coverslip. Hemolymph was obtained aseptically by puncturing a proleg of late larval stages of the insect. A volume of 0.1 ml of hemolymph, containing approximately 100,000 cells, was allowed to fall into the culture tube, which previously had received 0.9 ml of GRACE or SCHNEIDER medium (Grand Island Biological Company, Grand Island, New York). The growth medium had the following supplements: fetal calf serum, 10%; native homologous hemolymph, 1%; and cysteine, 5×10^{-3} M which prevented melanization of the native hemolymph (MAZZONE, 1968). The cells were incubated at 25° C, and the medium was changed every 5 days.

Many cells attached to the coverslip surface within 10 to 15 min after initiation of the cell cultures. The attached cells grew in clusters and occupied most of the vessel surface in 5 to 7 days (Fig. 121). Some of the cells remained floating in the medium as single ovoid types, which multiplied in a few days, forming spindle-shaped cells (Fig. 122). Some of these cells maintained a loose connection to the coverslip surface.

MAKINO and YOSHIDA (1949) reported the diploid number of chromosomes in the gypsy moth as 62. Mitosis was demonstrated in the cultures by the addition of colchicine, 0.001% final concentration, to the culture medium. After 24 hrs, analyses of squashed preparations revealed numerous

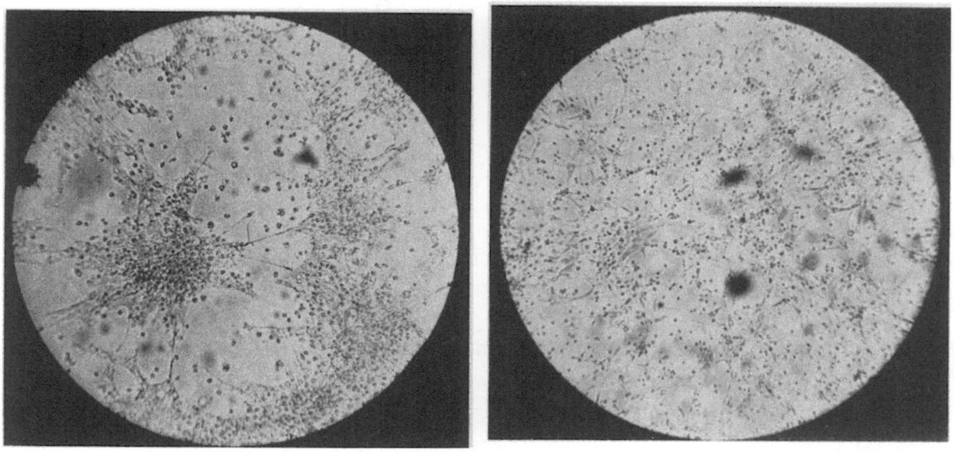

Fig. 121. Gypsy moth hemocytes in culture after 1 day (left) and 5 days (right). 100×

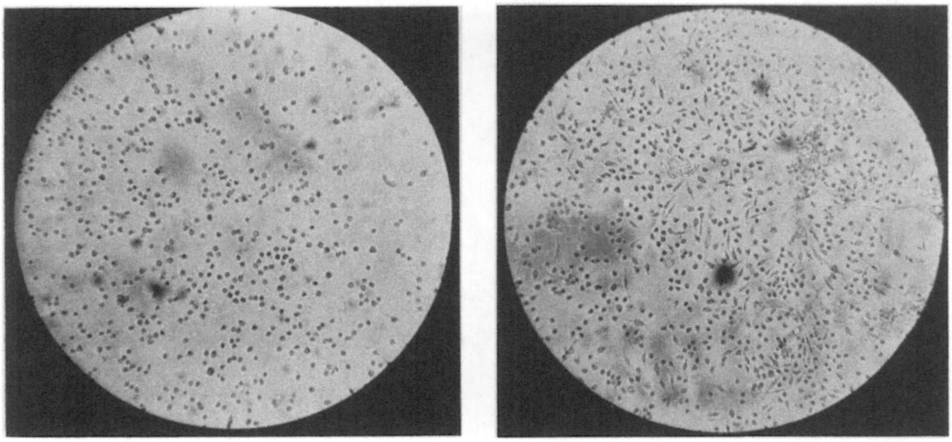

Fig. 122. Gypsy moth hemocytes floating in the medium after 1 day (left) and 5 days (right). 100×

cells in different phases of mitosis in diploid and polyploid numbers. The chromosomes are small and are seen in different focal planes. Cells in prophase and pre-metaphase are shown in Fig. 123.

B. Subculturing of Cells

In attempts to detach the cells trypsin (0.25 %) and trypsin-EDTA (0.25 %, 0.1 %) were first used. Very few cells were liberated, and these failed to attach and proliferate in passage. Pronase (Weinstein, 1966), at the same concentration as trypsin and the same exposure time, released more cells than trypsin or trypsin combined with EDTA. However, with each kind of enzymatic reagent most of the cells suffered irreparable damage. In the case of Pronase some cells attached to the new vessel surface and grew slowly.

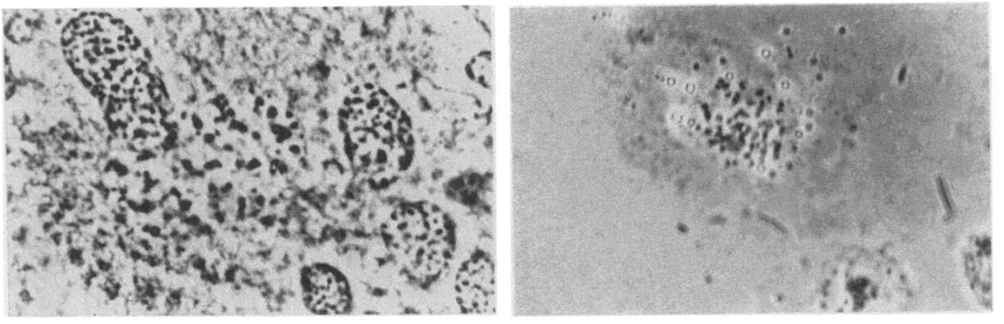

Fig. 123. Chromosomes of gypsy moth hemocytes, prophase (left) and pre-metaphase (right). 650×

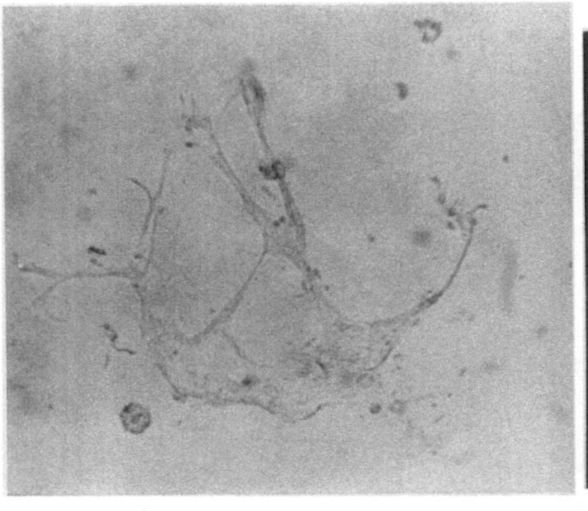

Fig. 124

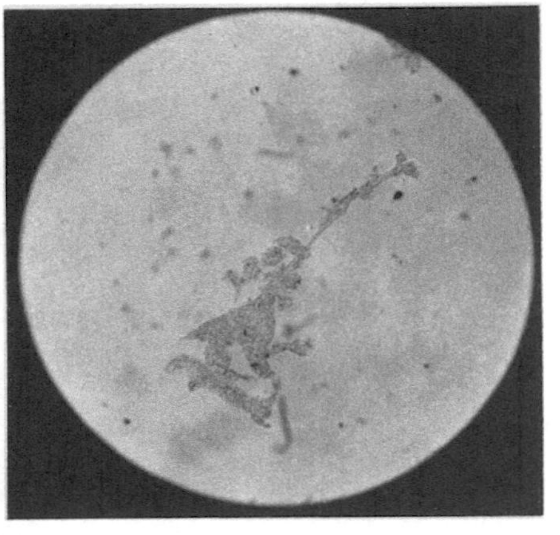

Fig. 125

Fig. 124. A cell mass that grew outward from the coverslips after placing it cell-side down in a Petri plate. 100×

Fig. 125. A cell mass that adhered to the Petri plate surface after removal of the coverslip. 100×

A somewhat better redistribution of cells was achieved by the following procedures. The coverslip, containing a monolayer of cells, was removed from the cell culture and placed cell-side down in a Petri plate. Growth medium was added, and after a few days cells began to migrate outward from the coverslip (Fig. 124). Alternatively, after the same period of time the coverslip was lifted from its location on the plate surface. This procedure effected a stripping of cells from the coverslip, and they remained attached to the Petri plate surface (Fig. 125). Cells subcultured by the coverslip method appear to be growing better in their first passage than cells detached enzymatically. Thus far it has been impossible to subculture the cells that were observed to multiply as floating cells in the culture medium.

C. Viral Infection of Primary Cultures

A comparison was made between the infections resulting from two sources of the virus: hemolymph containing viral material and purified polyhedra. For the present it will be assumed that polyhedra are incapable of infecting tissue culture cells and must be lysed, liberating the virus rods, which do produce infection in such systems (VAUGHN, 1968). In the first case larvae were allowed to feed on synthetic diets (VANDERZANT et al., 1962; ODELL and ROLLINSON, 1966) containing polyhedra purified with zonal rotors, provided by the MAN Program at the Oak Ridge National Laboratory, Oak Ridge, Tennessee.

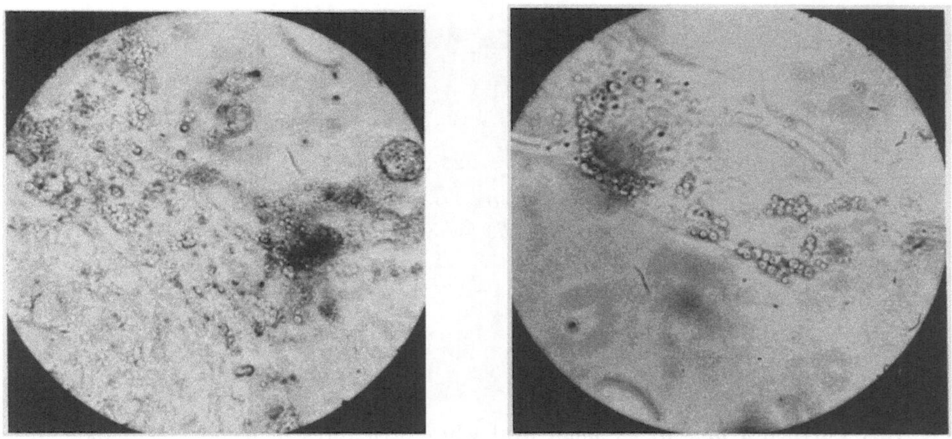

Fig. 126. Polyhedra present in hemocytes 7 days after exposing virus infected hemolymph to cell culture. 1250 ×

After the larvae were in the advanced stages of polyhedrosis, hemolymph was obtained as described above. The culture fluid of several Leighton tubes was poured off, and 0.1 ml volumes of infected hemolymph were added to each tube and allowed to run down the entire cell surface area. After 20 min growth medium was added to each tube as a rinse. The fluid was poured off, and this rinsing was repeated 4 times. Fresh growth medium was then added, and the cultures were incubated for 7 days. At this time the cells showed cytopathic effects. Analyses also revealed a general presence of polyhedra as shown in Fig. 126.

For the second source of virus, polyhedra were lysed according to BERGOLD'S alkaline pH method (BERGOLD, 1953), yielding a fraction rich in virus rods. An aliquot was added to each of several cultures. The same exposure time and rinsing procedure were followed. After 7 days the cells were observed to suffer damage but not to the same degree observed in the case of infectious hemolymph. Fewer polyhedra were observed, and they were smaller, indicating that they were not fully developed (Fig. 127). Under the conditions imposed, the differences in infectivity characteristics and polyhedra production in the two cases appear to follow the observations made by other investigators (TRAGER, 1935; AIZAWA and VAGO, 1959; VAUGHN and FAULKNER, 1963).

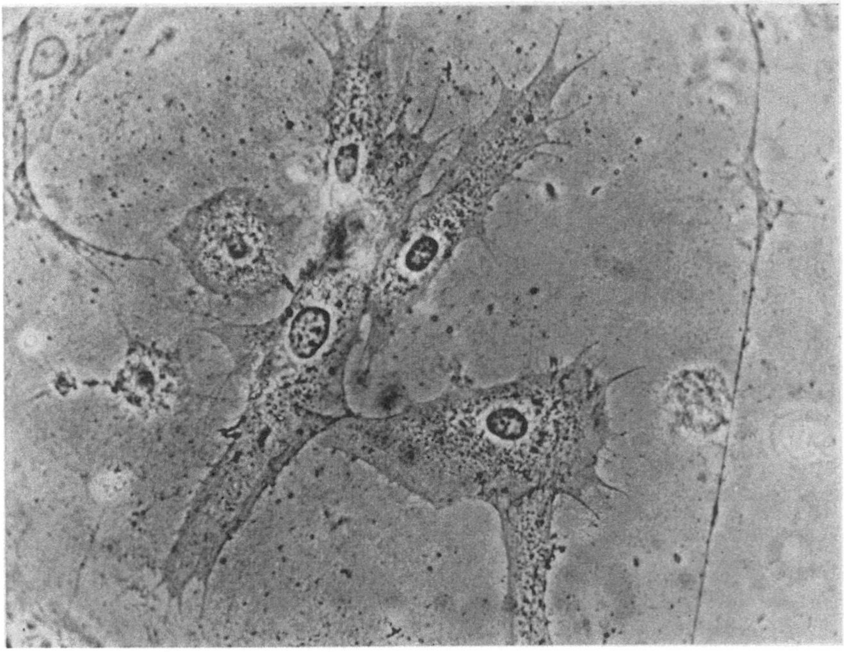

Fig. 127. Polyhedra present as small oval white structures in hemocytes 7 days after exposing virus rods, isolated from purified polyhedra, to cell culture. 1250 ×

II. Detection of Viral Antigens of the Densonucleosis Virus by Antibodies Labeled with Peroxidase Enzyme

E. KURSTAK, S. BELLONCIK, and S. GARZON

A. Introduction

This report describes a relatively new technique for detecting and localizing viral antigens within infected cells. The method involves the attachment of horseradish peroxidase to viral antibody and treatment of the infected cells with this complex (KURSTAK et al., 1969; LEDUC et al., 1969; WICKER and AVRAMEAS, 1969). The antibody-peroxidase complex binds specifically with the viral antigen within the cell and its site can be visualized by adding 3,3′-diamino-benzidine, which reacts with the peroxidase, resulting in the precipitation of a colored compound. The location of the viral antigen can then be observed using a standard light microscope or an electron microscope.

The events occurring within cultured cells of *Galleria mellonella* (L.) infected with densonucleosis virus (KURSTAK and CÔTÉ, 1969) were analyzed by the immunoperoxidase technique. In addition, this method was applied for electron microscopic studies.

B. Materials and Methods

Hemocytes and ovary cells of *G. mellonella* were cultivated *in vitro* at 32° C in Leighton tubes. GRACE'S modified medium supplemented with 30% fetal calf serum and 10% whole egg ultrafiltrate was used. Cells were inoculated with purified densonucleosis virus ($10^{5.4}LD_{50}$/Leighton tube). The techniques for virus purification and for obtaining anti-DNV specific antibodies in rabbits were described elsewhere (KURSTAK and STANISLAWSKI, 1968; KURSTAK and CÔTÉ, 1969).

The specific virus antibodies were labeled with horseradish peroxidase enzyme. For this purpose, 15 mg of peroxidase (Aldrich Co., Wisconsin) were dissolved in 1 ml of tris acetate 0.05 M buffer at pH 7.2; 5 mg of anti-virus antibody protein were added, combined with 40 mg of cyclohexyl -3,2 morpholino ethyl carboimid methaparatoluen sulfonate (MCDI), dissolved in 1 ml of tris acetate buffer. The disclosure of peroxidase activity by the 3,3-diamino-benzidine or by Nadi's reaction, was achieved using standard techniques (PEARSE, 1960; GRAHAM and KARNOVSKY, 1966).

For electron microscopy, 12 mg of peroxidase and 5 mg of specific virus antibody were dissolved in 1 ml of 0.1 M phosphate buffer (PBS), pH 6.8, with 0.05 ml of 1% glutaraldehyde (AVRAMEAS, 1969; LEDUC et al., 1969).

Every 2 hrs virus infected cells were prefixed *in situ* in Leighton tubes for 10 min at room temperature in 1% formaldehyde in PBS, pH 7.2, and then exposed for 4 hrs to enzyme labeled antibodies. After washing with PBS at pH 7.2 and 1 hr post-fixation with 2% osmic tetroxide, the thin sections were treated with lead citrate and observed with an electron microscope. In order to eliminate the excess non-labeled oxidase, and also to ensure that the results obtained were due to the fixation of labeled antibodies and not to a non-specific adsorption of the enzyme, 0.1 ml of cell extracts were mixed with enzyme and this mixture was allowed to stand for 45 min at 26° C. Then 8 mg of MCDI dissolved in 0.2 ml of the same buffer were added. This mixture was then treated at the time of the labeling with specific antisera, as described previously. With this preparation, the marking of cells was negligible or at times barely discernible. It would seem, therefore, that the non-labeled peroxidase does not interfere in a significant manner with the specificity of the method and does not alter the interpretation of the results for intracellular localization of viral antigens.

C. Results and Discussion

Hemocytes cultivated *in vitro* were used for the localization of the densonucleosis virus. Fig. 128A illustrates the appearance of the normal unstained cells. Such monolayers were maintained easily for 6 weeks.

Treatment of the cells 5 to 7 hrs after infection with anti-densonucleosis virus antibody-peroxidase complex permits the detection of virus antigen within the cytoplasm and in the perinuclear region of only a few cells. At 10–11 hrs, the viral antigen appears to be concentrated within the nucleus (Fig. 129A). Subsequently, the whole nucleus becomes evenly stained and the same is observed for the cytoplasm near the nuclear membrane. Under electron microscopy the immuno-peroxidase method reveals viral antigenic sites within the infected cell (NAKANE and PIERCE, 1967; KURSTAK et al., 1969; LEDUC et al., 1969). The dense network shown in Fig. 129B is

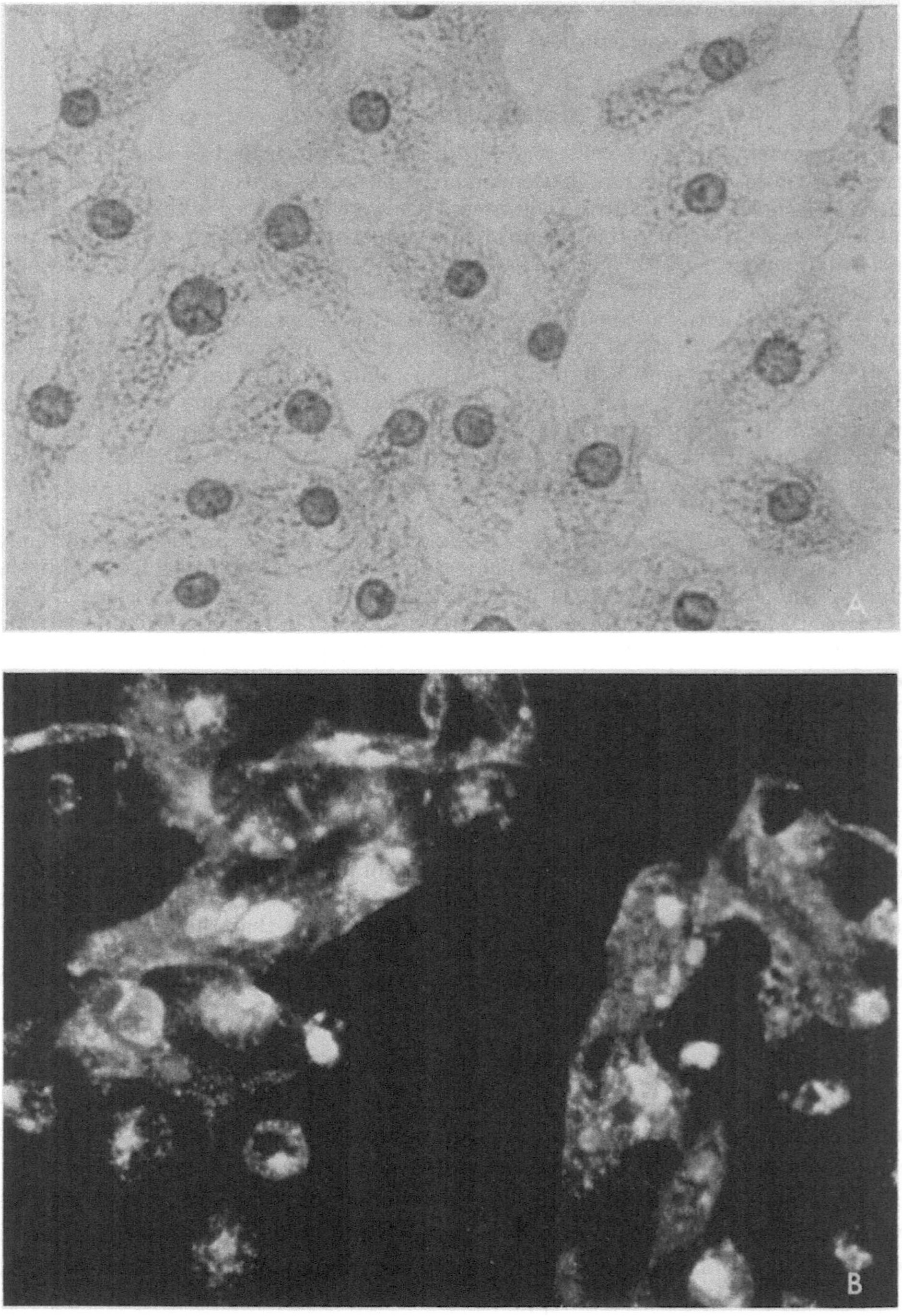

Fig. 128 A and B. *In vitro culture* of hemocytes of *Galleria melonella*. A Non-infected unstained hemocytes. Magnification 850×. B Hemocytes infected with densonucleosis virus. Note localization of the virus antigenic material in the nucleus and cytoplasm, revealed by the immunofluorescence reaction. Magnification 850×

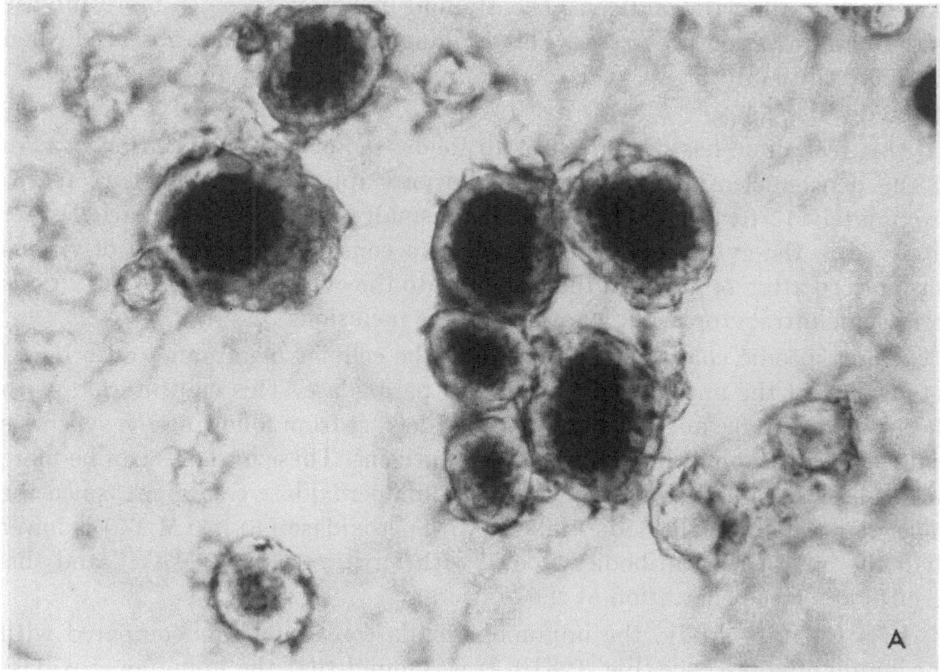

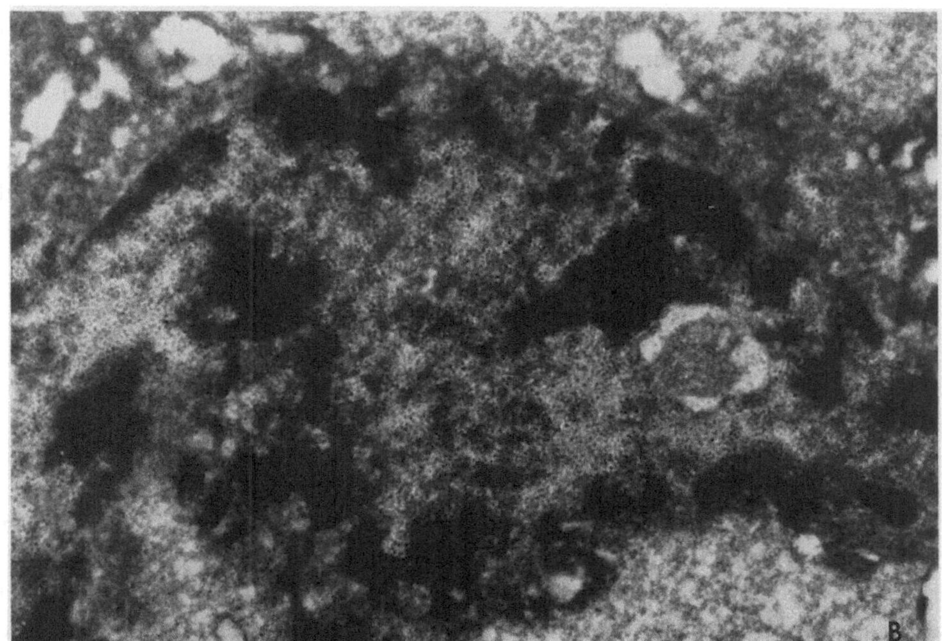

Fig. 129A and B. Intracellular detection of the densonucleosis virus antigens by immunoperoxidase reaction during late infection. A Intracellular specific marking of the virus structure antigens. Enzyme activity revealed by Nadi's reaction. Magnification 1600 ×. B An electron micrograph of the dense network, representing viral antigenic material. Note virions marked by immunoperoxidase. Lead citrate staining. Magnification 35,700 ×

identified as the viral antigen. The staining reaction becomes more intense as the viral infection progresses. This staining reaction was not observed when anti-viral antibody or peroxidase were applied separately to infected or uninfected cells.

This technique has permitted us to follow the cycle of replication *in vitro* of the densonucleosis virus. The first virions appear in the nucleus at approximately 12 to 15 hrs, at which time a small number of virus particles are observed in the cytoplasm. The passage of a considerable quantity of virions was noticed after 24 hrs from the nucleus into the cytoplasm, where the virions form some intracytoplasmic para-crystalline inclusions.

A non-specific coloration results when the cellular membranes adsorb and retain part of the antibodies labeled with peroxidase. This can interfere with the study of surface antigens, but does not detract from following the synthesis and the localization of intracellular viral antigens. These antigens can be more easily disclosed in the cell during the immunoperoxidase treatment, since the molecular weight of antibodies labeled with peroxidase (40,000 M.W.) is lower than the one of the antibodies labeled with ferritin (650,000 M.W.), and this ensures a better penetration of the antisera.

In a separate study, the immunoperoxidase reaction was compared with the immunofluorescence (Fig. 128B). It was found that the immunoperoxidase method eliminates the problem of auto-fluorescence and the sensitivity of the enzymatic reaction permits a more precise localization of the site of viral antigen. The preparations are permanent and suitable for examination under the light microscope. In addition, the immunoperoxidase reaction can also be used as a specific staining technique for electron microscopic studies.

Acknowledgment. This work was supported by research grants MT-2385 from the Medical Research Council of Canada and A-3746 from the National Research Council of Canada.

III. Growth of a Microsporidian Parasite in Cultured Cells of Tent Caterpillars *(Malacosoma)*

T. J. KURTTI and MARION A. BROOKS

A. Introduction

The microsporidia associated with insects, like other obligate intracellular protozoa, are refractory to cultivation when isolated from the host. One way that this difficulty may be circumvented is by growing cells *in vitro* which will serve as a substrate for these organisms. There are several reports that the microsporidian parasites will grow to the spore stage in cultured cells (TRAGER, 1937; GUPTA, 1964; ISHIHARA and SOHI, 1966; ISHIHARA, 1968; SHADDUCK, 1969).

This report presents the results of our studies on the primary cultivation of cells from two species of tent caterpillars, *Malacosoma disstria* HÜBNER and *Malacosoma americanum* (FABRICIUS), and describes the successful intracellular cultivation of a microsporidian parasite associated with *M. disstria*.

B. Materials and Methods

The *M. disstria* and *M. americanum* larvae were reared in the laboratory, from field-collected eggs, on an artificial diet.

Larval and pupal tissues were used as sources of cells for cultivation. The techniques for preparation and culture of larval hemocytes and imaginal wing tissues are described elsewhere (KURTTI and BROOKS, 1970). Gonadal tissues were extirpated from surface-sterilized 3-day old pupae and placed into culture medium where they were minced into small fragments.

A suitable medium for these cells is a modification of GRACE'S medium (YUNKER, VAUGHN and CORY, 1967). Antibiotics were not added to the medium because of their possible interference with the growth and metabolism of the microsporidian. Cultures were incubated at 25° C.

The microsporidian was isolated from naturally infected *M. disstria* larvae, and had the characteristics of the protozoan, *Glugea disstriae* (Thomson) (THOMSON, 1959). Spore inocula used for infecting cultured cells were obtained from organ cultures of infected silk glands. Glands were removed from larvae, placed in roller tubes containing 3 ml of culture medium, and incubated at 25° C for several days. During this time many spores were released into the culture medium. The resulting suspension of spores was harvested by centrifugation.

Since the *M. disstria* were infected, the cells cultured *in vitro* were also infected. An attempt was made, therefore, to see if we could cross infect hemocytes of *M. americanum* and *Galleria mellonella* (Linnaeus) larvae with the *M. disstria* parasite. Two basic problems are encountered when attempting to use uninfected cultured cells as a substrate for the cultivation of an intracellular microorganism. One is finding a technique for infecting these cells with a viable microbe capable of vegetative growth, and the other is maintaining the cells in a way that will support the growth and development of the parasite. TRAGER (1937) succeeded in infecting primary cultures of *Bombyx mori* ovarian cells with blood of larvae infected with *Nosema bombycis*. Recently ISHIHARA and SOHI (1966) have used OHSHIMA'S (1937) alkaline method to infect *B. mori* ovarian cultures with *N. bombycis* spores. ISHIHARA (1968) has found that this treatment will also infect primary cultures of mammalian and chick embryonic cells with *N. bombycis*.

Alkaline treatment acts on the microsporidian infecting *M. disstria*, causing eversion of the polar filament and subsequent extrusion of the sporoplasm from the spore, when the spores are placed in culture medium. We used two variations of the alkaline treatment, and both succeeded in infecting cultured *M. americanum* and *G. mellonella* hemocytes. In both cases the spore suspension from the salivary gland cultures were washed with sterile distilled water and then suspended for forty minutes in sterile 0.1 M KOH. In one method the KOH solution was adjusted to a pH of 11 with 0.1 M HCl and an aliquot of this suspension was introduced directly into the hemocyte culture containing 2 ml of medium. In the other method, the KOH suspension was neutralized with 0.1 M HCl, centrifuged, resuspended in fresh culture medium, and then this solution was added to a culture vessel containing hemocytes. Hemocyte cultures from the two insect species were set up to have an initial density of $5\text{–}7.5 \times 10^5$ cells/ml. The cultures were incubated for 4 days and studied with phase microscopy to be certain that they were free of microsporidia. On the fourth day of cultivation the spore suspensions were added to give a final suspension of 10 spores for every hemocyte. The *M. americanum* hemocytes were incubated at 25° C and the *G. mellonella* hemocytes at 25° C and at 30° C. The temperature of incubation for the latter hemocytes was not found to influence the results.

C. Results and Discussion

Studies by various workers (Hirumi and Maramorosch, 1964; Mitsuhashi, 1965, 1967a; Kurtti and Brooks, 1970) have indicated that not all larval tissues from lepidopterous insects can be used for primary cell cultures capable of growth. We have found that for *M. disstria* and *M. americanum* the best results were obtained when larval hemocytes and imaginal tissues were used. The hemocyte cultures were composed of four major types of cells: prohemocytes, plasmatocytes, spherule cells, and granular hemocytes. Of these, only

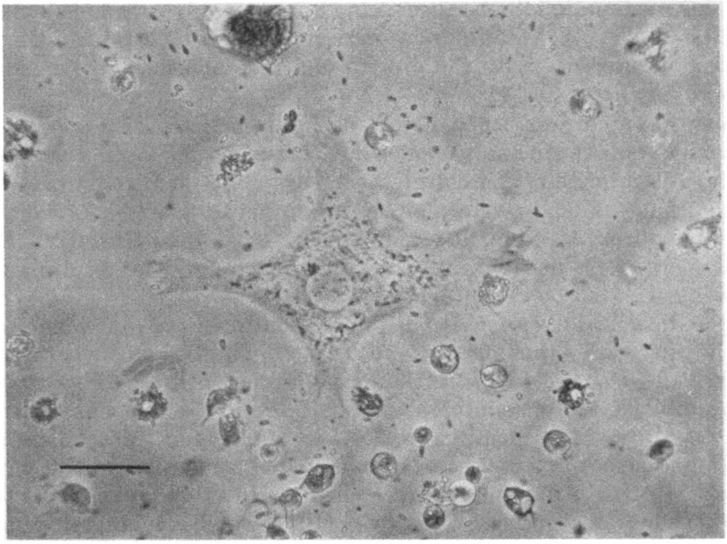

Fig. 130. Cultured ovarian cell of *Malacosoma disstria* naturally infected with microsporidia. Phase contrast. Bar = 100 μ

the prohemocytes showed mitotic activity. Fibroblast-like cells and epitheliocytes were obtained from imaginal tissues; both were active in mitosis.

Intracellular replication and development of the microsporidian parasite was observed in primary cultures of naturally infected hemocytes of *M. disstria*. These cells, when introduced into culture, were often found to be "infected" with spores but no vegetative forms, i.e., no schizonts were seen. However, after 3–4 weeks *in vitro*, large intracellular clusters of both schizonts and spores were found scattered throughout the monolayer of hemocytes. These foci within the large plasmatocytes often contained several hundred microsporidia. The cells from imaginal tissues also had spores in their cytoplasm but unlike the hemocytes there was never any evidence for the growth of the protozoan. In contrast to larval tissues, where the level of infection was low, the pupal gonadal cells were heavily infected with both schizonts and spores. The cells heavily infected with schizonts were able to attach to the culture vessel and were also capable of migration (Fig. 130). The vegetative

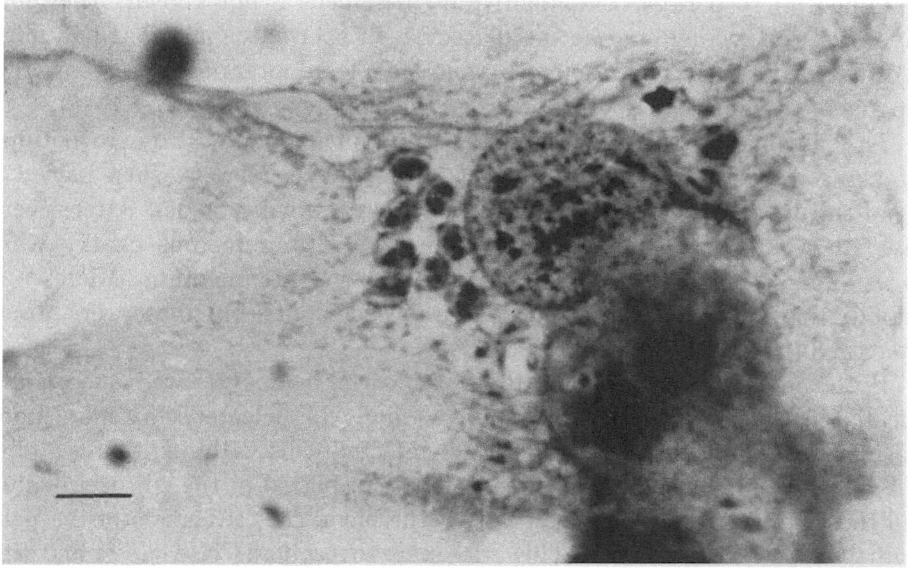

Fig. 131. Hemocyte of *Malacosoma americanum* infected in culture with microsporidia from *M. disstria*. The parasites are in the binucleate schizont stage. Wright-Giemsa stain. Bar = 10 μ

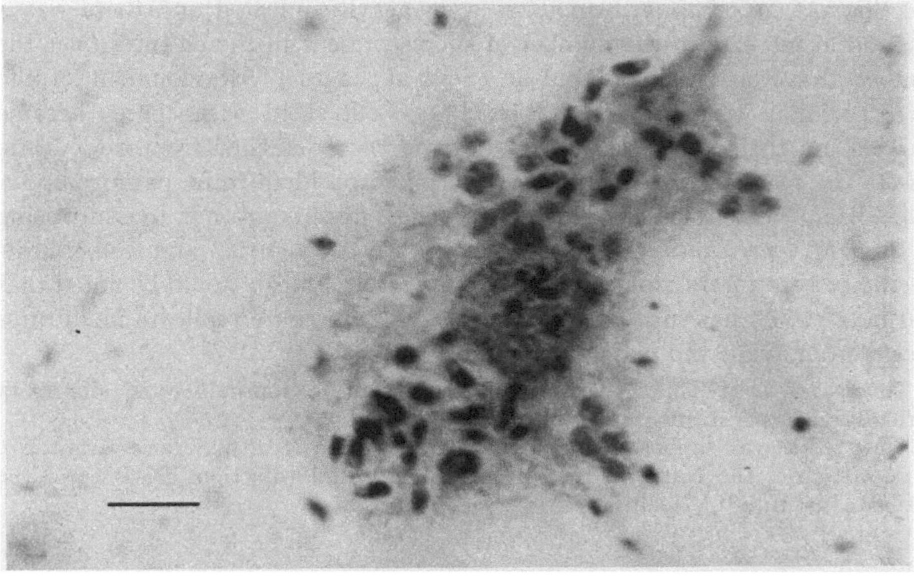

Fig. 132. Hemocyte of *M. americanum* infected with *M. disstria* parasite as in Fig. 131. Parasites are in a later stage, with schizonts and sporonts present. Wright-Giemsa stain. Bar = 10 μ

forms of the parasite were observed to sporulate *in vitro*. After 14 days most of the cultured cells were filled with spores and many became rounded up and detached from the culture vessel. Cells migrating from spermatocytes often contained microsporidia. The identity of these cells is not known: They may have been cells which line the spermatocysts or spermatogonia that had failed to differentiate because of their infection. Male lepidopterans have been shown capable of transmitting microsporidia to their progeny (THOMSON, 1958).

The silk glands are the principal site of infection of the microsporidian attacking *M. disstria* (THOMSON, 1959). The infected glands when placed in organ culture were generally translucent with a few white nodes. After several days the glands became quite hypertrophied with numerous chalky white nodes. Mainly spores were released into the culture medium. Microscopic examination did reveal the presence of a few schizonts, but these were absent after the 0.1 M KOH treatment.

Both the *M. americanum* and the *G. mellonella* hemocytes were successfully infected with the KOH-treated spores. Single binucleate schizonts (sporoplasms) were found in the cytoplasm of infected cells within the first 24 hrs. The majority of the spores failed to evert when placed in the culture medium and these were actively phagocytized by the blood cells. Eversion of the polar filament was often delayed. This was seen to occur as late as 24 hrs after inoculation into the medium. While both species of insects became infected, active growth and development of the microsporidian was found only within the *M. americanum* hemocytes. Schizonts were not seen in *G. mellonella* hemocytes after the seventh day of cultivation. On the thirteenth day post-infection, *M. americanum* hemocytes were heavily infected and there was a substantial increase in the number of spores present in the cultures over the number that was originally put in. Several stages of development of the microsporidian were seen (Figs. 131, 132). Cells filled with spores became rounded up, their nuclei were pycnotic, and they detached from the culture vessel. The reasons for the failure of the microsporidian to be pathogenic for *G. mellonella* hemocytes are not known. These results are not too surprising in that *M. americanum* is closely related to *M. disstria*, and some host-specificity type interactions may be playing a role. But it should be noted that ISHIHARA (1968) has succeeded in getting *N. bombycis* to replicate in cultures of cells from avian and mammalian embryos.

Acknowledgment. This is paper No. 7221, Scientific Journal Series, Minnesota Agricultural Experiment Station, St. Paul.

This work was supported in part by U.S. Public Health Service Grant No. AI 00961, from the National Institute of Allergy and Infectious Diseases of the National Institutes of Health.

Chapter 10

Drosophila Cell Culture and Its Application for the Study of Genetics and Virology

I. *Drosophila* Cells *in Vitro*: Behavior and Utilization for Genetic Purpose

CLAUDIO BARIGOZZI

The utilization of the technique of culturing tissues and cells *in vitro* has offered a new tool for investigating the genotype of undifferentiated and differentiated cells. The main phenomena which can be investigated by means of this technique are: the fine analysis of the somatic chromosomes, the construction of maps, the study of DNA replication and of the gene action in normal diploid condition or in cells in which extrachromosomes or haploidization have changed the basic gene dosage. In fact, in populations of cultured cells three phenomena occur, which can be exploited for the goals outlined above: mitosis, formation of several types of heteroploid and polyploid cells, and fusion between cells, leading to the formation of hybrid diploid or tetraploid genomes.

The totality of findings in this field is concerned with vertebrate cells, and, the great majority, with mammalian cells. These cells offer a great opportunity because they are easily cultured whether they are undifferentiated or incompletely or completely differentiated. Short term cultures are frequently a sufficient means to investigate the fine chromosome structure and DNA replication. For other problems established cell lines are more useful. Vertebrate cells, however, also have serious limitations: formal genetics (i.e., a genetic map) is a largely unknown territory, with only the partial exception of the mouse; well mapped gene markers detectable on single cells are rare; the chromosomes are generally numerous and nearly always difficult to classify by pairs in an unambiguous idiogram; these conditions make it difficult to define a morphological aberration in a given chromosome pair. For this reason, there is more to expect from culturing invertebrate cells, and especially insect cells, whose genetics is well known and the chromosomal conditions favorable. These considerations lead to the choice of *Drosophila melanogaster* as the best material for the main problems previously mentioned. Unfortunately the low number of easily classifiable chromosomes (8), the easily distinguishable distribution of hetero- and euchromatin, and the exceptional number of markers (including some very useful biochemical characters, identifiable on single cells) are not associated with cells that can be easily cultured. So far, apart from some successful cultures of imaginal discs (Schneider, 1966), one technique has been fully published for culturing embryonic cells (Horikawa and Fox, 1964; Horikawa et al., 1966). This technique has proved very useful for particular purposes. Established cell lines are now also maintained by Echalier and Ohanessian (1969, 1970, and this chapter).

The foregoing is an attempt to evaluate the results of genetic relevance obtained thus far, and the lines of research which can be envisaged using both short and long term cultures of *Drosophila* cells.

A. Short Term Cultures

Analysis of the karyotype. Cells grown in medium H-5 of Horikawa and Fox (1964) are gathered from a population of about 6 hrs old embryos, both males and females. The age of the single cell is, however, rather heterogenous, and this may be responsible for the different size of the cells in culture. According to Horikawa and Fox (1964), cells belong to only two types, large and small, but this conclusion was not confirmed by Dolfini and Tiepolo (1968), who found a contiunous size variation in another wild stock. Autoradiography has definitely demonstrated that embryonic cells, cultured according to the technique of Horikawa and Fox, multiply. According to Ottaviano-Gottardi (1968) mitoses are frequent during the first week and result in a considerable increase in the number of cells. Although stock differences have been found, mitoses are frequent in all cultures. Metaphases are often very clear and, when hypotonic solution treatment and air drying are used, spreading allows unambiguous recognition of each chromosome pair. Heteropycnosis is also strong (Figs. 133, 134), especially when the chromosomes are well stretched.

Examination of metaphases gives complete information on the chromosome number of embryonic cells. The first account was given by Dolfini and Gottardi (1965) who found that, 12 hrs after cultivation, 88.1 % of 135 metaphases of Varese wild stock cells were normal (8 chromosomes). The remainder comprised numbers ranging from 7 to 11. These numerical aberrations, however, referred to the 1st and to the 4th pair only: the 2nd and the 3rd remained normal. Incidently, it should be remembered that in *D. melanogaster* aberrations of the 2nd and of the 3rd pair are always lethal. In Varese stock, polyploid cells are rare. In other stocks, with information restricted to a couple of stocks carrying a Y; 3 translocation, polyploidy is frequent (Figs. 135, 136; stock P-80; Halfer, unpublished data). Whether hetero- and polyploid cells were already present in the embryo or whether they were produced in the culture, must be questioned. The answer is furnished by the calculation of the duration of the mitotic cycle (Dolfini and Tiepolo, 1968). Using pulse labeling with ^{3}H-thymidine, it was found that the entire cycle lasts over 24 hrs, because the S phase takes on the average 21 hrs and 25 min. Since the obser-

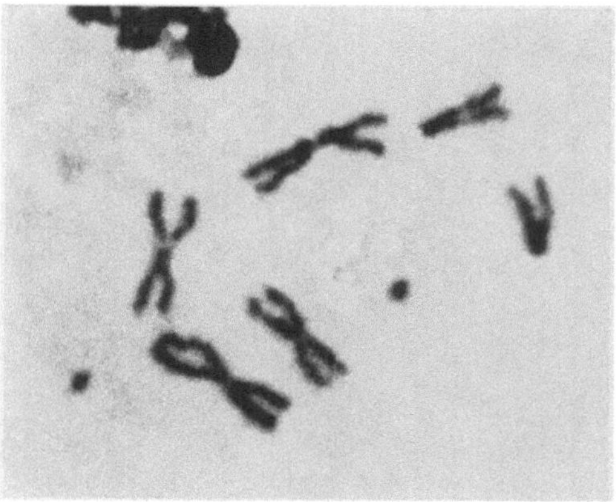

Fig. 133

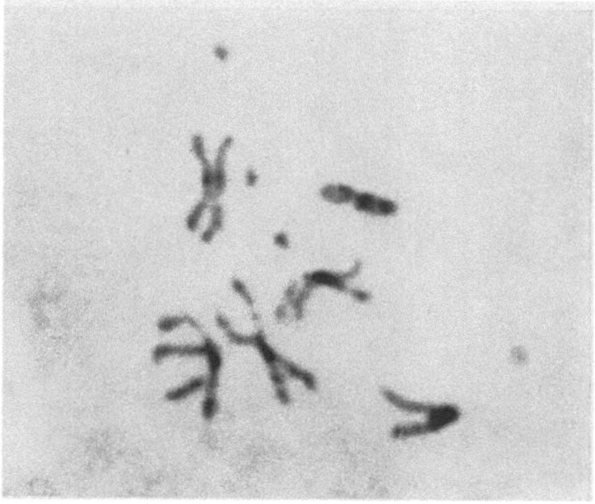

Fig. 134

Fig. 133. Metaphase of a female cell from a wild stock (Varese). Squash preparation stained with acetic orcein, treated with hypotonic solution and air dried. 2600 ×

Fig. 134. Metaphase of a male cell (Varese) showing positive heteropycnosis of the heterochromatin. 2600 ×

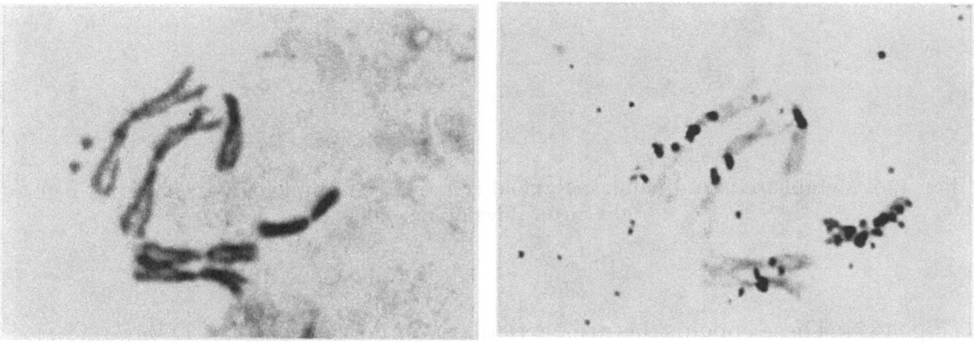

Fig. 135. Metaphase of male cell. Left, before developing the autoradiograph; right autoradiograph showing late DNA replication of heterochromatic portions. Continuous labeling with ^{3}H-thymidine. 2600 ×

vations were made within the first 24 hrs of cultivation, it must be concluded that some aberrations including polyploidy were present before the culture was started. The occurrence of the polyploidy throws new light on the very obscure origin of polyteny.

Euchromatin and heterochromatin. Hetero- and euchromatin can be clearly distinguished in embryonic cells. The first problem examined by Barigozzi et al. (1966) was to determine whether heterochromatin replicates later than euchromatin, as is the case in mammals, insects and other groups (Lima-De-Faria and Jaworska, 1968) and in other species. The technique used was again autoradiography based on incorporation of ^{3}H-thymidine. The results were clear: all heterochromatic sections and Y replicate later than euchromatin

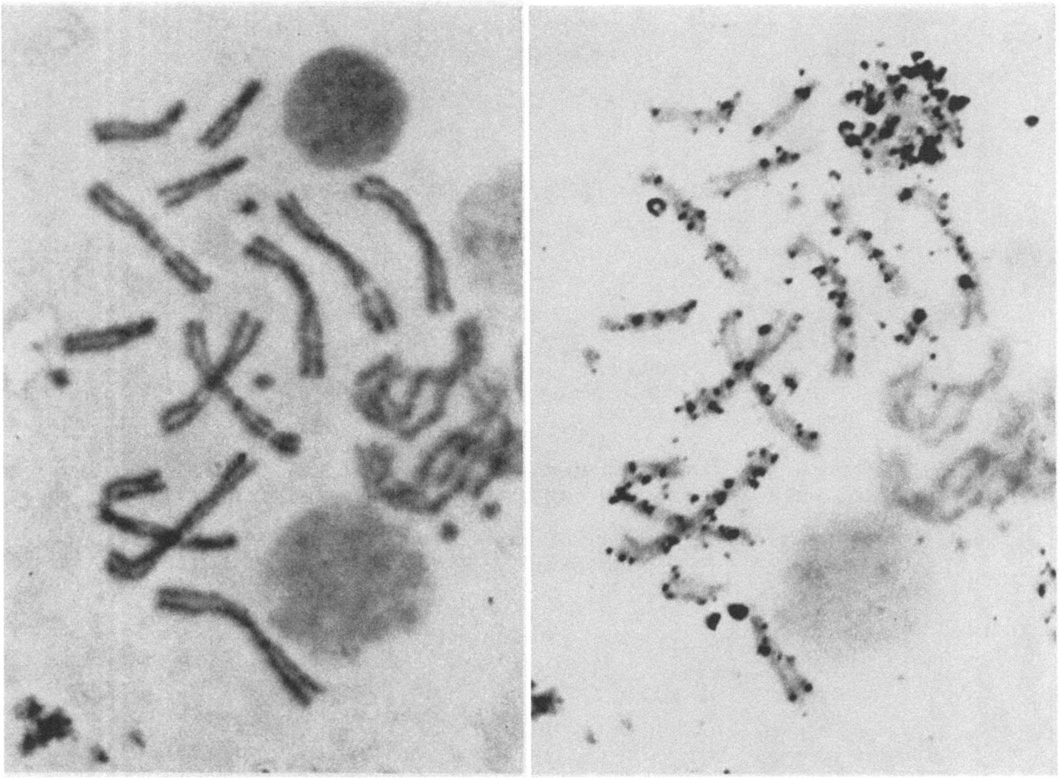

Fig. 136. Metaphase of a dividing polyploid cell, nearly tetraploid (Varese). Left, before, right, after developing. 2600 ×

(Fig. 137). These conclusions are particularly important because *D. melanogaster* represents the animal species in which heterochromatin has been most thoroughly analyzed. Late replication of heterochromatin can be better studied in *Drosophila* than in the majority of other organisms, because this species exhibits many chromosomal rearrangements. This poses the question of whether the relative replication time is the same when the heterochromatic chromosome (Y) remains intact or when it is split into two sections, forming one centric and one acentric fragment. There are different possibilities, translocation to heterochromatin or to euchromatin of X of the 2nd, 3rd or 4th chromosome, which may provide us with information on DNA replication control within the chromosome and on the role of position effect in DNA replication.

This unique opportunity has been exploited by our group (Halfer et al., 1969 and unpublished data) with the following results: three translocations, involving in all instances the Y and respectively, the X, the 2nd and 3rd chromosomes, have been analyzed completely. Recently, Dolfini (personal communication) supplied a fourth case of possible Y-X translocation occurring

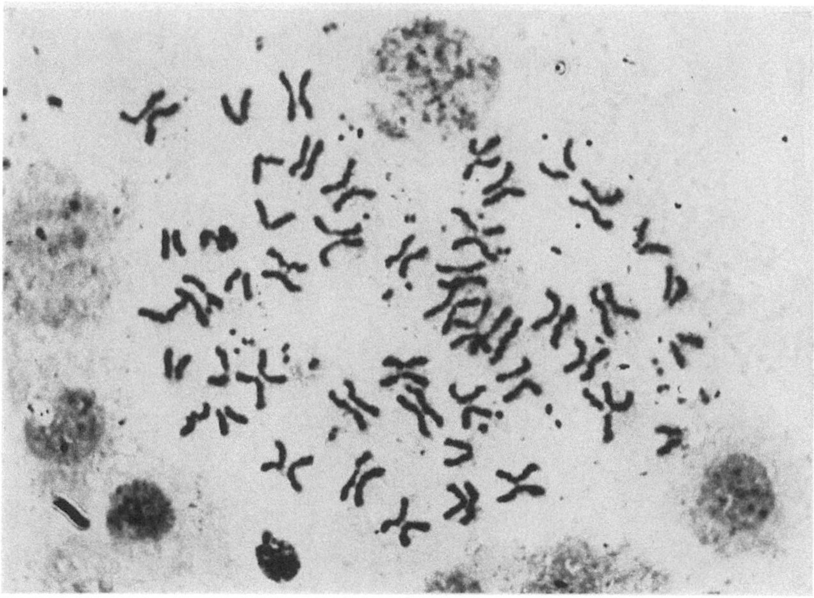

Fig. 137. Early anaphase of a 19-ploid cell (Varese) with 64 chromosomes. 1400×

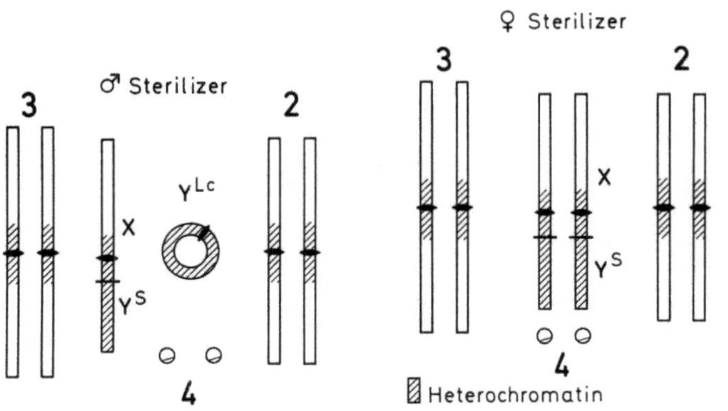

Fig. 138. Scheme of "Sterilizer" T(Y; X) translocation

in embryonic cells of the established cell line K_C (see ECHALIER, this chapter). This translocation is still under investigation. The breaking points of the four cases are summarized schematically in Figs. 138 and 140. The results (Figs. 139, 141, 142) lead to the conclusion that the Y chromosome heterochromatin is always the last one to replicate in the set; the two portions, however, become asynchronous, while the intact Y does no timing differentiation along its length. The timing asynchrony takes on different aspects in each case, thus showing that retardation seems to be dependent on the presence of a section called C of Y (Fig. 143). This should be considered, then, as a controlling

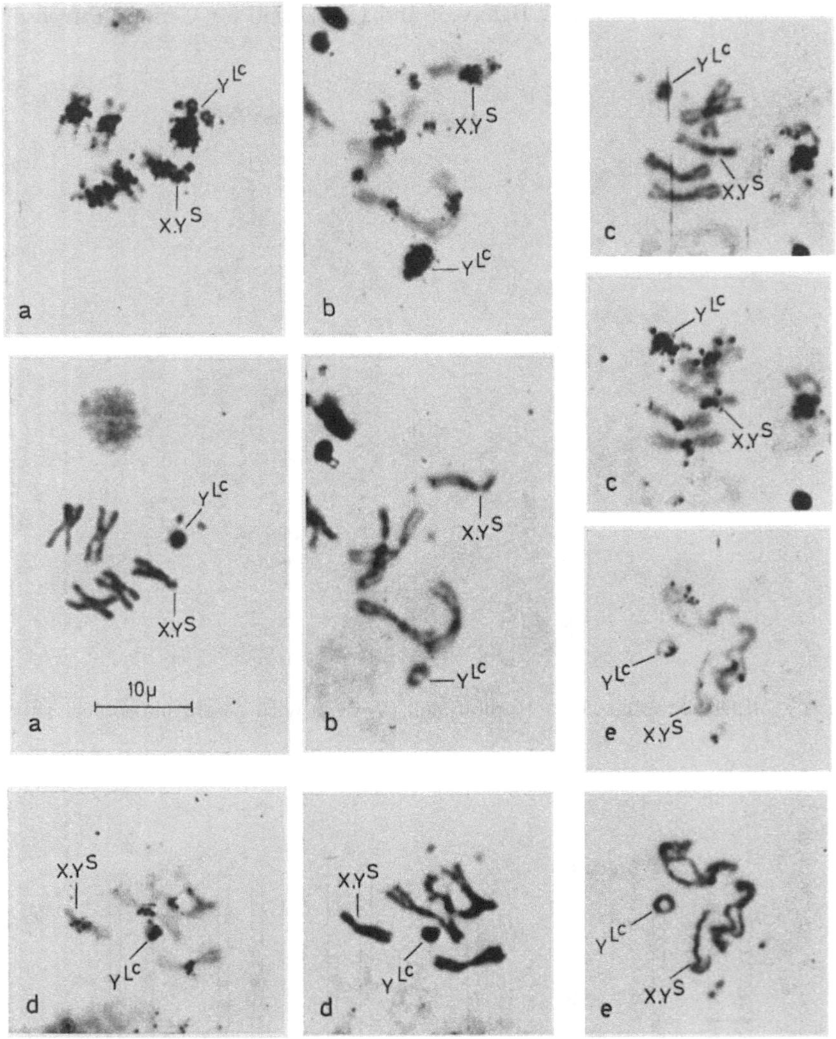

Fig. 139. "Sterilizer". Series of metaphases with the respective autoradiographs showing early replication of the translocation Y portion

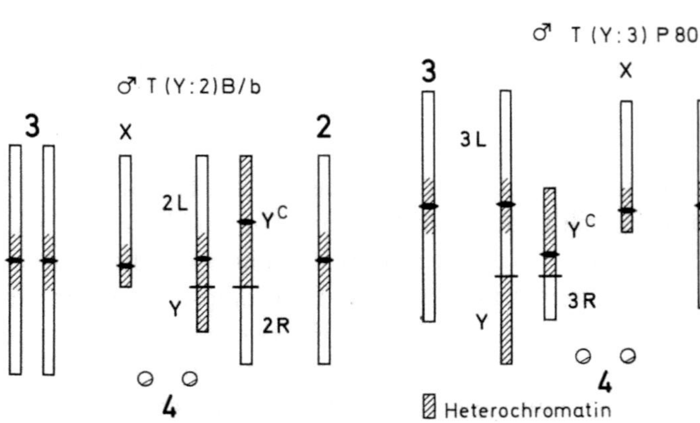

Fig. 140. Scheme of two translocations between the Y chromosome and the autosome, T (Y; 2) and T (Y; 3)

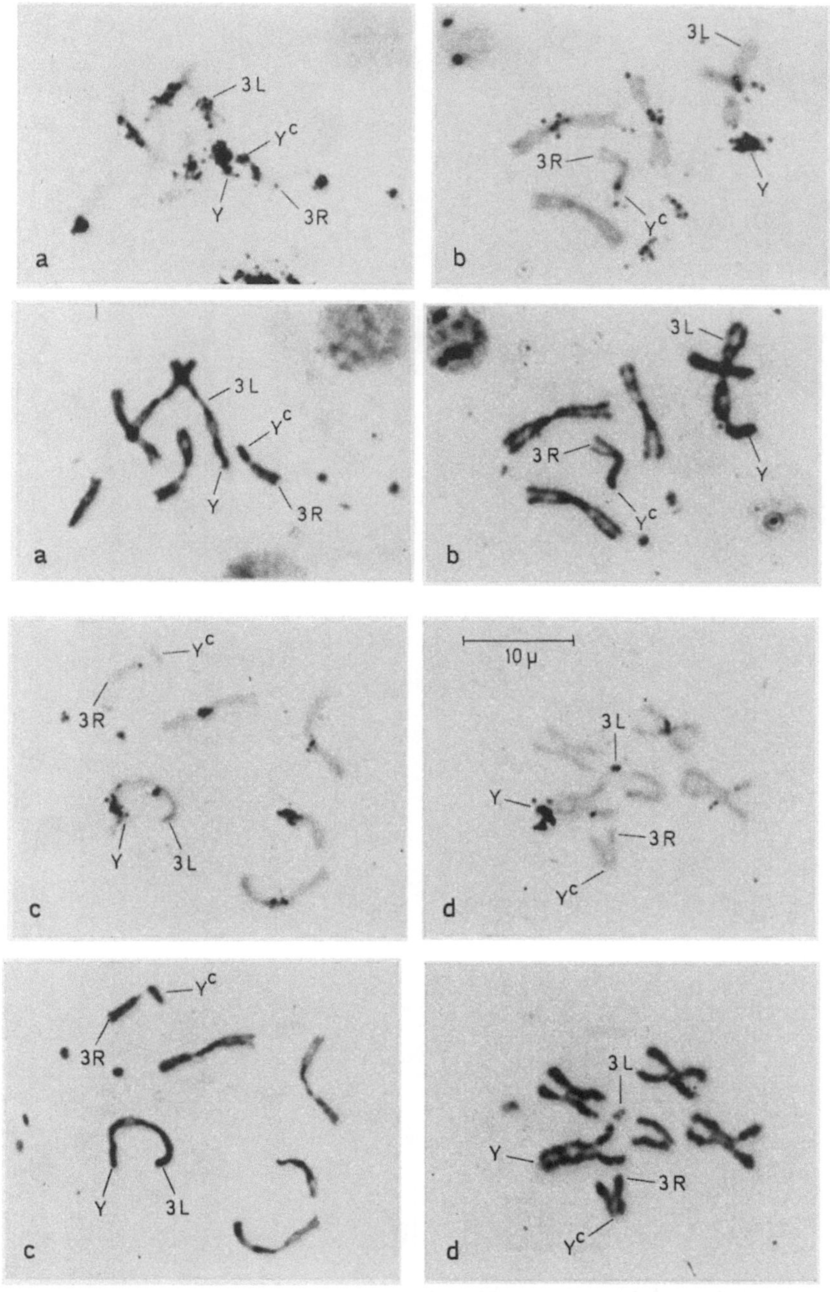

Fig. 141. Series of metaphases and the respective autoradiographs of T (Y; 3), showing late replication of the acentric Y chromosome portion translocated to the 3rd chromosome

center of DNA replication of Y, which seems to act by exerting an influence on the whole chromosome, which is interrupted when a section of Y becomes separated. This is an example of cytological work and wealth of problems,

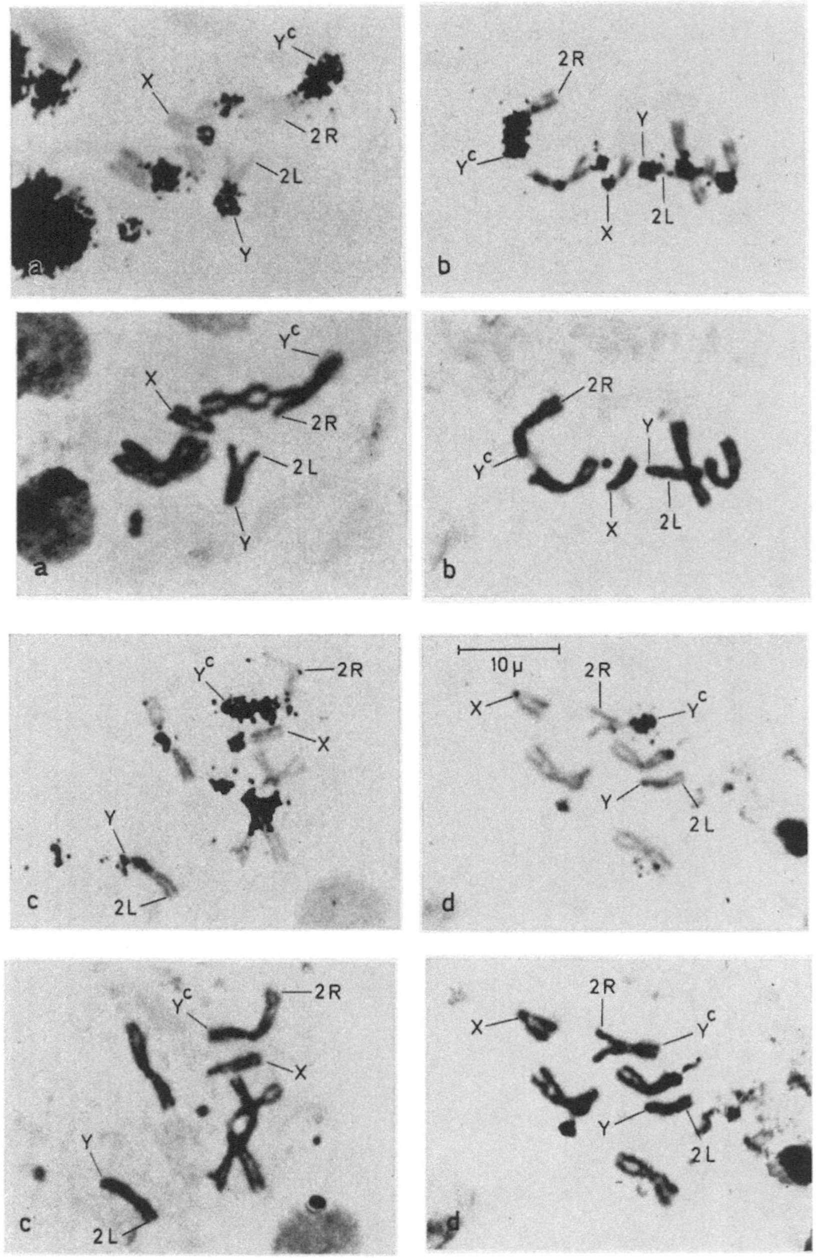

Fig. 142. Series of metaphases and the respective autoradiographs of I (Y; 2), showing late replication of the centric portion of the Y chromosome

that can be analyzed only when the resolving power of the chromosomal structures is maximal, as in cultured cells. Dividing cells *in situ*, treated in similar manner with ^{3}H-thymidine by injection into the body, are far from rendering high quality preparations.

Behavior of cultured cells. There are two procedures for growing *Drosophila* cells *in vitro*; (1) suspended in a flask and (2) on a cover glass at the bottom of a Carrel flask. Both conditions can be obtained using the H-5 medium of HORIKAWA and FOX (1966) or the medium of ECHALIER and OHANESSIAN (1969, 1970). In the technique of ECHALIER and co-workers (1965, 1969), the cultures are not started from isolated cells, but from small fragments of embryo. Cells adhere to the surface and tend to form a network of spindle-like elements. Pulsations or rhythmic movements are noticeable. Later on, this first network disappears, and residual cells form a secondary network, which may evolve into a homogenous layer of epithelioid cells. The first stage can also be obtained using H-5 medium (DOLFINI, 1969).

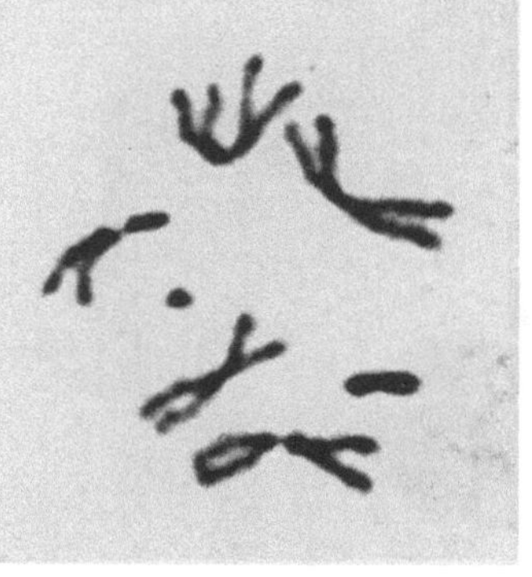

Fig. 143. Metaphase of a translocation found in the established cell line K_c: Y seems to have lost a portion of a long arm, probably translocated to the X. 2600 ×

B. Long Term Cultures

The results first obtained by HORIKAWA and FOX (1964) and later by ECHALIER and OHANESSIAN (1969) are doubtless the most promising for a utilization of cell culture for genetic purposes. However, the former authors have apparently discontinued their work and the results gathered by ECHALIER and co-workers are only now ready for discussion from several standpoints. In this paper only some information on the cytological aspects of line K_C will be considered, partly carried out by Miss S. DOLFINI in Milan, in close collaboration with Dr. ECHALIER'S group. Other points are discussed by ECHALIER in this chapter.

Duration of the mitotic cycle. These new experimental conditions are certainly biologically more "normal" than those represented by the short term cultures. In fact, short term cultures did not survive for long periods, while the last data on a long-term cell line were obtained 23 months after stabilization. According to DOLFINI et al. (1970), the duration of S is considerably shorter than in the former case, 10 instead of over 21 hrs.

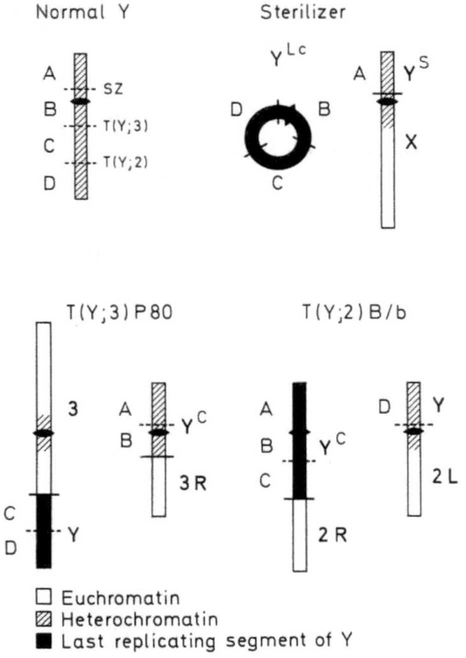

Fig. 144. Interpretation of the DNA replication in the Y chromosome translocations; the last replicating portions in black

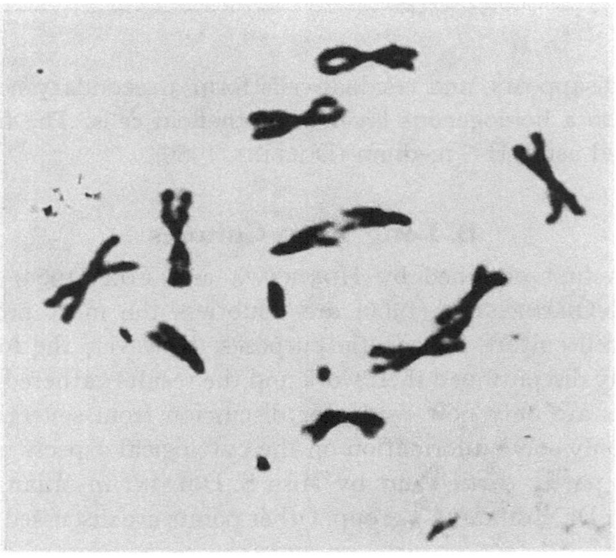

Fig. 145. The result of a fusion between a male and a female cell (metaphase) in cell line K_c. 2600×

Dynamics of genomes in stabilized cells. Systematic checking of cultured cells has generally shown changes in chromosome number. In the case of the K_C *Drosophila* established cell line submitted to karyology, differences were found when preparations made 17 and 23 months after the line had been established were compared. In the first check tetraploid cells were rare while the later check gave (in nearly 900 metaphases) 5–7% of tetraploids. This observation seems to prove that polyploidization occurs in culture. Among the genomes present, some new variants were found. Beside normal female cells with 2 X's there were XO cells and cells with one short heterochromatic body (the Y?) and one of the X's showing an extra piece. This latter cell type (Fig. 144) seems to be the result of the translocation between X and Y (DOLFINI, personal communication).

All these data, preliminary though they are, prove that chromosomal markers can be found in cultured cells and the importance of this finding can not be overemphasized.

Fusion between cells. The abundance of male and female cells in the same culture is by itself a sufficient prerequisite to detect the products of cell fusion. The presence of a short heterochromatic chromosome and of an X with an extra piece makes this task even easier. Dividing cells showing tetraploid sets, including one short heterochromatic chromosome (the Y?), two normal X's, one with an extra piece, four 2nd and four 3rd chromosomes are present although very rare. The 4th pair is not considered here, because it requires a more detailed presentation of data. Cells of this type are more difficult to explain as the result of simple polyploidization plus the presence of an extra X, than as the result of fusion between two cells of different sex (Fig. 145).

C. Conclusion

The results discussed in this paper show that the utilization of cell cultures *in vitro* for resolving genetic problems is now a concrete possibility.

The advantages offered by insect cells are clearly demonstrated by present techniques. Short term cultures supply all the possibilities provided by mammalian cells. The clear-cut identification of individual chromosomes and the distribution pattern of eu- and heterochromatin, together with the large amount of genetic information available and the abundance of chromosomal structural aberrations, permit the investigation in *Drosophila* of basic problems such as mitosis, DNA-replication, and heterochromatin functions with a resolving power unparalleled in other systems. Among insects, not only *Drosophila*, but other Diptera and Orthoptera, among others, are likely to offer excellent material. The choice of a species for cell cultivation directed towards genetic aims should take into consideration the following cytological prerequisites: low chromosome number, ease of identification of the different chromosome pairs and, possibly, a detectable amount of heterochromatin. There is no doubt that the importance of insect cell cultivation will increase enormously if the general use of established cell lines becomes possible. The preliminary results discussed in this volume are sufficient to demonstrate the large spectrum of possibilities opened by this technique for genetic purposes.

II. Established Diploid Cell Lines of *Drosophila melanogaster* as Potential Material for the Study of Genetics of Somatic Cells

GUY ECHALIER

It can be predicted that the study of genetics of somatic cells will develop considerably in the near future. After the spectacular breakthrough of molecular genetics, the time has come to apply its main results, relating to the function and regulation of the genome of bacteria and viruses, to eukaryotic cells. Many leading biologists have already undertaken a reconversion to different eukaryotic systems, but, in their search for the proper material, all the laboratories have used mammalian or, at least, vertebrate cells. The established lines of diploid cells that ECHALIER and OHANESSIAN (1969) have recently grown from embryos of *Drosophila melanogaster* may offer many advantages in this new field.

Drosophila cell lines have now been cultivated for two years and their karyotypes are fairly constant and predominantly euploid (Figs. 146, 147 and 148). They can be easily grown in our medium (ECHALIER et al., 1965, 1968, 1970) supplemented with fetal calf serum. Cell division time at 26° C was estimated to be about 19 hrs (DOLFINI et al., 1970 and BARIGOZZI, this chapter). Because of the role of the fly *Drosophila* in the history of genetics, we have in our possession massive data on its genome. Moreover, the well-known chromosomal set of this species is relatively simple.

Before these *Drosophila* cell lines become standard material for genetic studies, however, several technical obstacles must be overcome. Most of the work here reported is still in progress.

A. Isolation of Clones

It is always difficult to grow a single cell of a metazoan organism, when isolated in a culture vessel. Several techniques were devised to overcome this difficulty. One method consists of confining one cell in a very small volume of medium, either in a microdrop under liquid paraffin (LWOFF et al., 1955), or in a sealed capillary tube (SANFORD et al., 1948). Using the second technique, SUITOR et al. (1966) succeeded in cloning an insect cell line for the first time. Another technique developed by PUCK et al. (1956) consists of inoculating a culture flask with a very dilute suspension of cells and leaving it undisturbed for two weeks. If the conditions are suitable, each cell will form an isolated colony. The chances of contamination between the clones are reduced when the medium is solidified with agar. A useful procedure for collecting the clones is to cover the bottom of the flask with small fragments of coverslips on which the clones will grow. Only the dilution technique has been successful with *Drosophila* cells. But its results are still extremely variable and the "plating

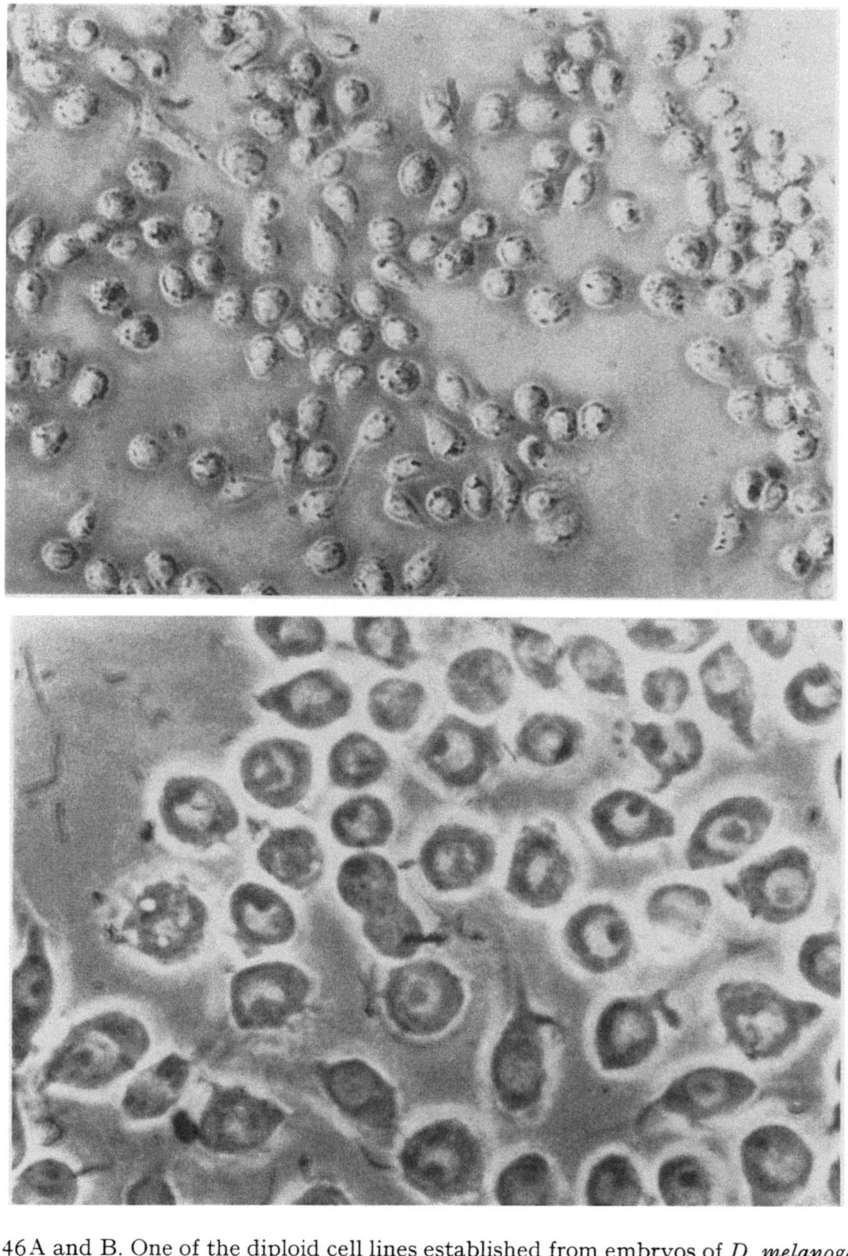

A

B

Fig. 146A and B. One of the diploid cell lines established from embryos of *D. melanogaster* (line K). Cells of this line adhere loosely to the culture vessel, are approximately round and somewhat isolated from each other. At higher magnification the clear spherical nucleus with one distinct nucleolus is apparent. (Living cells, phase contrast. A, 450×; B, 1000×)

efficiency" remains tremendously low: a Falcon plastic flask must be inoculated with as many as 10,000 cells to yield some 50 colonies. Our difficulties are probably due to the fact that we have to pipette the cell suspension

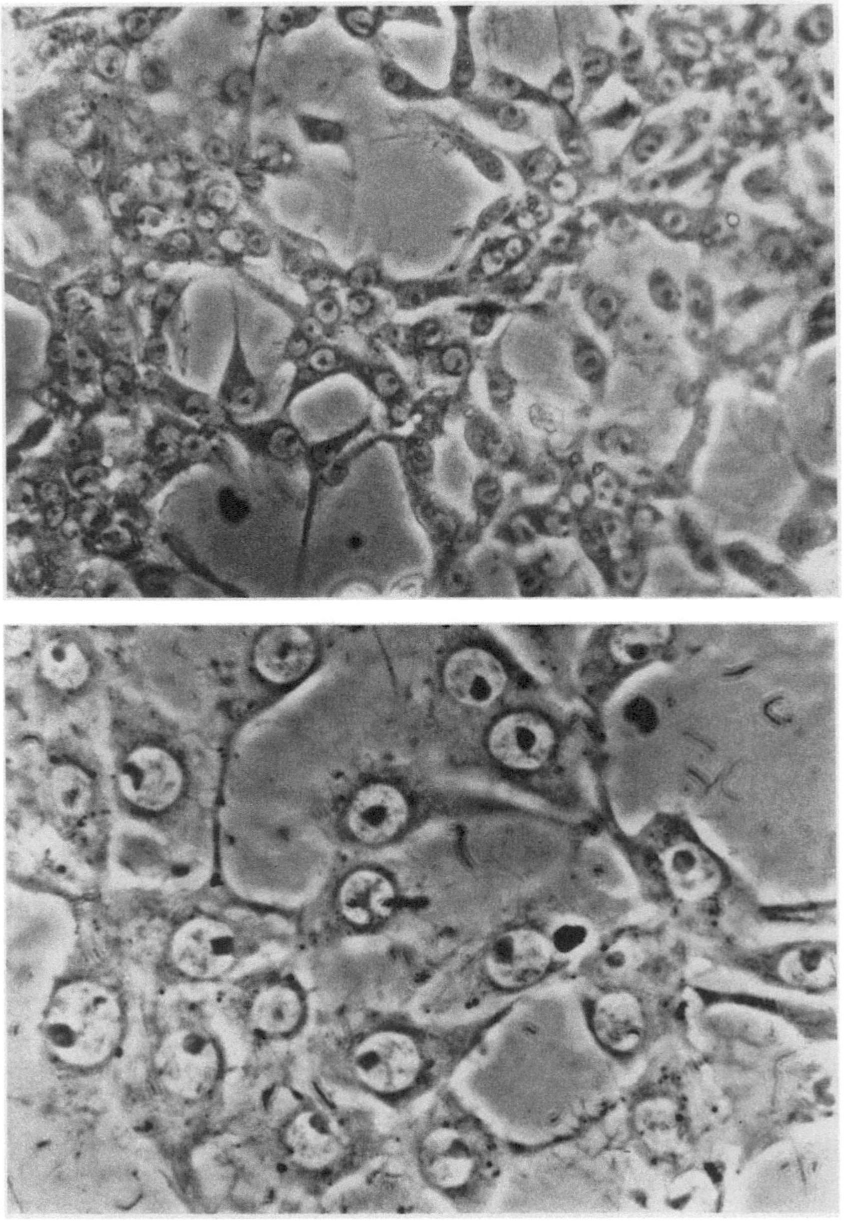

A

B

Fig. 147 A and B. Diploid cell line C, established from embryos of *D. melanogaster*. These cells are firmly attached to the glass, are fibroblast-like and form a net-like structure. (Living cells, phase contrast. A, 450 ×; B, 1000 ×)

violently, to make sure that the cells are isolated, since *Drosophila* cells have low tolerance for trypsin. Moreover, our medium may still be inadequate. Nevertheless, 7 clonal sublines have been established from our cell line K.

B. Selection of Suitable Genetic Markers

Of the thousand of mutations described in *Drosophila*, very few may be used in cultivated cells. Mutations disturbing one of the main biochemical pathways that any undifferentiated cell requires for its growth and multiplication must be sought. Moreover, at the present time the establishment of new cell lines from a *Drosophila* strain with a specific genotype is still laborious. The following is a rapid survey of the different kinds of markers that may be isolated from our existing cell lines.

Chromosomal markers. This type of marker, restricted to eukaryotic cells, has been widely used in genetics of somatic mammalian cells. In most cases, the cells of two distinct species are crossed to facilitate the identification of their chromosomal sets.

The relative simplicity of the karyotype of *D. melanogaster* is an obvious asset. An extensive karyologic analysis of our cell lines is being done by Miss S. DOLFINI (BARIGOZZI, this chapter). Two types of potentially interesting chromosomal markers need to be mentioned here. The sex chromosomes, easy to recognize in *Drosophila*, are excellent natural markers. Miss DOLFINI (personal communication) found that in line K most of the cells have a female chromosomal set (Fig. 148A) while the opposite is true in line C (Fig. 148B). Likewise, the loss of a specific chromosome, or even of a small fragment of it—not difficult to spot in the karyotype of *Drosophila*—may be very useful. It might be possible to locate a gene when the absence of one definite chromosome is correlated with the loss of some gene product. For instance, Miss DOLFINI observed that in many cells of the line K one and sometimes both chromosomes of the 4th pair were lost. POULSON (1940) noticed that "during the embryogenesis of *Drosophila*, in the absence of the K chromosomal genes, cellular proliferation can continue, but differentiation is adversely affected".

It would be of great interest to isolate clones with deficient chromosomal sets. This may be accomplished by the use of *para*-fluorophenylalanine, which induces haploidization in several fungi (LHOAS, 1961) and possibly in a mammalian cell line (GILLOIS et al., 1969). The frequency of aneuploid metaphase figures in *Drosophila* cell lines, 24 hrs after treatment with this compound, was raised up to 60% (unpublished results). So, we can greatly increase our chances of selecting an aneuploid cell from the population and growing a clone from it.

Morphological features. The individual shape of the cells as well as the general growth pattern of the various lines are fairly distinct (Figs. 146 and 147), but the morphology of cells in culture is inconstant. Such characteristics are not reliable markers.

Nutritional mutants. PUCK and KAO (1967) recently adapted to somatic mammalian cells a method for selecting auxotrophic mutants from a large population. This technique requires that the cells be grown in a medium of definite composition and this has not yet been possible in the case of *Drosophila* cell lines.

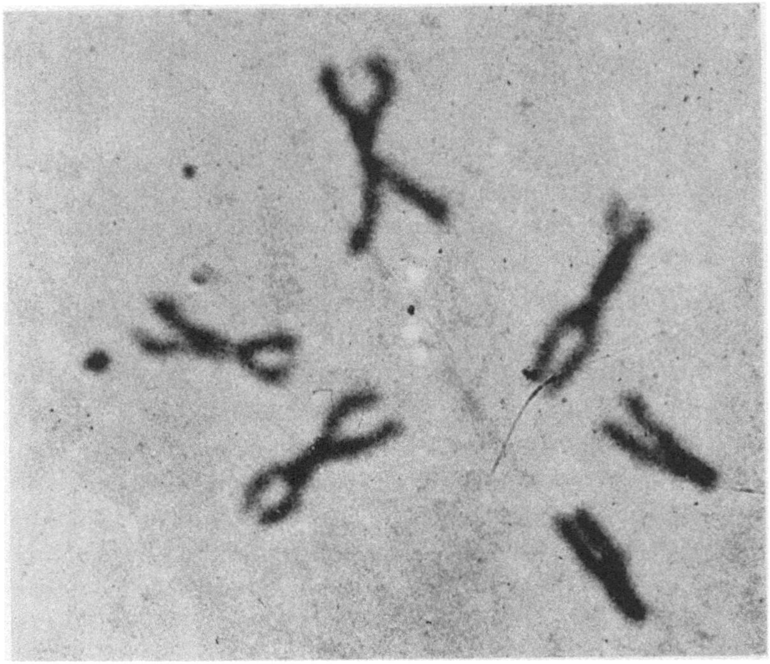

A

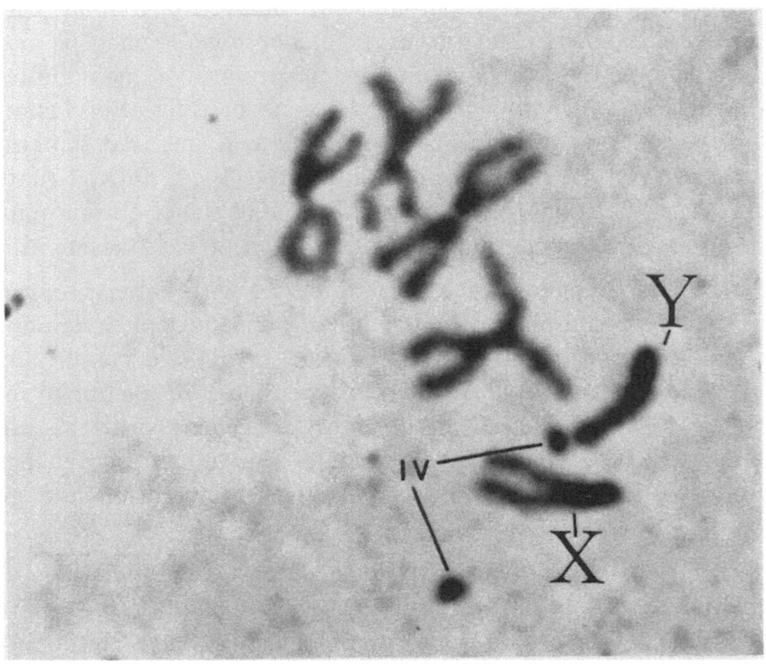

B

Fig. 148 A and B. Karyotypes of two established lines of *D. melanogaster* cells. A Line K. These cells have a female chromosome set, and sometimes they lose one small chromosome of the 4th pair. B Line C. The majority of the cells of this line are of the male type. Orcein-stained preparations. (Courtesy of Miss S. DOLFINI)

Our culture medium (ECHALIER and OHANESSIAN, 1970) contains large amounts of a lactalbumin hydrolysate and yeast extract. We have made progress, however, towards a quasi-synthetic medium. One subline multiplies rapidly in a medium with a known concentration of amino acids. Another line does fairly well when some specific vitamins are substituted for yeast extract. But, thus far, the cells fail to multiply if they are cultivated without both lactalbumin hydrolysate and yeast extract. If this can be done, we will be able to formulate an *optimum medium* for the growth of new *Drosophila* cell lines, and a *minimum medium* for the selection of auxotrophic mutants.

Drug-resistant markers. Markers of this kind, especially resistance to analogues of purines or pyrimidines, such as 8-azaguanine and 5-bromodeoxy-uridine, have been frequently used in genetics of somatic cells. They are of value because the enzymatic deficiencies which account for such resistance are located on one of the essential biosynthetic pathways leading to nucleic acids and because a very selective technique was devised (SZYBALSKA and SZYBALSKI, 1962; LITTLEFIELD, 1966) for detecting the few hybrid cells originating from a fusion between two cells resistant to the two different drugs.

One of my collaborators, J. L. BECKER, is investigating the frequency and mechanism of resistance to several purine and pyrimidine analogues in *Drosophila* cells (unpublished work). The frequencies of mutation in a population of the line K, have been estimated at about 10^{-5} for resistance to 3 γ/ml of 8-azaguanine and at about 10^{-6} for resistance to 30 γ/ml of 5-bromodeoxy-uridine. These results are very similar to those published for mammalian cells. We have already isolated one clone of azaguanine resistant cells. The frequencies of reversion have not yet been established. BECKER can now assay, in the resistant lines, the activities of the enzymes involved, guanine pyrophosphoryl-ase or thymidine kinase.

Isozyme pattern variations. During the last few years, such isozyme variations have been extensively studied in the *Drosophila* fly. For instance, several specific esterases were identified by their different electrophoretic mobilities, and their genetic control determined (WRIGHT, 1963; BECKMAN and JOHNSON, 1964; JOHNSON, 1964).

Dr. MARTOJA, in our laboratory, has recently established by cytochemical tests, the presence of esterases in our cell line K (personal communication). These enzymes act on substrates, such as α-naphthylacetate used in *Drosophila* genetics.

Virus susceptibility. This topic is discussed by OHANESSIAN (this chapter). A worker in her group, Miss RICHARD-MOLARD, succeeded in growing new cell lines from a strain of *Drosophila* which is genetically "refractory" to sigma virus. One of the genes controlling this character was previously located on the second chromosome (OHANESSIAN-GUILLEMAIN, 1963). These special cell lines, in contrast to the others, do not support the multiplication of the sigma virus (unpublished work).

C. Cell Fusion

Cell fusion is a "must" for classic genetic research. Since the discovery by Barski et al. (1960) that somatic hybridization can occur spontaneously in a mixed population of two mammalian cell lines, a technique was devised (Harris and Watkins, 1965; Yerganian and Nell, 1966) using specific strains of viruses, to considerably increase the frequency of cell fusion.

I have tried to induce polykaryocytosis by this technique in *Drosophila* cells, by a treatment with the Sendai strain of parainfluenza virus, the simian parainfluenza SV5, or the Aujeski virus of the herpes group. These three viruses are most commonly used to fuse mammalian cells. Up to now, there has been no apparent success. Of great interest is the report of Conover (Chapter 3) announcing the achievement of hybridization between human HeLa and mosquito cells. However, this interphylum hybridization, interesting though it is, may involve different conditions.

Another way worth exploring is the possible fusion effect of some arboviruses on insect cells. Paul et al. (1969) and Suitor and Paul (1969) while reporting the cytopathic effects of some arboviruses of serogroup B on Singh's *Aedes* cell line, describe the formation of large syncytia by fusion of infected cells. In collaboration with Dr. Hannoun the multiplication of arboviruses in *Drosophila* cell lines and their effect are being investigated (Hannoun and Echalier, this chapter).

Although we still need a fusion technique as simple and rapid as the one used with mammalian cells, spontaneous cell fusion does take place in *Drosophila* cell lines, as Dr. Barigozzi reports in this chapter. Miss Dolfini was able to observe several karyologic figures which are reasonable evidence of such an event.

D. Somatic Pairing and Crossing Over

In this particular field, *Drosophila* cells, or dipteran cell lines, will gain decisive advantage over the other cells. An accurate genetic analysis implies that some recombination, that is, some crossing-over, takes place between homologous chromosomes.

D. melanogaster is one of the very few organisms and, as far as I know, the only metazoon in which somatic crossing-over was duly established (Stern, 1936). This phenomenon requires the intimate association of homologous prochromosomes during interphase. Somatic pairing might even be the rule in dipteran cells, as is suggested by the formation, in many tissues, of giant polytene chromosomes. Cytogeneticists noticed (Kaufmann, 1934) when making "squash" preparations of dipteran embryos or larvae, that in metaphase the two chromosomes of the same pair stayed very often in close proximity. A careful karyologic examination of our *Drosophila* cell lines seems to confirm that such proximity is not merely fortuitous. Nichols (Chapter 2) illustrated the same phenomenon in his photographs of the karyotypes of Singh's *Aedes* cell lines.

E. Conclusion

This report is not exclusively speculative. Progress is being made in all the lines of approach discussed above, and it is interesting to note that new *Drosophila* embryo cell lines have just been established by KAKPAKOV et al. (1969).

III. Arbovirus Multiplication in an Established Diploid Cell Line of *Drosophila melanogaster*

C. HANNOUN and G. ECHALIER

The important applications of insect cell culture to virology became evident from the onset, with the pioneering work of TRAGER (1938). Since then, various insect tissues and cells were grown *in vitro* and their susceptibility to certain arboviruses was tested (see Chapters 4–7). Recently, we have studied the conditions of multiplication of several arboviruses in established cell lines of *Drosophila* (ECHALIER and OHANESSIAN, 1969). The preliminary results of this work are here reported.

A. Materials and Methods

Cell cultures. The cells belonged to the line K, one of the diploid cell lines established from *D. melanogaster* embryos (ECHALIER and OHANESSIAN, 1969). At the beginning of the two experimental series, they had been cultivated *in vitro* for 20 months (54 passages) and 2 years (67 passages), respectively. The cells were grown in Falcon plastic flasks (30 ml), at 26° C, with 4 ml of the medium devised for these cells (ECHALIER and OHANESSIAN, 1970) and supplemented with 2% (1st experiment) or 20% (2nd experiment) fetal calf serum. The maximum number of cells in a bottle increased to 10^7. The cultures were transferred every 8 to 10 days.

Virus strains. Prototype virus strains were used, except in the case of the following: Sindbis virus: "large plaque" (G) and "small plaque" (p) variants (HANNOUN et al., 1964); West Nile virus: In addition to the prototype strain Egypt 101, strain *Halima,* recently isolated from a horse (PANTHIER et al., 1966) was tested at the 4th passage level; Tahyna virus: strain *Theobaldia,* isolated in Czechoslovakia by DANIELOVA (personal communication) from the mosquito *Theobaldia annulata* and the sister strains *488 V* and *488 N,* isolated in France from the blood of a rabbit by inoculation into infant mice by visceral and neural routes (HANNOUN, unpublished data).

Infection of the cultures and virus titrations. The initial inoculum was 0.2 ml of a 10% suspension of suckling mouse brain, having a titer of 10^7 to 10^8 mouse LD_{50}/ml. The titer was determined in BHK 21 cells at the time of inoculation into the cultures. After one day, the medium was removed, washed, and fresh medium was introduced. The cell culture fluids were titrated by inoculating tube cultures of BHK 21 cells with each of a series of dilutions of each fluid sample. The ID_{50} titer was determined on the 5th and 10th day after inoculation.

Cytological examinations. Microscopic examinations were made on Giemsa and Hematoxylin-stained preparations of infected and uninfected cells.

B. Experimental Results

Long term experiments: Establishment of a carrier culture state in Drosophila cells. Dengue-2 and Tahyna virus multiply in *Drosophila* cells, as shown in Table 46, and virus is released for at least 107 days. The experiment is still in progress. Control experiments demonstrated that survival of infectious particles of the two viruses in culture medium without cells never exceeded 3 days at 26° C. In the case of Dengue-2 virus, release was slow, since it was not regularly recovered before the 13th day. Even then and at later intervals the titers in the culture medium remained low and fluctuated significantly, on the same day, from one bottle to another. However, no culture remained permanently negative. These fluctuations can be attributed to the relatively low sensitivity to this virus of the cell system used for titration. It was noticed that the average titers became significantly higher at the end of the third

Table 46. *Multiplication of Dengue-2 and Tahyna viruses in Drosophila line K*

Days after inoculation	Virus titers[a]	
	Dengue-2	Tahyna
2	N	2.3
3	1.4	2.4
5	N	2.3
6	N	1.8
7	2.2; 1.3	2.6; 1.7
8	N	2.7
9	N	2.7
12	N	2.5; 1.4
13	1.8	2.2; 1.3
14	2.2	2.4; 2.3
15	1.3	2.4; 2.3
16	N	2.7
19	2.1; N; N	2.9; 1.9; 1.7
21	N; N; N	3.7; 2.8; 2.1
23	2.2; N; N	3.1; 2.4; 2.4
25	N	2.7
29	N; N; N	2.8; 2.7; 2.4; 2.2; 2.2
33	N; N; N	4.4; 3.2; 3.2; 2.7; 2.4; 2.4
36	N; N	3.2; 3.1; 2.7; 2.6; 2.3
54	2.2; 2.2; 1.9	2.7; 2.7; 2.3; 2.2; 2.2; 2.2
57	2.2; 1.7; N	2.7; 2.3; 2.2; 2.2; 2.2; 2.2
79	2.2; N	1.8; 1.6; N; N; N; N
82	N; N; N; N	1.8; N; N; N
85	2.9; 2.9; 2.6; 2.4; 2.2	1.9; 1.7; 1.7; 1.3; 1.3; N; N
90	2.7; 2.4; 1.9	2.2; 2.2; 1.8; 1.6; N
107	2.4; 2.2; 1.9; 1.9; 1.9	2.6; 2.6; 2.2; 2.1; 1.9; 1.9; 1.9; 1.7

[a] Expressed as the logarithm of the ID_{50}/ml of culture fluid, as determined in BHK 21 cells. N means <1.2. The initial inocula for the Dengue-2 and Tahyna experiments were 2.5 and 4.5, respectively. The culture medium was changed at the end of the first day and at intervals of 8–10 days, thereafter.

month. For Tahyna virus, multiplication was much more rapid and virus was constantly found in the culture fluid from the beginning of the experiment, at least until the 57th day. The titers were higher than in the case of Dengue-2, but it should be remembered that BHK 21 cells are more susceptible to Tahyna than to Dengue-2 virus. At the end of the third month, the average titers decreased slightly.

Susceptibility of Drosophila cells to different groups of arboviruses. Eighteen viruses belonging to several major groups were tested and the results are presented in Table 47. With group A viruses, multiplication was rapid and the titers of the culture fluids reached high values. The large plaque variant of Sindbis virus grew more rapidly than the small plaque variant. The multiplication of Sindbis virus (prototype strain) in cultures of *Drosophila* cells, both in "primary cultures" and in continuous lines was carefully investigated by F. BRAS, in A. OHANESSIAN's group. She observed a fairly high multiplication rate and the establishment of a carrier culture state (unpublished data).

Table 47. *Susceptibility of Drosophila cell line K to arboviruses*

Viruses	Logarithm ID_{50} introduced/culture	Virus titers days after inoculation[a]			
		4th	7th	11th	14th
A group					
Chikungunya	6.9	4.3	>5.2	>5.2	>5.2
Sindbis (G)	6.2	4.4	>5.2	>5.2	4.7
Sindbis (p)	6.1	3.4	not done	3.3	3.3
B group					
Dengue-1	2.5	N	1.7 ?	N	2.4
Dengue-2	2.5	N	2.2 ?	N	2.4
Dengue-3	2.7	N	N	N	1.9
Dengue-4	2.5	N	N	N	N
Yellow Fever FN	2.5	N	N	N	N
Saint Louis encephalitis	4.7	2.2	2.2	N	1.7
West Nile (Egypt 101)	5.0	>5.2	>5.2	>5.2	>5.2
West Nile (Halima)	4.5	4.9	not done	4.7	>5.2
Japanese encephalitis	5.9	N	2.9	3.1	2.3
Central European encephalitis	4.5	2.3	2.8	2.2	2.2
Other groups					
Tahyna (Theobaldia)	5.5	2.7	4.5	4.2	3.7
Tahyna 488 V	4.5	not done	not done	not done	4.8
Tahyna 488 N	4.5	3.2	4.8	4.5	3.4
Ilesha	5.4	N	N	N	N
Sandfly fever (Sicily)	3.5	N	N	N	2.2

[a] See Table 46.

Among group B viruses, West Nile multiplied exceptionally well, reaching high titers quickly. No significant difference could be observed between two viruses of different origin and passage level. The other group B viruses either grew to low titers or not at all. *Drosophila* cells are not completely unsusceptible, however, to Central European Encephalitis (CEE), an exclusively tick-borne virus.

With regard to other groups, the results obtained with Tahyna virus (Table 46) were confirmed on several strains (Table 47). The cells did not seem to be susceptible to Ilesha or Sandfly fever (Sicilian, SFS) viruses, except for recovery of SFS virus at 14 days, but this needs to be further evaluated.

Cytological observations. Systematic examinations of stained slides, with special attention to the possibility of induction of cell fusion by the infecting viruses, failed to reveal any significant morphological alterations of the cells in any of the 18 viruses investigated.

IV. Sigma Virus Multiplication in *Drosophila* Cell Lines of Different Genotypes

Annie Ohanessian

Sigma virus has a hereditary association with the *Drosophila* fly. This association, called "stabilized condition", is regularly transmitted through maternal inheritance (L'Heritier, 1958; Brun, 1963). Diploid cell lines of *Drosophila melanogaster*, obtained by Echalier and Ohanessian (1969) were used to further investigate the relationships between the virus and its host cells. Sigma multiplication was previously observed in "primary" cultures of embryonic cells of *Drosophila* (Ohanessian and Echalier, 1967). Three aspects of virus multiplication in cell lines were studied: (a) virus multiplication for short periods; (b) virus persistance for periods exceeding one year; (c) virus multiplication in cell lines of different genotypes.

A. Short Term Multiplication

The yield of virus from two cultures is represented in Fig. 149. One culture was treated with anti-sigma virus antiserum after adsorption. No virus was isolated from the supernatant fluid after the treatment with antiserum. In the two cultures, the maximum yield of sigma virus was attained on the third day. At that time, the titer was considerably higher than the original inoculum. This rapid growth was followed by a period of reduced virus production. About 10^6 infectious units were obtained 22 days after infection. Thermal inactivation of the virus is depicted in Fig. 150: Without cells, no virus could be detected after 13 days. That the virus actually multiplied in cell cultures is demonstrated by the comparison between Figs. 149 and 150. These results have been confirmed by several other experiments of the same kind.

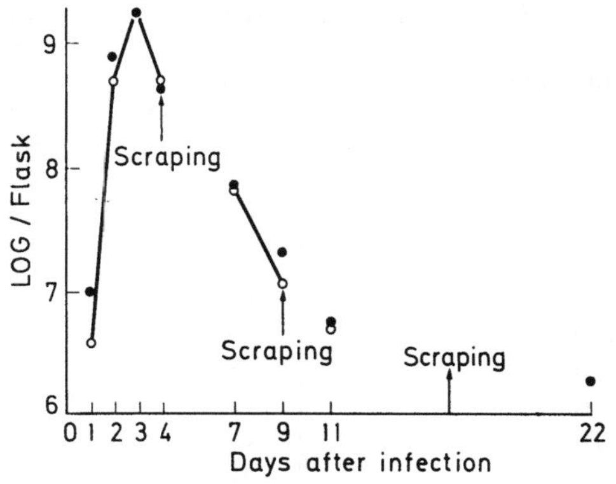

Fig. 149. Multiplication of sigma virus in *Drosophila* cell cultures. Some cells were removed by scraping and the medium changed on days 4, 9, and 16. The virus was titrated as described by BRUN (1963). The titers obtained from the culture treated with anti-sigma virus antiserum after adsorption are identified by open circles

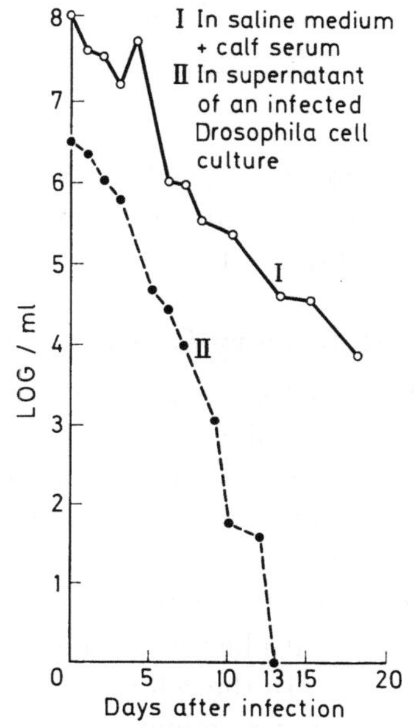

Fig. 150. Thermal inactivation of sigma virus at 25° C. *I* Virus suspended in saline solution with calf serum. *II* Virus suspended in the supernatant of an infected cell culture, but without cells

B. Long Term Infection

One culture was infected and maintained as in previous experiments. The culture medium was changed about once a week and titrated at intervals. When necessary, some of the cells were scraped and transferred to subcultures. Only the virus yields of the initial culture are shown in Table 48. For a period of over a year the viral titer remained approximately constant, 10^5–10^6 infectious units per flask. This suggests that there was continuous and regular virus production. No cytopathic effect was observed: the cells appeared quite

Table 48. *Yield of sigma virus from infected Drosophila cell culture*

Day after infection	Log infectious units/flask	Day after infection	Log infectious units/flask
0	4.38	114	5.57
7	6.02	124	5.24
14	5.92	128	5.50
21	5.92	135	5.80
28	5.79	143	5.26
35	5.70	156	6.54
52	5.00	177	6.00
58	5.76	191	5.00
65	5.23	210	5.97
72	5.55	219	3.90
79	4.71	300	6.24
86	5.50	311	6.24
94	5.58	318	5.86
100	4.35	360	6.44
103	5.46	388	5.82
107	5.70	437	6.72

normal and their multiplication proceeded at the usual rate. Similar results were obtained in the different sub-cultures. Thus, there is a typical carrier state between sigma virus and its host cells.

C. Infection of Refractory Cell Lines of *Drosophila*

Recently, RICHARD-MOLARD (personal communication) of my group has succeeded in isolating several cell lines from *Drosophila* homozygous for the refractory gene "ref". Sigma virus variant P^- can not multiply in *Drosophila* homozygous for this gene (OHANESSIAN-GUILLEMAIN, 1963). Virus variant P^+, on the other hand, can multiply in this line. RICHARD-MOLARD inoculated refractory and non-refractory cell lines with P^- and P^+. The results are shown in Table 49. Both P^+ and P^- multiplied in the non-refractory cell line and the titer of P^- was particularly high. In the cell line derived from "ref" *Drosophila* only P^+ multiplied. Thus, in this preliminary experiment the results obtained in the *Drosophila* fly were reproduced in cell culture.

D. Conclusion

Drosophila cell cultures have proven their usefulness for the study of sigma virus infection and sigma virus genetics. BRAS and PRINTZ (personal communications) have shown that other viruses such as Sindbis and vesicular stomatitis, which multiply in the fly, also multiply in *Drosophila* cell lines. A carrier state is also obtained with Sindbis virus. It will be interesting to compare these results with those of BANERJEE and SINGH (1968), PELEG (1969b) and REHACEK (1968c) who have established persistent virus infection with other insect cell lines.

Table 49. *Growth of sigma virus in Drosophila cell cultures of different genotype*

Day after infection	Log infectious units/flask			
	P^+ virus		P^- virus	
	$\frac{+}{+}$ cells	$\frac{\text{ref}}{\text{ref}}$ cells	$\frac{+}{+}$ cells	$\frac{\text{ref}}{\text{ref}}$ cells
0	6.04	6.04	6.20	6.20
8	4.9	6.25	7.30	N[a]
12	4.17	5.74	8.27	N
20	6	6.79	7.97	N

[a] No virus demonstrated.

V. The *SR* Condition in *Drosophila bifasciata*: *In Vivo* and *in Vitro* Growth

Elaine A. Leventhal

A cursory examination of this volume indicates the breadth of interest in pathogens that prey on or are transmitted by insects. Even "*Drosophila* geneticists" are looking at "sick fruit flies" and characterizing "cytoplasmically inherited" agents which produce abnormalities that mimic lethal or sublethal mutants. The prototype infection in *Drosophila* is sensitivity to carbon dioxide caused by sigma virus (L'Heritier, 1958). Plus and Duthoit (1969) has described a new virus *P* that causes a cessation in egg production, and Akai et al. (1967), Filshie et al. (1967), and Perotti and Bairati (1968) have all described uniformly shaped polyhedral viral particles in several types of tissues from a variety of life stages in *Drosophila melanogaster*.

The *SR* (sex ratio) strain of *D. bifasciata* carries an infectious agent that can be artificially transmitted both inter- and intraspecifically to normal females (Leventhal, 1968). After a sufficient incubation period the injected females and their daughters produce abnormal sex ratios that reflect male embryonic mortality. Whole fly, embryonic squashes and sections, as well as hemolymph smears have been examined for the presence of pathogens and none have been observed. Therefore, *SR* was presumed to be submicroscopic.

The test for the presence of the *SR* agent has been biological; egg mortality and sex ratio distortion; thus, the major limiting factor in the study of the *SR* condition has been the lack of rapid *in vivo* or *in vitro* assay systems.

The *D. bifasciata* sex ratio condition is similar to the spirochete infections in *D. willistoni* studied by Poulson (1963) and Poulson and Sakaguchi (1961 a and b). The kinetics of the treponeme infection have been described by Williamson (1965), and a reliable assay devised: the presence of spirochetes in the hemolymph.

The work presented here attempts to further characterize the properties of the *D. bifasciata SR* agent in two ways. First, in order to correlate infectivity with a transmissible agent, it is necessary to demonstrate serial passage of *SR*. Effecting interspecific serial passage would also allow maintenance of *SR* in a vigorous host that could be genetically manipulated to clarify the possible chromosomal components associated with *SR*. Second, two sets of preliminary experiments utilizing vertebrate and insect tissue culture techniques are reported. These studies are aimed toward the development of an efficient assay procedure.

A. Methods

Serial passage. D. bifasciata SR females were ground in a Potter type glass homogenizer with 0.25 M sucrose diluent containing 1.5% of 0.05 M Tris buffer. The homogenate was frozen (−70° C) and thawed (+15° C) five times to rupture the cells. The brei was clarified at 1500 rpm for 10 min, the pellet discarded and the supernatant centrifuged at 10,000 rpm for 30 min. The second pellet was discarded and the undiluted supernatant was injected in 0.5 µl doses into the abdomens of host females. All the progeny of the injected females were examined, sample F_2 matings were set up and the remainder of the flies were frozen. Controls were either untreated or injected with 0.5 µl doses of diluent. LEVENTHAL (1968) found no differences in progeny ratios in these types of "controls", females inoculated with normal (uninfected) *D. bifasciata* fly homogenate, diluent alone, or untreated. Therefore, only untreated or diluent injected flies were inoculated in these experiments.

Tissue culture. Since stable *Drosophila* cell lines were not available, the first experiments made use of vertebrate cell cultures that support arbovirus growth (HORSFALL and TAMM, 1965). These were chick embryo (maintained in Hanks' medium), green monkey kidney (Vero, grown in modified Eagle's medium), and baby hamster kidney (BHK-21, grown in 973 K medium) cell cultures. Twelve tubes of chick and BHK-21 cells were maintained as control (untreated) tubes and 12 were inoculated with the *SR* preparation. Of 40 tubes of Vero cells (transfer 125), 20 were inoculated with the *SR* preparation, 10 were inoculated with uninfected fly extract and 10 were kept as untreated blanks. The cultures were initially seeded with 10^5 cells, and allowed to multiply for 5 days at which time a confluent cell sheet was formed. The inocula, prepared from frozen serial passage flies were subjected to pretreatment either by sonication or freeze-thawing and clarification of the homogenate by centrifugation at 10,000 rpm. The supernatant was centrifuged at 40,000 rpm for one and one-half hrs in a Beckman-L centrifuge, and the pellet was resuspended in 0.5 ml of diluent. Initially, 1 gram of flies was homogenized in 24 ml of diluent; thus, the final pellet preparation constituted a 50-fold concentration of *SR* particles. One tenth ml of this preparation was added to the cultures with 1 ml of appropriate medium and allowed to adsorb for 1 hr. The infectious fluid was removed, the cells were rinsed in 1 ml of fresh medium, and 2 ml of maintenance medium added. Cultures were kept at 37° C and observed daily.

Primary cell cultures from *Drosophila* embryos were prepared according to the method of HORIKAWA and FOX (1964). Three separate sets of 8 cultures were made, 4 were left untreated and 4 were inoculated with 0.1 ml of undiluted *SR* preparation, in 1 ml of H-5 medium, and left to adsorb for 1 hr. The cells were gently pelleted at 1000 rpm in an IEC clinical centrifuge. The *SR*-containing medium was aspirated, the cell pellet rinsed with 1 ml of H-5 and the cells centrifuged again. Wash medium was removed and the cells suspended in H-5 supplemented with 10% calf serum. Cultures were kept at 30° C and watched for 4 days.

The possible growth of the *SR* particle in the various tissue cultures was assayed by inoculating uninfected hosts with 0.5 μl doses of tissue culture extracts and observing the progeny ratios.

B. Results and Discussion

Serial passage in Drosophila. The results of these experiments are presented in Table 50. Passage 1 represents the initial transfer of infective material from *D. bifasciata* SR breis into normal *D. melanogaster* hosts. The decrease in the number of male progeny, the *SR* effect, was similar to that reported in earlier studies (LEVENTHAL, 1968). Although not seen in F_1, persistence of the *SR* effect into the F_2 generation is clear. The same sex ratio distortion was seen in F_3, but the data are not included in Table 50. The passage 1 progeny were collected, frozen and used to inoculate the second passage. The *SR* distortion was absent in the F_1 generation, but was clearly evident in the F_2 progeny of the inoculated *D. melanogaster* and *D. bifasciata* females.

Experiments with vertebrate cell cultures. One group of culture tubes received sonicated inoculum and the other the freeze-thawed material. One tenth ml doses of *SR* preparations were allowed to adsorb to the cells for 1 hr and then the cells were washed and fresh maintenance medium added to the tubes. The inoculated cell lines were examined daily for 2 weeks and the media were changed regularly. No cytopathic effects (CPE) were noted in any of these lines tested, though there seemed to be some retardation of growth in the Vero cells by day 7. However, by day 9, regeneration of the cell sheet appeared to have taken place. The *SR* inoculated lines were indistinguishable from control tubes by day 12.

Table 51 gives the results of the bioassay that tested the infectivity of the primary inocula used in the vertebrate cell cultures. Two pretreatments of *SR* material gave different results. Using the sonicated material, no *SR* effect was seen in the *D. melanogaster* F_1 progeny, though a decrease in the number of males did occur in the F_2 flies tested. *D. bifasciata* hosts showed the effect immediately and there was maternal transmission to the next generation. The freeze-thawing treatment also produced an inoculum infectious for *D. melanogaster* females, though ineffective in *D. bifasciata* hosts. These results could be interpreted as indicating an augmentation of infectivity mediated by treatment superimposed on differences in species specificity for the *SR* infection (LEVENTHAL, 1968). However, because the number of flies tested was small and since *D. bifasciata* is not a vigorous species and females lay very small broods, conclusions about the efficacy of the pretreatments are premature. It is clear, nevertheless, that the initial inoculum contained infective particles.

To determine whether maintenance of *SR* had occurred *in vitro*, in the absence of CPE, the cultures that showed some aberrant growth, the Vero cells, were ruptured by alternate freezing and thawing and the cell contents centrifuged for 2 hrs at 35,000 rpm in the Beckman-L centrifuge. The pellet material was mixed with 0.5 ml of diluent and injected in 0.5 μl doses into virgin females.

Table 50. *Sex ratios produced by primary passage (passage 1) of SR into D. melanogaster and second passage into D. melanogaster and D. bifasciata females*

	F_1 generation							F_2 generation					
	No.		mean male ratio	total off-spring	significance[a]			No. lines	mean male ratio	total off-spring	significance[a]		
	inoc.	fer-tile			t	df	P				t	df	P
I. *D. melanogaster*													
Passage 1													
Experimental	15	13	0.483	2912				27	0.440	10368			
Control	10	9	0.495	1288	0.896	20	N.s.	9	0.494	1654	3.269	34	<0.005
II. *D. melanogaster*													
Passage 2													
Experimental	12	12	0.471	2867				15	0.449	3825			
Control	9	7	0.492	1380	1.314	17	<0.10	5	0.490	692	1.949	18	<0.05
III. *D. bifasciata*													
Passage 2													
Experimental	12	6	0.491	450				7	0.371	630			
Control	12	6	0.500	604	0.867	10	N.s.	7	0.496	1542	8.110	12	<0.005

[a] t = Student's t value; df = Degrees of freedom; P = Probability of random distribution; N.s. = Not significant ($P > 0.05$).

Table 51. *Sex ratios from D. melanogaster and D. bifasciata females receiving pretreated inocula*

	F_1 generation							F_2 generation					
	No.		mean male ratio	total off-spring	significance[a]			No. lines	mean male ratio	total off-spring	significance[a]		
	inoc.	fer-tile			t	df	P				l	df	P
I. *D. melanogaster*													
Sonicated homo-genate	5	4	0.498	1473	0.843	7	N.s.	4	0.414	466	2.735	7	<0.025
Freeze-thawed homogenate	5	5	0.420	1993	2.527	8	<0.025	5	0.431	959	1.912	8	<0.05
Control (diluent)	5	5	0.506	676				5	0.503	656			
II. *D. bifasciata*													
Sonicated homo-genate	15	10	0.428	897	2.199	13	<0.025	6	0.407	946	3.622	9	<0.005
Freeze-thawed homogenate	5	3	0.443	230	2.183	6	<0.05	5	0.499	747	0.170	8	N.s.
Control (diluent)	5	5	0.496	560				5	0.500	560			

[a] See Table 50.

Table 52. *Sex ratio effect produced in normal females in bioassay for replication of SR in Vero cell cultures*

	No.		mean male ratio	total off-spring	significance[a]		
	inoc.	fert.			*t*	df	p
I. *D. melanogaster*							
Experimental	10	10	0.406	938	1.344	14	<0.10
Control	10	6	0.464	543			
II. *D. bifasciata*							
Experimental	10	5	0.414	401	2.114	8	<0.05
Control	10	5	0.500	578			

[a] See Table 50.

Actual replication of *SR in vitro* could not be demonstrated because of the absence of a cell-free control. The results of this experiment are seen in Table 52. The *SR* effect was clearly induced in 3 of the 10 injected *D. melanogaster* females. The sex ratios were also the most extreme obtained by artificial transfer and approach those reported from wild populations (MAGNI, 1956). Of the 5 *D. bifasciata* females tested, 1 produced a significantly reduced number of sons and the sex ratios of the 4 remaining females were depressed.

Experiments with primary Drosophila cell cultures. A 0.05 ml quantity of undiluted *SR* inoculum in 0.95 ml of the culture medium was allowed to remain in contact with the cell suspension for 1 hr. The cells were then centrifuged at 1000 rpm for 5 min and the medium removed and replaced with fresh culture fluid supplemented with 10% calf serum. The cultures were observed daily. Fig. 151a is a photograph of a noninfected culture. Fig. 151b is a 1-day infected culture. Note the granular appearance of the cytoplasm. By day 3 (Fig. 151c) vacuolation was present in the larger cells. By day 4 (Fig. 151d)

Table 53. *Bioassay in D. melanogaster and D. bifasciata females of SR grown in Drosophila embryonic cell culture*

No. inoculated	No. *SR* in F_1	Mean male ratio from *SR* lines	No. *SR* in F_2	Mean male ratio from *SR* lines	No. *SR* in F_3	Mean male ratio from *SR* lines
D. melanogaster						
50	17/46	0.368	4/25	0.364	7/15	0.329
Control:						
10	0/10	0.501	0/10	0.496	0/10	0.499
D. bifasciata:						
25	2/13	0.354	2/5	0.392	3/7	0.354
Control:						
25	0/20	0.499	0/10	0.500	0/10	0.500

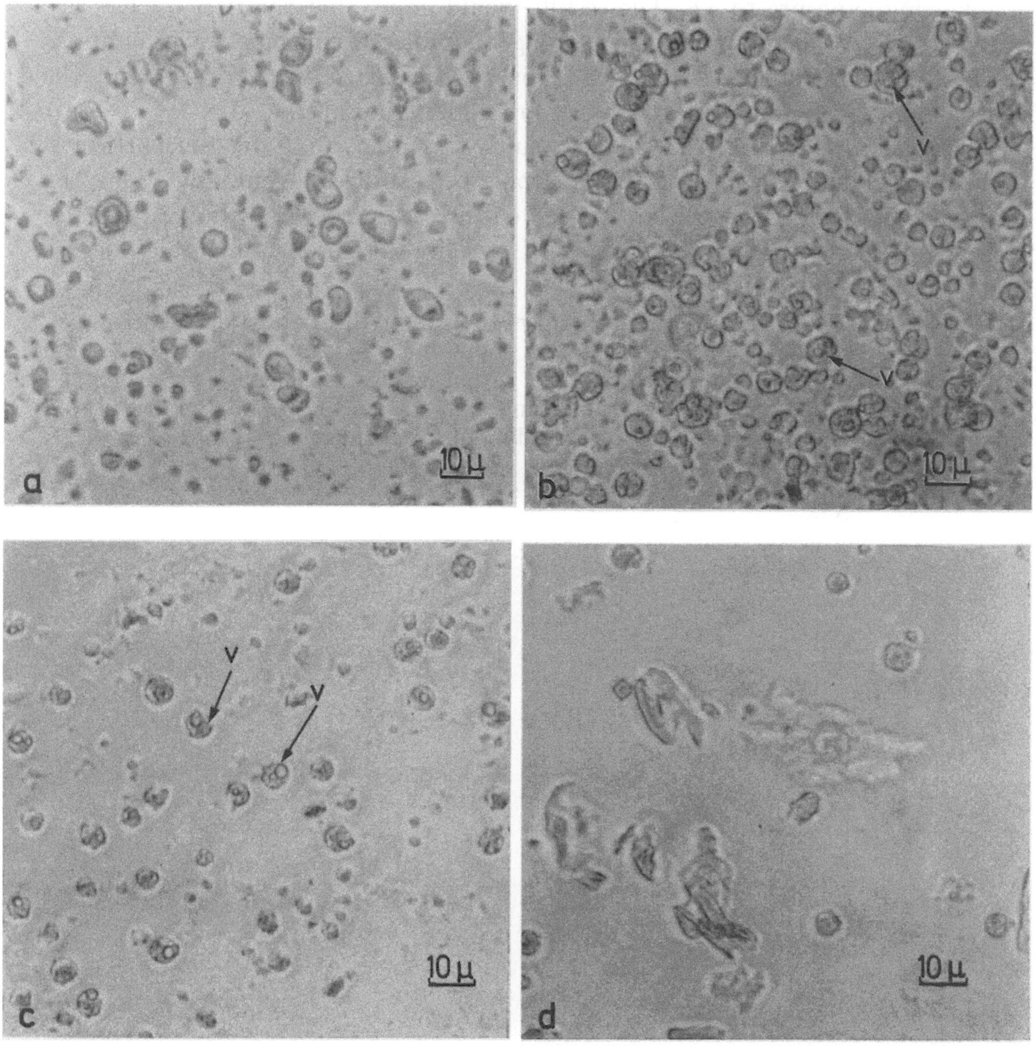

Fig. 151 a–d. Photomicrographs of *Drosophila* embryonic cell cultures: a Control non-infected culture. b 1 day post *SR* inoculation. Note granular cytoplasm with cytoplasmic vacuoles (*v*). c Three days post *SR* inoculation. Note extensive cytoplasmic vacuolations. d Four days post *SR* inoculation. Note debris and few intact cells. 495 ×

there were no viable cells. The medium and cell debris of the 4 day cultures were used to inoculate a small sample of *D. melanogaster* and *D. bifasciata* 1-day old females. The fluid was first passed through a 0.22 µ millipore filter and then injected into the flies in 0.5 µl doses. The *SR* effect appeared (Table 53) in the progeny of the injected females of both tester species and was maintained at a stable level through the F_3 generation. The *SR* effect was not as pronounced as that produced by *SR* grown in Vero cells (Table 52). The Vero *SR* was maintained and possibly allowed to replicate for 2 weeks before being assayed while *Drosophila* cells could survive no longer than 4 days following exposure to *SR*. The results would imply that the titer of the Vero inoculum was higher, producing a stronger *SR* effect in the bioassay. Conditions allowing more efficient SR replication in the *Drosophila* cultures and the development of a quantitative assay are under study.

It can be concluded from the results presented, that the *SR* particle is pathogenic for *D. melanogaster* embryonic cell cultures, eventually lysing the host cells. The pathogen from the cell cultures is infective for normal females resulting in the production of aberrant progeny ratios reflecting a high degree of male mortality.

Acknowledgment. Publication Number 1441 from the Laboratory of Genetics, University of Wisconsin. This study was supported by Grant No. GM 11777 from the National Institutes of Health. The author wishes to thank Dr. S. BUCKLEY for providing the vertebrate cell lines and Drs. JAMES CROW, CARTER DEMISTON, and HOWARD TEMIN for their critical reading of the manuscript.

Chapter 11

New Opportunities in Biological Research Offered by Arthropod Cell Cultures

I. Some Speculations on the Possible Role of Arthropods in the Evolution of Arboviruses

R. WALTER SCHLESINGER

Although we really know nothing about the evolutionary origin of any group of viruses, those that are transmitted in nature by arthropods to plants or higher animals could provide uniquely stimulating approaches to the problem. They are the only viruses known to bridge the evolutionary gaps between the different kingdoms (animal and plant) and between different phyla within the animal kingdom (arthropods and vertebrates). Here Nature itself demonstrates, in a very practical sense, the universality of the genetic code and of mechanisms involved in gene transcription or translation.

Surely the newly-won ability to switch such viruses back and forth from vector cells to vertebrate cells, both cultured under controlled conditions, broadens tremendously the range of experimental attacks available to us.

My remarks will be restricted to a consideration of those arthropod-borne animal viruses which suggest, on various grounds, that they may possess some evolutionary relatedness. They are represented by group A and B and possibly other arboviruses which may fit into a unique taxonomic niche in the Lwoff-Tournier scheme of viruses, the "togaviruses" (LWOFF and TOURNIER, 1966). These are enveloped viruses of diameters in the range 400–550 A, containing single-stranded RNA genomes with sedimentation coefficients 40–45 S and molecular weights variously estimated in the order of $2–4 \times 10^6$, of possible, but not yet proven, icosahedral nucleocapsid symmetry. Those that have been adequately analyzed contain at least two, but more often three constituent polypeptides of which one, associated with the envelope, is a glycopeptide. Some of them have been shown to replicate their RNA by mechanisms consistent with those well established for other RNA viruses. The demonstration of double-stranded replicative form (RF) and of replicative intermediate (RI) RNA species prescribes the existence of RNA replicase(s) which can fulfill dual functions independently of host cell DNA: the laying-down on the parental (+) strand of a (−) strand which, in turn, serves as a functional template for the synthesis of progeny viral RNA. Preliminary data presented by STEVENS (1970) and STOLLAR et al. (1971) of our Department suggest operation of similar mechanisms in *A. albopictus* cells experimentally infected with Sindbis virus. RNA replicases have been isolated from vertebrate cells infected with Sindbis virus, but as far as I know, not yet from arthropod cells.

Of the 204 viruses listed in the recently published "Catalogue of Arthropod-borne Viruses of the World" (TAYLOR, 1967), 42 belong to group B and 20 to group A on the basis of common group-specific antigenic determinants. Several other antigenically defined subgroups or ungrouped arboviruses as well as certain viruses not classified as arboviruses may have fundamental properties compatible with those of the togaviruses. Whether or not this turns out to be so, the steadily increasing number of agents assigned to group A and B in itself poses some perplexing problems: (1) How many more will the continued survey of wild-caught arthropods or of potential vertebrate hosts bring to light? (2) Can we assume that structural, chemical, and antigenic likeness within each group implies evolutionary relatedness? (3) To the extent that this assumption is justified, what are the selective pressures that lead to differentiation of types and subtypes? (4) Do they have their primordial origin in their arthropod vectors or in some vertebrate host? (5) What are the phenotypic or genotypic characteristics that determine the biological behavior of a given member of these groups—principally, pathogenicity for human beings?

It seems to me that the history of discovery during the past two decades strongly suggests that continued search will yield additional, hitherto unrecognized type-specific variants within each group—all the more since many initial isolations were essentially accidental byproducts of blind surveys. They yielded viruses unrelated to known diseases of vertebrate hosts in nature. Hence, there certainly was no obvious incentive to look for them in a directed fashion. By the same token, there is no *a priori* reason for thinking that the field investigations thus far conducted have exhausted or even scratched the surface of Nature's reservoir of similar agents.

Logic and the analogy to other groups of viruses—notably influenza A and the avian or murine RNA tumor viruses—present persuasive arguments for the notion that the members of each group share some evolutionary links. It stands to reason that, the more complex the ecological sphere in which a putative ancestral virus can operate, the more varied and unpredictable would be the nature and intensity of selective pressures to which it would have to adapt itself. It is hard to imagine a biological cycle more conducive to mutational and phenotypic changes than one in which the transmission, survival, and replication of genetic material is at the mercy of the haphazard mutual attraction of diverse blood-sucking arthropods and their vertebrate victims.

Despite the large number of togavirus species that have been isolated from arthropods or vertebrates in regions clearly constituting the "right" ecological environments, no member of these groups has, as far as I know, ever been known to establish itself outside of such regions. That is not to say that some of these viruses cannot spread without the intervention of transmitting arthropods or may not have the potential for adapting themselves to alternative modes of vertebrate-to-vertebrate transmission. Venezuelan equine encephalitis (VEE) seems to be a good example for the alternative of aerosol infection of man. But it seems that identifiable *index* infections of man or other vertebrates in nature can invariably be traced to conditions favorable for arthropod transmission. Perhaps rubella virus represents the extreme that has completely lost its dependence on arthropod vectors.

This combination of circumstances impresses me as the most potent argument in favor of the proposition that these viruses have their evolutionary origin in arthropods. Other arguments are more teleological in nature: they involve the lifelong, usually harmless viral persistence and replication in arthropods, contrasted by evanescent viremia in most vertebrates; the transovarian transmission of some, especially in ticks; the fact that, for many of these viruses, the range of competent arthropod vectors is much narrower than that of infectible vertebrate species.

Despite their teleological undertones, these considerations do permit us, however, to place the question of arboviral evolution in perspective relative to the origin of viruses in general. Let us postulate, for the sake of simplistic argument, that the theory of the derivation of viruses from cellular genes or gene product has some intrinsic merit. It is supported by the functional similarities to episomes, plasmids, and the mechanism of gene-controlled replication of such endosymbionts as the killing factors of *Paramecium*. A case can be made for the notion that most DNA bacteriophages, when first isolated from natural sources, may be temperate, i.e., that they can maintain themselves in their indigenous host bacteria in a lysogenic relationship as excellently discussed by BRADLEY (1967). This condition would demand at least some molecular homology with bacterial chromosomal DNA. Although the T phages of *E. coli* B have for long been the preferred prototype for the study of productive phage replication, we should not overlook the possibility that their extreme degree of virulence may represent the ultimate in laboratory-bred degeneracy of a virus. It is the necessary habit of us virologists to select those viral clones which can be readily assayed by virtue of their capacity for efficient cell destruction and whose replication is ideally close to synchrony. Even the recently discovered RNA phages—more germane as a model for our discussion—may, in their laboratory derivatives, represent highly selected populations. This possibility is strengthened by the fact that even these strains have little or no inhibitory effects on cellular macromolecular syntheses. Indeed ZINDER (1965) states "it is not impossible that their natural host is other than *E. coli*".

In the field of animal virology, HUEBNER and TODARO (1969) have provided an exciting model for these deliberations in the form of their recently published hypothesis concerning the genesis of RNA tumor viruses. This model, to be sure, involves a direct role for cellular DNA—the "switch-on" of postulated virogenes or oncogenes—a concept supported by the apparent dependence of replication of these RNA viruses on DNA synthesis.

No comparable mechanism can be suggested for arboviruses. Their replication—both in vertebrate and, as shown by STEVENS (1970) and STOLLAR et al. (1971), in mosquito cells—is not inhibited by Actinomycin D, and involves synthesis of double-stranded RF-RNA. Therefore, replication presumably does not require a DNA-dependent step. Instead, the *de novo* "creation" of an independently replicating RNA species would require the inclusion in the initial template molecule of information for RNA replicase. To qualify as a candidate for becoming a *viral genome*, such a molecule or its descendants would also have to code for structural protein(s). Increasing evidence points to the replication of the togaviruses on membrane-bound ribosomes or certainly in close proximity to cellular membrane structures. What choice does a cell have to rid itself of any aberrant RNA-protein complexes formed in this way? One could easily imagine a process that might be termed "reverse pinocytosis" or "exocytosis", i.e., the shedding of modified membrane "blebs" some containing, by chance or by design, the protein-associated RNA molecule which is endowed with the capacity for independent replication. Such "blebs", would be recognizable as "viruses" when they are taken up by foreign cells in which their independent replication may interfere with normal cellular function.

In the case of arboviruses, Nature provides this transfer to foreign host cells through the agency of arthropod transmission to vertebrates. The complexity of the ensuing cycles of vertebrate-arthropod-vertebrate-arthropod transmission, continuing for eons, would impose an unending variety of selective pressures. These could result in the emergence of an infinite number of variants, all derived from one ancestral RNA molecule that arose in one arthropod cell. Again, we in the laboratory accentuate or add to that natural selective process by (a) concentrating on those viruses that cause known diseases, (b) by forcing presumably innocuous viruses to be pathogenic by injecting them into mouse brains or cell cultures.

We all know that nothing is easier than to mount (or bend) arguments in favor of a biased working hypothesis as long as one is protected by total lack of hard evidence for or against it. The sole purpose of my remarks is to focus attention on experimental approaches that the ready availability of cultured cell lines derived from various vector species opens up, regardless of whether the resulting evidence will tend to support or demolish my bias.

(1) Do "normal" cells produce togavirus-like particles or proteins (antigens) or can they be "induced" to do so by various experimental interventions, as suggested earlier in this volume by SHINEDLING and GREENBERG (1971)?

(2) Are they intrinsically unable to make interferon or do they fail to produce it after infection with certain togaviruses (PELEG, 1969b) because they recognize their RNA as "self", i.e., closely related to some normal cellular RNA species?

(3) Do arthropod cells contain RNA species (RNA plasmids?) capable of autonomous (i.e., DNA-independent) replication? In particular, can one demonstrate RNA-dependent RNA replicases or a spectrum of double-

stranded RNA molecules beyond that found in supposedly "normal" vertebrate cell cultures?

(4) Can one demonstrate, by hybridization, some degree of base sequence homology between viral RNA (or its complementary strand) and cellular DNA or RNA species derived from arthropod cells?

(5) What is the role of cellular membrane composition—as reflected in the viral envelope—in determining phenotypic expressions of viruses capable of alternating between arthropod and vertebrate cells? Characteristic differences between the glycoproteins of these types of cells (e.g., in their sialic acid contents) have been recognized, and JENKIN et al. (1971) in this volume, reported on significant differences in lipid composition.

It seems to me that the ability to carry out controlled experiments on these and similar questions may prove to be the most important contribution of this new research tool to fundamental virology.

II. Future Possibilities of Arthropod Cell Culture in Parasitology and in Developmental Biology[1]

WILLIAM TRAGER

I deeply appreciate the honor of having been asked to give this lecture dedicated to the memory of Earl C. SUITOR, Jr. Dr. SUITOR had all too few years in which to work. Yet the stimulating effect of his research on the whole field of arthropod cell culture has been apparent throughout the course of this symposium and testifies to the burgeoning activity of this field—an activity that Dr. SUITOR himself helped to create. He published several papers that pointed out some of the important lines of investigation now being followed (SUITOR, 1966a, b, 1969). Among these was a paper by WOOD and SUITOR (1966) on the development of a microfilaria in mosquito cell cultures, an illustration of the application of arthropod cell culture to parasitology, one of the subjects I would like to discuss.

The possible application to parasitology was actually among the reasons for my own early work on insect tissue culture. I hope you will forgive me for beginning a discussion of the future by going back into the distant past. It was just thirty-five years ago that I was fortunate enough to obtain encouraging initial outgrowths of cells from ovarian tissue of full-grown silkworms (*Bombyx mori* larvae). It was a simple matter to grow the silkworm nuclear polyhedrosis virus in such cultures (TRAGER, 1935) and to demonstrate a devastating cytopathic effect, although at that time the term had not been coined. But when I tried to use these silkworm tissue cultures to grow the protozoan parasite *Nosema bombycis*, cause of the serious disease of silkworms, *pébrine*, I encountered difficulties. Although the resistant spores of the organism hatch out in the digestive tract of the silkworm, they would not hatch

[1] Earl C. SUITOR, Jr. Memorial Lecture.

in a tissue culture. However, it was possible to infect a tissue culture by adding to it hemolymph from a caterpillar infected two days previously, and evidently containing infectious forms of the parasite (TRAGER, 1937). At that time OHSHIMA's method (1937) of pretreating microsporidian spores with dilute alkali had not been published. This method permits the ready infection of appropriate tissue cultures by microsporidia, as described for the tent caterpillar (*Malacosoma*) by KURTTI and BROOKS (1971) in this volume. The same method had been used earlier by ISHIHARA and SOHI (1966) with *Nosema bombycis*. Spores of this parasite were exposed to 0.1 M KOH for 40 min and then inoculated into 4- or 5-day old cultures of ovarian tissue of *Bombyx mori*. The cells of the outgrowth became infected.

All of these studies represent only a beginning. The microsporidia are remarkable parasites. A simple ameboid cell injected into the cytoplasm of the host cell (ISHIHARA, 1968a) multiplies and then gives rise to spores of unusually complex structure. The host-parasite relationships likewise are of exceptional fnterest. Growth of microsporidia in insect tissue culture provides opportunities for many different kinds of studies bearing on intracellular parasitism. A few, such as investigations of the fine structure of both the parasites and the infected cells, are already under way. But we would like to know how these parasites obtain nutrients from their host cells, what happens at the interface between host cell and parasite cell, how the parasites produce hypertrophy of the host cell, and whether messenger RNA passes from one to the other. I look forward to fascinating results with microsporidia in arthropod tissue cultures.

Other sporozoan parasites of insects might also be expected to develop in appropriate insect cell cultures. One thinks especially of the mosquito stages of malaria. With the abundance and diversity of mosquito cell lines now available it seems reasonable to hope that suitable conditions will be found for the complete *in vitro* development of the sporogonic cycle of malaria.This cycle is not necessarily an intracellular one (WEATHERSBY, 1952), and metabolites from a cell culture might support development of the parasites.

Certainly this seems to be the case in the growth of trypanosomes as observed by me some years ago in tsetse fly tissue cultures. Cell outgrowths were obtained from various pupal tissues of *Glossina palpalis* in a culture medium containing sheep serum, one of the first uses of vertebrate serum for insect tissue culture. *Trypanosoma brucei*, *T. congolense* and *T. vivax* all grew profusely in initial cell or organ cultures from tsetse flies, but they did not grow in the culture medium alone (TRAGER, 1959). It is of interest that they often invaded islands of fly tissue and formed seething masses of protozoa. With all three species of trypanosomes the cultures showed forms morphologically resembling all the stages in the developmental cycle in the fly, including metacyclic trypanosomes as seen in salivary glands or proboscis. Only with *T. vivax*, however, could infectivity be demonstrated. It is regrettable that over 10 years have passed without serious attempts to extend this work. This remains an attractive area for future research in attempts at understanding what it is that induces non-infective midgut forms of the African trypanosomes to transform into the infective metacyclic forms.

Symbiosis or mutualism is an aspect of parasitism that is only beginning to receive the attention it has long deserved. Insects present a bewilderingly rich array of intracellular microorganisms presumed to be symbiotic, and in a few cases actually shown to be essential to the life of their host (KOCH, 1967). Among these are the bacteroids of roaches, the bacteria-like and yeast-like symbionts inhabiting the mycetome of leafhoppers and other plant-feeding Hemiptera, the bacteria-like symbionts of the sucking lice, of tsetse flies, and of other insects feeding exclusively on blood. These intracellular symbionts have not been cultured. With future developments in *in vitro* culture of the host cells it may be possible also to grow these symbionts. This would open the way to detailed investigations of physiological relationships between symbionts and their host cells. When I was working with *Glossina* I often noticed the intracellular bacteria in alimentary tract explants. Some of them would become extracellular, but they were not seen to multiply in the medium, nor did the cells containing them (the mycetocytes) ever grow. On the other hand, MITSUHASHI and MARAMOROSCH (1964) have observed *in vitro* mitotic division of infected mycetocytes of leafhoppers and have noted motile intracellular bacteria in old leafhopper tissue cultures. The symbiotic bacteria are readily seen in electron micrographs of sections of cultured embryonic leafhopper cells (MITSUHASHI, 1967b).

Investigations are only now beginning on events at the interface between an intracellular parasite and its host cell. We would like to know what materials go from the host to the parasite, from the parasite to the host, and how their passage across the cell membranes is controlled. Insect tissue cultures infected with pathogenic or symbiotic intracellular organisms should provide rich material for comparative studies.

I would like now to discuss briefly some possible applications of insect tissue culture methods to work in developmental biology. There is no need to document the importance of *Drosophila* in genetics and cytogenetics. Work of the past 10 years on activation by ecdysone of specific regions of the giant polytene chromosomes of larval Diptera is basic to modern attempts at understanding the control of gene expression (KRUEGER and LEZZI, 1966). We know that the giant chromosomes develop in larval tissues that grow throughout larval life only by increase in cell size (TRAGER, 1935) and that most of these tissues are histolyzed during the pupal stage. Yet no one has seen anything suggesting *in vitro* growth by increase in cell size of strictly larval tissues of Diptera, starting with either embryonic or early larval material. One might suppose that a good culture medium ought to support growth by increase in cell size as readily as it supports cell multiplication. Of course, we must recognize that *in vitro*, even with vertebrate tissues, it is easier to obtain uncontrolled propagation of certain kinds of cells than to elicit controlled, differentiated development analogous to that occurring during ontogeny *in vivo*.

Early workers with insect tissue culture noted partial differentiation of adult structures from imaginal disks explanted *in vitro*. In *Glossina* (TRAGER, 1959) I described the development *in vitro* of the eyes, with formation, in material explanted from 6-day old pupae (hence $^1/_5$ to $^1/_4$ of the way through the pupal period). A more detailed recent study is that of SCHNEIDER (1964) who explanted cephalic ganglia with eye-antennal disks from 96-hr *Drosophila* larvae and found differentiation of antennae in 70% of the explants and of the eyes in about 10%. Development was considerably slower than *in vivo*, and there was no development if the disks were taken from slightly younger larvae. Much more work is needed before imaginal disk explants can become fully useful in studies on differentiation and its control.

Still another approach to such problems is suggested by the fascinating *in vivo* experiments of HADORN (1966) and GEHRING (1968). The disks from larval *Drosophila* can be implanted into adult flies. Here, since there is no ecdysone, they do not differentiate but merely increase in mass. By cutting such disks in pieces they can be, in a sense, "subcultured" and maintained for a long time in successive adult fly hosts. If then they are finally implanted into a larval host about to pupate, they will differentiate, along with the host tissues, into an adult structure. Usually, but not always, they form the structure they would have formed in the first place. If dissociated cells from a disk are allowed to reaggregate before implantation into an adult host they resume their growth within the host. Growth of such cells should be a major aim of future work on insect tissue culture in relation to developmental biology. It would be interesting to see what they would do if subsequently implanted into a larva about to pupate, or if supplied *in vitro* with appropriate hormones.

Perhaps ten years from now some of us will enjoy a discussion of progress along some of the lines I have suggested.

References

AGARWAL, H. C., RAO, K. D. P.: Variations in the neutral lipids of *Dysdercus koenigii* (Hemiptera) during its life cycle. J. Insect Physiol. **15**, 161–166 (1969).

AGUI, N., YAGI, S., FUKAYA, M.: Induction of moulting of cultivated integument taken from a diapausing rice stem borer larva in the presence of ecdysterone. Appl. entomol. Zool. **4**, 156–157 (1969).

AHMED, M. E., BLACK, L. M., PERKINS, E. G., WALKER, B. L., KUMMEROW, F. A.: Lipid in potato yellow dwarf virus. Biochem. biophys. Res. Commun. **17**, 103–107 (1964).

AIZAWA, K., SATO, F.: Culture de tissus de ver à soie, *Bombyx mori* L., dans un milieu sans hemolymph. Ann. Epiphyties **14**, 125–126 (1963).

— — MURAKAMI, A.: Role of hemolymph in tissue culture of the silkworm, *Bombyx mori* [in Japanese, English summary]. J. Sericult. Sci. Japan **30**, 445–450 (1961).

— VAGO, C.: Sur l'infection à Borrelinavirus en culture de tissus d'insectes. Ann. Inst. Pasteur **96**, 455–460 (1959).

AKAI, H., GATEFF, E., DAVIS, L. E., SCHNEIDERMAN, H. A.: Virus-like particles in normal and tumorous tissues of *Drosophila*. Science **157**, 810–813 (1967).

AKSTEIN, E.: The chromosomes of *Aedes aegypti* and of some other species of mosquitoes. Bull. Res. Council Israel **11**B, 146–155 (1962).

ANDERSON, R. E., CUMMING, R. B., WALTON, M., SNYDER, F.: Lipid metabolism in cells grown in tissue culture: O-alkyl, O-alk-1-enyl, and acyl moieties of L-M cells. Biochim. biophys. Acta (Amst.) **176**, 491–501 (1969).

APPLEBAUM, S. W., EBSTEIN, R. P., WYATT, G. R.: Dissociation of ribosomal ribonucleic acid from silkmoth pupae by heat and dimethylsulfoxide: evidence for specific cleavage points. J. molec. Biol. **21**, 29–41 (1966).

ARVY, L., GABE, M.: Sur la multiplication *in vitro* des cellules sanguines de *Forticula auricularia*. C. R. Acad. Sci. (Paris) **140**, 787–789 (1946).

AVRAMEAS, S.: Coupling of enzymes to proteins with glutaraldehyde. Use of the conjugates for the detection of antigens and antibodies. Immunochemistry **6**, 43–52 (1969).

BAILEY, J. M.: Lipid metabolism in cultured cells. I. Factors affecting cholesterol uptake. Proc. Soc. exp. Biol. (N.Y.) **107**, 30–35 (1961).

— Cellular lipid nutrition and lipid transport. In: Lipid metabolism in tissue culture cells (G. H. ROTHBLAT and D. KRITCHEVSKY eds.). Philadelphia: The Wistar Institute Press 1967.

— MENTER, J.: Lipid metabolism in cultured cells. VII. Linoleic acid content of cells grown in lipid-free synthetic medium. Proc. Soc. exp. Biol. (N.Y.) **125**, 101–105 (1967).

BANERJEE, K., SINGH, K. R. P.: Establishment of carrier cultures of *Aedes albopictus* cell line infected with arboviruses. Indian J. med. Res. **56**, 812–814 (1968).

— — Loss of mouse virulence in chikungunya virus from the carrier culture of *Aedes albopictus* cell line. Indian J. med. Res. **57**, 1003–1005 (1969).

BARIGOZZI, C., DOLFINI, S., FRACCARO, M., RAIMONDI, G. R., TIEPOLO, L.: *In vitro* study of the DNA replication pattern of somatic chromosomes of *Drosophila melanogaster*. Exp. Cell Res. **43**, 231–234 (1966).

BARILE, M. F.: Mycoplasma and cell cultures. Cell cultures for virus vaccine production. Nat. Cancer Inst. Monogr. **29**, 201–203 (1968).

BARROSO, C., MUNICIO, A. M., RIBERA, A.: Biochemistry of the development of the insects *Dacus oleae* and *Ceratitis capitata*—evolution of total fatty acids. Comp. biochem. Physiol. **28**, 239–244 (1969).

BARSKI, G., SORIEUL, S., CORNEFERT, F.: Production dans des cultures *in vitro* de deux souches cellulaires en association, de cellules de caractère "hybride". C. R. Acad. Sci. (Paris) **251**, 1825–1827 (1960).

BARTH, L. G., BARTH, L. J.: Control of differentiation of cells from *Rana pipiens* gastrulae. Anat. Rec. **131**, 530–531 (1958).

BARTH, R. H., BUNYARD, P. P., HAMILTON, T. H.: RNA metabolism in pupae of the oak silkworm, *Antheraea pernyi:* the effects of diapause, development and injury. Proc. nat. Acad. Sci. (Wash.) **52**, 1572–1580 (1964).

BECKER, Y., JOKLIK, W. K.: Messenger RNA in cells infected with vaccinia virus. Proc. nat. Acad. Sci. (Wash.) **51**, 577–585 (1964).

BECKMAN, L., JOHNSON, F. M.: Esterase variations in *Drosophila melanogaster*. Hereditas (Lund) **51**, 212–220 (1964).

BENSON, A. A.: Orientation of lipids in chloroplast and cell membranes. J. Amer. Oil Chemists' Soc. **43**, 265–270 (1966).

BERGOLD, G. M.: Insect viruses. In: Advances in virus research, vol. I (K. M. SMITH and M. A. LAUFFER, eds.). New York: Academic Press 1953.

BERRY, S. J., KRISHNAKUMARAN, A., OBERLANDER, H., SCHNEIDERMAN, H.: Effects of hormones and injury on RNA synthesis in saturniid moths. J. Insect Physiol. **13**, 1511–1537 (1967).

BHAT, U. K. M., SINGH, K. R. P.: Structure and development of vesicles in larval tissue culture of *Aedes aegypti* (L.). J. med. Entomol. **6**, 71–74 (1969).

BLACK, L. M.: Some viruses transmitted by agallian leafhoppers. Proc. Amer. phil. Soc. **88**, 132–144 (1944).

— Occasional transmission of some plant viruses through the eggs of their insect vectors. Phytopathology **43**, 9–10 (1953).

— Biological cycles of plant viruses in insect vectors. In: The viruses (F. M. BURNET, and W. M. STANLEY, eds.), vol. 2, p. 157–185. New York: Academic Press 1959.

— Some recent advances on leafhopper-borne viruses. In: Biological transmission of disease agents (K. MARAMOROSCH, ed.), p. 1–9. New York: Academic Press 1962.

— Insect tissue cultures as tools in plant virus research. Ann. Rev. Phytopath. **7**, 73–100 (1969).

— SMITH, K. M., HILLS, G. J., MARKHAM, R.: Ultrastructure of potato yellow-dwarf virus. Virology **27**, 446–449 (1965).

BRADLEY, D. E.: Ultrastructure of bacteriophages and bacteriocins. Bact. Rev. **31**, 230–314 (1967).

BRAKKE, M. K.: Density gradient centrifugation: a new separation technique. J. Amer. chem. Soc. **73**, 1847–1848 (1951).

— Stability of potato yellow-dwarf virus. Virology **2**, 463–476 (1956).

BRAND, T. VON: Biochemistry of parasites. New York: Academic Press 1966.

BRČÁK, J., KRÁLÍK, O.: Mycoplasma-like microorganism and virus particles in the leafhopper *Javesella pellucida* (F.) transmitting the oat sterile dwarf disease. Biol. Plantarum **11**, 95–96 (1969).

BRELAND, O. P.: Studies on the chromosomes of mosquitoes. Ann. entomol. Soc. Amer. **54**, 360–375 (1961).

BRUN, G.: Étude d'une association du virus σ et de son hôte la *Drosophile:* l'état stabilisé. Thesis: Lab. exp. Biol. Orsay, France (1963).

BUCKLEY, S. M.: Susceptibility of the *Aedes albopictus* and *Aedes aegypti* cell lines to infection with arboviruses. Proc. Soc. exp. Biol. (N.Y.) **131**, 625–630 (1969a).

— *In vitro* propagation of tick-borne viruses other than group B arboviruses. Internat. symposium on tick-borne arboviruses (excluding group B), Smolenice, Czechoslovakia (1969b).

BURDETT, W. J., HANLEY, E. W., GROSCH, H.: The effect of ecdysones on the maintenance and development of ocular imaginal discs *in vitro*. Tex. Rep. Biol. Med. **26**, 176–180 (1968).

CARLSON, J. G., GAULDEN, M. E.: Grasshopper neuroblast techniques. In: Methods in cell physiology (D. M. PRESCOTT, ed.), chap. 13. New York: Academic Press 1964.

Cell Culture Collection Committee: A culture collection of animal cell strains. Science **146**, 241–243 (1964).

Chamberlain, R. W., Sudia, W. D.: Mechanism of transmission of viruses by mosquitoes. Ann. Rev. Ent. **6**, 371–390 (1961).

Chao, J., Ball, G. H.: A cell line isolated from the hemocytes of *Samia cynthia* pupae. This volume, p. 28–32.

Charney, J., Coriell, L. L.: Demonstration, purification and partial characterization of abnormal (HSL) antigens in stable human cell lines. J. nat. Cancer Inst. **33**, 285–301 (1964).

Chirigos, M. A., Schwalb, E. D., Scott, D.: Friend leukemia virus spleen focus assay: Relationship of spleen foci to viremia, splenomegaly, and survival time. Cancer Res. **27**, 2249–2254 (1967).

Chiu, R.-J., Black, L. M.: Monolayer cultures of insect cell lines and their inoculation with a plant virus. Nature (Lond.) **215**, 1076–1078 (1967).

— — Assay of wound tumor virus by the fluorescent cell counting technique. Virology **37**, 667–677 (1969).

— Liu, H.-Y., MacLeod, R., Black, L. M.: Potato yellow dwarf virus in leafhopper cell culture. Virology **40**, 387–396 (1970).

— Reddy, D. V. R., Black, L. M.: Inoculation and infection of leafhopper tissue cultures with a plant virus. Virology **30**, 562–566 (1966).

Clark, A. J., Block, K.: The absence of sterol synthesis in insects. J. biol. Chem. **234**, 2578–2587 (1959).

Clark, K., Langley, P.: Studies on the initiation of growth and moulting in *Locusta migratoria*. I. The time and nature of the initiating stimulus. J. Insect Physiol. **9**, 387–392 (1963).

Clements, A. N.: The physiology of mosquitoes. New York: Macmillan Co. 1963.

Congote, L. F., Sekeris, C. E., Karlson, P.: On the mechanism of hormone action. XIII. Stimulating effects of ecdysone, juvenile hormone and ions on RNA synthesis in fat body cell nuclei from *Calliphora erythrocephala* isolated by a filtration technique. Exp. Cell Res. **56**, 338–346 (1969).

Conover, J. H., Zepp, H. D., Hirschhorn, K., Hodes, H. L.: Production of human-mosquito somatic cell hybrids and their response to virus infection. This volume, p. 85–92.

Converse, J. L., Nagle, S. C., Jr.: Multiplication of yellow fever virus in insect tissue cell cultures. J. Virol. **1**, 1096–1097 (1967).

Coriell, L. L.: Laminar flow system for sterile work. Cell cultures for virus vaccine production. Nat. Cancer Inst. Monogr. **29**, 569–572 (1968).

Courgeon, A. M.: L'activité mitotique, en culture organotypique dans les disques oculo-antennaires de larves de *Calliphora erythrocephala*, Meig. C. R. Acad. Sci. (Paris) **268**, 950–952 (1969).

Crowle, A. J.: A simplified micro double-diffusion agar precipitin technique. J. Lab. clin. Med. **52**, 784–787 (1958).

David-Ferreira, J. F., Manaker, R. A. S.: An electron microscope study of the development of a mouse hepatitis virus in tissue culture cells. J. Cell Biol. **24**, 57–78 (1965).

Davies, P., Weissmann, G.: Hydrolysis of histones at neutral pH by granulocyte lysosomes. J. Cell Biol. **43**, 29a (1969).

Davis, B., Dulbecco, R., Eisen, S., Ginsberg, S., Wood, A.: Microbiology, p. 1041–1044. New York: Hoeber 1968.

Day, M. F., Grace, T. D. C.: Culture of insect tissues. Ann. Rev. Entomol. **4**, 17–38 (1959).

DeDuve, C., Wattiaux, R.: Functions of lysosomes. Ann. Rev. Physiol. **28**, 435–492 (1966).

DHANDA, V., RODRIGUES, F. M., GHOSH, S. N.: Isolation of Chandipura virus from sandflies. Indian J. med. Res. **58**, 179–180 (1970).

DOGRA, G. S., TANDAN, B. K.: Adaptation of certain histological techniques for *in situ* demonstration of the neuroendocrine system of insects and other animals. Quart. J. micr. Sci. **105**, 455–466 (1964).

DOI, Y., TERANAKA, M., YORA, K., ASUYAMA, H.: Mycoplasma- or PLT group-like microorganisms found in the phloem elements of plants infected with mulberry dwarf, potato witches' broom, aster yellows, or paulownia witches' broom. Ann. phytopath. Soc. Japan **33**, 259–266 (1967).

DOLFINI, S.: Some comparative data on embryonic cells of *D. melanogaster* cultured *in vitro*. Drosophila Information Service **44**, 189 (1969).

— COURGEON, A. M., TIEPOLO, L.: The cell cycle of an established line of *Drosophila melanogaster* cells *in vitro*. Experientia (Basel) **26**, 1020–1021 (1970).

— GOTTARDI, A.: Changes of chromosome number in cells of *Drosophila melanogaster* cultured *in vitro*. Experientia (Basel) **22**, 1–7 (1966).

— TIEPOLO, L.: The cell cycle of embryonic cells of *Drosophila melanogaster* cultured *in vitro*. Second Int. Colloq. Invertebrate Tissue Culture, p. 182–188. Milano: Ist. Lombardo Sci. Lettr. (Baselli Foundation) 1968.

DUTKY, R. C., ROBBINS, W. E., KAPLANIS, J. N., SHORTINO, T. J.: The sterol esters of housefly eggs. Comp. Biochem. Physiol. **9**, 251–255 (1963).

DU TOIT, R. M.: Transmission of blue tongue and horse sickness by culicoides. Onderstepoort J. vet. Sci. **19**, 7–16 (1944).

EAGLE, H.: The specific amino acid requirements of a human carcinoma cell (strain HeLa) in tissue culture. J. exp. Med. **102**, 37–48 (1955).

— Amino acid metabolism in mammalian cell cultures. Science **130**, 432–437 (1959).

ECHALIER, G., OHANESSIAN, A.: Cultures *in vitro* de cellules de *Drosophile*. Obtention d'une souche à multiplication continue. Second Int. Colloq. Invertebrate Tissue Culture, p. 174–181. Milano: Ist. Lombardo Sci. Lettr. (Baselli Foundation) 1968.

— — Isolement, en cultures *in vitro* de lignées cellulaires diploides de *D. melanogaster*. C. R. Acad. Sci. (Paris) **268**, 1771–1773 (1969).

— — *In vitro* culture of *Drosophila melanogaster* embryonic cells. In Vitro **6**, 162–172 (1970).

— — BRUN, G.: Cultures "primaires" de cellules embryonnaires de *Drosophila melanogaster* (Insecte Diptère). C. R. Acad. Sci. (Paris) **261**, 3211–3213 (1965).

FAST, P. G.: A comparative study of the phospholipids and fatty acids of some insects. Lipids **1**, 209–215 (1966).

FILSHIE, B. K., REHACEK, J.: Studies of the morphology of MurrayValley encephalitis and Japanese encephalitis viruses growing in cultured mosquito cells. Virology **34**, 435–443 (1968).

— GRACE, T. D. C., POULSON, D. F., REHACEK, J.: Virus-like particles in insect cells of three types. J. invert. Path. **9**, 271–273 (1967).

FISCHINGER, P. J., O'CONNOR, T. E.: Tissue culture assay of helper activity of murine leukemia for immune sarcoma virus. J. nat. Cancer Inst. **40**, 1199–1212 (1968).

FLORKIN, M., JEUNIAUX, C. H.: Hemolymph composition. In: The physiology of insecta, vol. 3. (M. ROCKSTEIN, ed.), p. 109–152. New York: Academic Press 1964.

FRANKLIN, R. M.: Replication of bacteriophages RNA: some physical properties of single-stranded, double-stranded and branched viral RNA. J. Virol. **1**, 64–75 (1967).

FUJITA, N., YASUI, Y., KITAMURA, S., HOTTA, S.: Cultivation of Japanese encephalitis (JE) virus in primary *Culex* mosquito tissue cultures. Kobe J. med. Sci. **14**, 241–249 (1968).

FUKAMI, J.: Effect of rotenone on respiratory enzyme system of insect muscle. Bull. nat. Inst. agric. Sci. (Tokyo), Ser. C **13**, 33–45 (1961).

FUKAYA, M., NASU, S.: A *Chilo* iridescent virus (CIV) from the rice stem borer, *Chilo suppressalis* Walker (Lepidoptera: Pyralidae). Appl. entomol. Zool. **1**, 69–72 (1966).

FUKUSHI, T.: Transmission of the virus through the eggs of an insect vector. Proc. imp. Acad. (Japan) **9**, 457–460 (1933).

— Relationships between propagative rice viruses and their vectors. In: Viruses, vectors, and vegetation (K. MARAMOROSCH, ed.), p. 279–301. New York: Interscience Publ. 1969.

— SHIKATA, E.: Fine structure of rice dwarf virus. Virology **21**, 500–503 (1963).

— — KIMURA, I.: Some morphological characters of rice dwarf virus. Virology **18**, 192–205 (1962).

GAHAN, P. B.: Histochemistry of lysosomes. Int. Rev. Cytol. **21**, 1–63 (1967).

GÁMEZ, R., BLACK, L. M.: Application of particle-counting to a leafhopper-borne virus. Nature (Lond.) **215**, 173–174 (1967).

— — Particle counts of wound tumor virus during its peak concentration in leafhoppers. Virology **34**, 444–451 (1968).

— CHIU, R.-J.: The minimum concentration of a plant virus needed for infection of monolayers of vector cells. Virology **34**, 356–357 (1968).

GARTLER, S. M.: Genetic markers in cell culture. Decennial review conference on cell tissue and organ culture. Nat. Cancer Inst. Monogr. **26**, 167–176 (1967).

GAW, Z., LIU, N. T., ZIA, T. U.: Tissue culture methods for cultivation of virus grasserie. Acta virol. **3**, Suppl., 55–60 (1959).

GEHRING, W.: The stability of the determined state in cultures of imaginal disks in *Drosophila*. In: The stability of the differentiated state (H. URSPRUNG, ed.), p. 136–154. Berlin-Heidelberg-New York: Springer 1968.

GIANFELICI, E.: Differentiation *in vitro* du complexe cerebro-endocrinien chez *Calliphora erythrocephala*. Ann. Endocr. (Paris) **29**, 496–500 (1968a).

— Evolution en culture *in vitro* de la neurosecretion chez l'insect *Calliphora*. Second Intern. Colloq. Invert. Tissue Culture, p. 155–167. Milano: Instituto Lombardo Sci. Lettr. (Baselli Foundation) 1968b.

GIANNOTTI, J., DEVAUCHELLE, G., VAGO, C.: Micro-organismes de type mycoplasme chez une cicadelle et une plante infectées par la phyllodie. C. R. Acad. Sci. (Paris) **266**, 2168–2170 (1968a).

— VAGO, C., DEVAUCHELLE, G., MARCHOUX, G.: Recherches sur les microorganismes de type mycoplasme dans les cicadelles vectrices et dans les végétaux atteints de jaunisses. Entomol. exp. appl. **11**, 470–474 (1968b).

GILLOIS, M., ECHARD, G.: Premiers resultats concernant l'induction de pseudoméloses. C. R. Acad. Sci. (Paris) **268**, 708–710 (1969).

GLASER, R. W.: The growth of insect blood cells *in vitro*. Psyche (Stuttg.) **24**, 1–7 (1917).

GOLDSCHMIDT, R.: Some experiments on spermatogenesis in vitro. Proc. nat. Acad. Sci. (Wash.) **1**, 220–222 (1915).

GOODCHILD, A. J. P.: Culture of insect tissues. Nature (Lond.) **173**, 504–505 (1954).

GRACE, T. D. C.: Effects of various substances on growth of silkworm tissues grown *in vitro*. Austr. J. biol. Sci. **11**, 407–417 (1958).

— Tissue culture for arthropod viruses. Trans. N. Y. Acad. Sci., Ser. II, **21**, 237–241 (1959).

— Establishment of four strains of cells from insect tissues grown *in vitro*. Nature (Lond.) **195**, 788–789 (1962).

— Establishment of a line of mosquito (*Aedes aegypti* L) cells grown *in vitro*. Nature (Lond.) **211**, 366–367 (1966).

CRACE, T. D. C.: Establishment of a line of cells from the silkworm *Bombyx mori*. Nature (Lond.) **216**, 613 (1967).
— Insect cell culture and virus research. In Vitro **3**, 104–117 (1968).
— Insect tissue culture and its use in virus research. Advanc. Virus Res. **14**, 201–220 (1969).
— BRZOSTOWSKI, H. W.: Analysis of the amino acids and sugars in an insect cell culture medium during cell growth. J. Insect Physiol. **12**, 625–633 (1966).
GRAHAM, R. C., JR., KARNOVSKY, M. J.: The early stages of absorption of injected horseradish peroxidase in the proximal tubules of mouse kidney: Ultrastructural cytochemistry by a new technique. J. Histochem. Cytochem. **14**, 291–302 (1966).
GRANADOS, R. R., HIRUMI, H., MARAMOROSCH, K.: Electron microscopic evidence for wound-tumor virus accumulation in various organs of an inefficient leafhopper vector, *Agalliopsis novella*. J. invert. Path. **9**, 147–159 (1967).
— MARAMOROSCH, K., SHIKATA, E.: Mycoplasma: Suspected etiologic agent of corn stunt. Proc. nat. Acad. Sci. (Wash.) **60**, 841–844 (1968a).
— WARD, L. S., MARAMOROSCH, K.: Insect viremia caused by a plantpathogenic virus: electron microscopy of vector hemocytes. Virology **34**, 790–796 (1968b).
GREEN, D. E., TZAGOLOFF, A.: Role of lipids in the structure and function of biological membranes. J. Lipid Res. **7**, 587–602 (1966).
GREENBERG, B.: *In vitro* cultivation of *Musca domestica* L. and *Musca sorbens* Weidemann tissues. Exp. Cell Res. **54**, 284–287 (1969).
GREENBERG, J. R.: Synthesis and properties of ribosomal RNA in *Drosophila*. J. molec. Biol. **46**, 85–98 (1969).
GREENE, A. E., CORIELL, L. L., CHARNEY, J.: Species identification of poikilothermic tissue culture cells. In: Biology of amphibian tumors (M. MIZELL, ed.), p. 112–120. Berlin-Heidelberg-New York: Springer 1969.
— — LEHR, H. B.: Viability of cell cultures stored up to 8 years in liquid nitrogen. Cryobiology **6**, 289 (1969).
GUBLER, D. J.: A method for the *in vitro* cultivation of ovarian and midgut cells from the adult mosquito. Amer. J. Epidemiol. **87**, 502–508 (1968).
GUGGENHEIM, M. A., FRIEDMAN, R. M., RABSON, A. S.: Interferon action in heterokaryons. Proc. Soc. exp. Biol. (N.Y.) **130**, 1242–1245 (1969).
GUPTA, A. P., SUTHERLAND, D. J.: *In vitro* transformations of the insect plasmatocyte in some insects. J. Insect Physiol. **12**, 1369–1375 (1966).
GUPTA, K. S.: Cultivation of *Nosema mesnili* Paillot (Microsporidia) *in vitro*. Curr. Sci. **33**, 407–408 (1964).
HADORN, E.: Konstanz, Wechsel und Typus der Determination und Differenzierung in Zellen aus männlichen Genitalanlagen von *Drosophila melanogaster* nach Dauerkultur *in vivo*. Develop. Biol. **13**, 424–509 (1966).
HAINES, T. W.: Tissue culture of *Aedes aegypti* cells and its application in studies on the behavior of Eastern equine encephalomyelitis virus. Diss. Abstr. **20**, 31 (1959).
HALDANE, J. B. S.: 'Dex' or 'Order of magnitude'? Nature (Lond.) **187**, 879 (1960).
HALFER, C., TIEPOLO, L., BARIGOZZI, C., FRACCARO, M.: Timing of DNA replication of translocated Y chromosome sections in somatic cells of *Drosophila melanogaster*. Chromosoma (Berl.) **27**, 395–408 (1969).
HANKS, J. H., WALLACE, R. E.: Relation of oxygen and temperature in the preservation of tissues by refrigeration. Proc. Soc. exp. Biol. (N.Y.) **71**, 196–200 (1949).
HANNOUN, C., ASSO, J., ARDOIN, P.: Mutants à petites plages du virus Sindbis. Ann. Inst. Pasteur **107**, 598–603 (1964).
HARRIS, H., WATKINS, J. F.: Hybrid cells derived from mouse and man: Artificial heterokaryons of mammalian cells from different species. Nature (Lond.) **205**, 640–646 (1965).

HARTLEY, J. W., ROWE, W.P.: Production of altered cell foci in tissue culture by defective Moloney sarcoma virus particles. Proc. nat. Acad. Sci. (Wash.) **55**, 780–786 (1966).

HARTZELL, A.: Insect tissue culture: a review. Proc. Intern. Congr. Entomol., 10th. Montreal, 1956, **2**, 319–325 (1958).

HARVEY, E. B.: A review of the chromosome numbers in the Metazoa. Part 1 (p. 48). J. Morph. **28**, 1–63 (1916).

HAYFLICK, L.: Cell cultures and mycoplasmas. Texas Rep. Biol. Med. **23** (Suppl. 1), 285–303 (1965).

— Fundamental biology of the Class Mollicutes. Order Mycoplasmatales (L. HAYFLICK, ed.). In: The Mycoplasmatales and L-phase of bacteria. New York: Appleton-Century-Crofts 1969.

— Oncogenesis *in vitro*. Nat. Cancer Inst. Monogr. **26**, 355–377 (1967).

— MOORHEAD, P. S.: The serial cultivation of human diploid cell strains. Exp. Cell Res. **25**, 585–621 (1961).

HEINZE, K.: Phytopathogene Viren und ihre Überträger. Berlin: Duncker & Humblot 1959.

HERMAN, W. S.: The ecdysial glands of arthropods. Int. Rev. Cytol. **22**, 269–347 (1967).

HILLEMAN, M. R.: Toward control of viral infections of man. Science **164**, 506–514 (1969).

HINK, W. F.: On the source and nature of bactericidal factors in the hemolymph of normal and immune wax moth larvae, *Galleria mellonella* (L.). Ph. D. Thesis, Ohio State Univ. (1966).

— Established insect cell line from the cabbage looper, *Trichoplusia ni*. Nature (Lond.) **226**, 466–467 (1970).

— IGNOFFO, C. M.: Establishment of a new cell line (IMC-HZ-1) from ovaries of cotton bollworm moths, *Heliothis zea* (Boddie). Exp. Cell Res. **60**, 307–309 (1970).

HIRUMI, H.: An improved device for cultivating cells *in vitro* and for observations under high power phase magnification. Contr. Boyce Thompson Inst. **22**, 113–115 (1963).

— Study on insect tissue culture. Cultivation of embryonic leafhopper tissues *in vitro*. Wakayama Igaku **15**, 325–334 (1965).

— GRANADOS, R. R., MARAMOROSCH, K.: Electron microscopy of a plantpathogenic virus in the nervous system of its insect vector. J. Virol. **1**, 430–444 (1967).

— MARAMOROSCH, K.: Cultivation of leafhopper (Cicadellidae) tissues and organs *in vitro*. Ann. Epiphyties **14** (n° hors sér. III), 77–79 (1963a).

— — Recovery of aster yellows virus from various organs of the insect vector *Macrosteles fascifrons*. Contr. Boyce Thompson Inst. **22**, 141–151 (1963b).

— — Tissue culture of the monarch butterfly, *Anosia plexippus*. Contr. Boyce Thompson Inst. **22**, 259–268 (1964).

— — The *in vitro* cultivation of embryonic leafhopper tissues. Exp. Cell Res. **36**, 625–631 (1964a).

— — Insect tissue culture: use of blastokinetic stage of leafhopper embryo. Science **144**, 1465–1467 (1964b).

— — Insect tissue culture: further studies on the cultivation of embryonic leafhopper tissues *in vitro*. Contr. Boyce Thompson Inst. **22**, 343–352 (1964c).

— — Electron microscopy of wound tumor virus in cultured embryonic cells of the leafhopper *Macrosteles fascifrons*. Second Int. Colloq. Invertebrate Tissue Culture, p. 203–217. Milano: Ist. Lombardo Sci. Lettr. (Baselli Foundation) 1968.

— — Mycoplasma-like bodies in the salivary glands of insect vectors carrying the aster yellows agent. J. Virol. **3**, 82–84 (1969).

— — Cell culture of Hemiptera. In: Invertebrate tissue culture (C. VAGO, ed.). New York: Academic Press (in press).

HOFFMAN, G., KÖHLER, G.: Haltung von Zeckengeweben in künstlichen Nähremedien. Z. Parasitenk. **31**, 8 (1968).

HOLT, C. M.: Multiple complexes in the alimentary tract of *Culex pipiens*. J. Morph. **29**, 607–618 (1917).

HORIKAWA, M., FOX, A. S.: Culture of embryonic cells of *Drosophila melanogaster in vitro*. Science **145**, 1437–1439 (1964).

— KURODA, J.: *In vitro* cultivation of blood cells of *Drosophila melanogaster* in a synthetic medium. Nature (Lond.) **184**, 2017–2018 (1959).

— LING, L.-N., FOX, A. S.: Long term culture of embryonic cells of *Drosophila melanogaster*. Nature (Lond.) **210**, 183–185 (1966).

— SUGAHARA, M.: Studies on the effects of radiation on living cells in tissue culture. I. Radiosensitivity of various imaginal discs and organs in larvae of *Drosophila melanogaster*. Radiat. Res. **12**, 266–275 (1960).

HORSFALL, F., TAMM, I.: Viral and rickettsial infections of man, ed. 4, p. 1282. Philadelphia: J. P. Lippincott 1965.

HSU, S. H., LIU, H. H., SUITOR, E. C., JR.: Further description of a subline of Grace's mosquito (*Aedes aegypti* L.) cells adapted to hemolymph free medium. Mosquito News **29**, 439–446 (1969).

HUEBNER, R. J., HARTLEY, J. W., ROWE, W. P., LANE, W. T., CAPPS, W. I.: Rescue of the defective genome of Moloney sarcoma virus from a noninfectious hamster tumor and the production of pseudotype sarcoma viruses with various murine leukemia viruses. Proc. nat. Acad. Sci. (Wash.) **56**, 1164–1169 (1966).

— TODARO, G. J.: Oncogenes of RNA tumor viruses as determinants of cancer. Proc. nat. Acad. Sci. (Wash.) **64**, 1087–1094 (1969).

HULL, R. N., CHERRY, W. R., THRITCH, O. J.: Growth characteristics of monkey kidney cell strains LLC-MK-1, LLC-MK-2 and LLC-MK-2 (NCTC-3196) and their utility in virus research. J. exp. Med. **115**, 903–917 (1962).

ISHIHARA, R.: Growth of *Nosema bombycis* in primary cell cultures of mammalian and chicken embryos. J. invert. Path. **11**, 328–329 (1968).

— Some observations on the fine structure of sporoplasm discharged from spores of a microsporidian, *Nosema bombycis*. J. invert. Path. **12**, 245–258 (1968 a).

— SOHI, S. S.: Infection of ovarian tissue culture of *Bombyx mori* by *Nosema bombycis* spores. J. invert. Path. **8**, 538–540 (1966).

ISHIHARA, T.: Families and genera of leafhopper vectors. In: Viruses, vectors, and vegetation (K. MARAMOROSCH, ed.), p. 235–254. New York: Interscience Publ. 1969.

ITO, S., LOEWENSTEIN, W. R.: Permeability of a nuclear membrane: Changes during normal development and changes induced by growth hormone. Science **150**, 909–910 (1965).

ITTYCHERIAH, P. I., STEPHANOS, S.: *In vitro* culture of ovary of the plant bug, *Iphita limbata* Stal. Indian J. exp. Biol. **7**, 17–19 (1968).

JENKIN, H., TOWNSEND, D., MAKINO, S., YANG, T.: Comparative lipid analysis of *Aedes aegypti* and monkey kidney cells (MK-2) cultivated *in vitro*. This volume, p. 97–102.

JOHNSON, F. M.: A recessive esterase deficiency in *Drosophila*. J. Hered. **54**, 76–78 (1964).

JOHNSON, J. W.: Growth of Venezuelan, and Eastern, equine encephalomyelitis viruses in tissue cultures of minced *Aedes aegypti* larvae. Amer. J. trop. Med. Hyg. **18**, 103–114 (1969).

JONES, B. M.: The cultivation of insect cells and tissues. Biol. Rev. **37**, 512–536 (1962).

— Invertebrate tissue and organ culture in cell research. In: Cells and tissues in culture, vol. 3 (E. N. WILMER, ed.), p. 397–457. New York: Academic Press 1966.

JONES, B. M., CUNNINGHAM, I.: Growth by cell division in insect tissue culture. Exp. Cell Res. **23**, 386–401 (1961).
JONES, J. C.: The hemocytes of *Sarcophaga bullata* Parker. J. Morph. **99**, 223–257 (1956).
KAKPAKOV, V. T., GVOZDEV, V. A., PLATOVA, T. P., POLUKAROVA, L. G.: *In vitro* establishment of embryonic cell lines of *Drosophila melanogaster* [in Russian]. Genetika **5** (12), 67–75 (1969).
KARABATSOS, N., BUCKLEY, S. M.: Susceptibility of the baby-hamster kidney-cell line (BHK-21) to infection with arboviruses. Amer. J. trop. Med. Hyg. **16**, 99–105 (1967).
KARLSON, P., PETERS, G.: Zum Wirkungsmechanismus der Hormone. IV. Der Einfluß des Ecdysons auf den Nucleinsäurestoffwechsel von *Calliphora*-Larven. Gen. comp. Endocr. **5**, 252–259 (1965).
— SHAAYA, E.: Ecdysone titers during insect development. Gen. comp. Endocr. **5**, 690 (1965).
KAUFMANN, B. P.: Somatic mitosis of *Drosophila melanogaster*. J. Morph. **56**, 125–155 (1934).
KIMURA, I., SHIKATA, E.: Structural model of rice dwarf virus. Proc. Japan Acad. **44**, 538–543 (1968).
KITAMURA, S.: The *in vitro* cultivation of tissues from the mosquito, *Culex pipiens* var. *molestus*. I. Cultivation of ovary tissues *in vitro*. Kobe J. med. Sci. **10**, 85–94 (1964).
— The *in vitro* cultivation of tissues from the mosquito *Culex pipiens* var. *molestus*. II. An improved culture medium useful for ovarian tissue culture. Kobe J. med. Sci. **11**, 23–30 (1965).
— The *in vitro* cultivation of tissues from the mosquitoes. III. Further studies on the cultivation of ovarian tissues of three mosquito species and the examination of the origin of cells grown *in vitro*. Kobe J. med. Sci. **12**, 63–70 (1966).
KOCH, A.: Insects and their endosymbionts. In: Symbiosis, vol. II, (S. M. HENRY, ed.), p. 1–106. New York: Academic Press 1967.
KORDOVA, N., REHACEK, J.: Development of *Coxiella burneti* from their ultrafilterable particles in ticks *in vivo*, their organs *in vitro* and in tick tissue cultures. Med. Vet. Entomol., Proc. XII Int. Congr. Ent. London, p. 777 (1965).
KRAUSE, G., KRAUSE, J., GEISLER, M.: Beobachtungen *in vitro* an den Zellen der larvelen Ovariolenhülle von *Bombyx mori* L (Lepidoptera) in verschiedenen Kulturmedien. Z. Zellforsch. **70**, 393–418 (1966).
KRUEGER, H., LEZZI, M.: Regulation of gene action in insect development. Ann. Rev. Entomol. **11**, 1–22 (1966).
KUNKEL, L. O.: Studies on aster yellows. Amer. J. Bot. **13**, 646–705 (1926).
KURSTAK, E., CÔTÉ, J. R.: Proposition de classification du virus de la densonucléose (VDN) basée sur l'étude de la structure moleculaire et les propriétés physico-chimiques. C. R. Acad. Sci. (Paris) **268**, 616–619 (1969).
— — BELLONCIK, S.: Etude de la synthèse et de la localisation des antigènes du virus de la densonucléose (VDN) à l'aide d'anticorps conjugués à l'enzyme peroxydase (nouvelle méthode d'immunoperoxydase). C. R. Acad. Sci. (Paris) **268**, 2309–2312 (1969).
— STANISLAWSKI, M.: Localisation du matériel antigènique du virus de la densonucléose au cours de l'infection chez *Galleria mellonella*. Canad. J. Microbiol. **14**, 1350–1352 (1968).
KURTTI, T. J., BROOKS, M. A.: Growth of lepidopteran epithelial cells and hemocytes in primary cultures. J. invert. Path. **15**, 341–350 (1970).
— — Growth of a microsporidian parasite in cultured cells of tent caterpillars (*Malacosoma*). This volume, p. 204–208.
LAMPEN, J. O., MCLELLAN, W. L., JR., EL-NAKEEB, M. A.: Antibiotics and fungal physiology. Antimicrobial Agents Chemotherapy. **5**, 1006–1015 (1965).

LANE, N. J.: Lipochondria, neutral red granules and lysosomes: synonymous terms? p. 169–182 In: Cell structure and its interpretation: essays presented to John Randel Baker, I.R.S. (S. M. MCGEE-RUSSELL and K. F. A. ROSS, eds.). London: E. ARNOLD, Ltd. 1968.
LARSEN, W. P.: Growth in an insect organ culture. J. Insect Physiol. **13**, 613–619 (1967).
LEDUC, E. H., WICKER, R., AVRAMEAS, S., BERNHARD, W.: Ultrastructural localization of SV40 T antigen with enzyme-labelled antibody. J. gen. Virol. **4**, 609–614 (1969).
LEIDY, G., SPRUNT, K., REDMAN, W., ALEXANDER, H. E.: Sensitivity of populations of clonal lines of HeLa cells to polio viruses. Proc. Soc. exp. Biol. (N.Y.) **102**, 81–85 (1959).
LELOUP, A. M., GIANFELICI, E.: The persistence of neurosecretion in organotypic culture by a dipterous insect, *Calliphora erythrocephala*. Ann. Endocr. (Paris) **27**, 506–508 (1966).
LENDER, T., LAVERDURE, A. M.: Culture *in vitro* des ovaires de *Tenebrio molitor*. Croissance et vitellogenèse. C. R. Acad. Sci. (Paris), **265**, 451–454 (1967).
LESSEPS, R. J.: Culture of dissociated *Drosophila* embryos: aggregated cells differentiate and sort out. Science **148**, 502–503 (1965).
LEVENTHAL, E. A.: The sex ratio condition in *Drosophila bifasciata;* its experimental transmission to several species of *Drosophila*. J. invert. Path. **11**, 170–183 (1968).
L'HERITIER, P.: The hereditary virus of *Drosophila*. Advanc. Virus Res. **5**, 195–245 (1958).
LHOAS, P.: Mitotic haploidization by treatment of *Aspergillus niger* diploids with *para*-fluorophenylalanine. Nature (Lond.) **190**, 744 (1961).
LIMA-DE-FARIA, A., JAWORSKA, H.: Late DNA synthesis in heterochromatin. Nature (Lond.) **217**, 138–142 (1968).
LITTLEFIELD, J. W.: Selection of hybrids from matings of fibroblasts *in vitro* and their presumed recombinants. Science **145**, 709–710 (1964).
— The use of drug-resistant markers to study the hybridization of mouse fibroblasts. Exp. Cell Res. **41**, 190–196 (1966).
LOCKHART, R. Z., JR., EAGLE, H.: Requirements for growth of single human cells. Science **129**, 252–254 (1959).
LOCKSHIN, R. A.: Programmed cell death. Activation of lysis by a mechanism involving the synthesis of protein. J. Insect Physiol. **15**, 1505–1516 (1969).
— WILLIAMS, C. M.: Programmed cell death. II. Endocrine potentiation of the breakdown of the intersegmental muscles of silkmoths. J. Insect Physiol. **10**, 643–649 (1964).
— — Programmed cell death. I. Cytology of degeneration in the intersegmental muscles of the *Pernyi* silkmoth. J. Insect Physiol. **11**, 123–133 (1965).
LWOFF, A., DULBECCO, R., VOGT, M., LWOFF, M.: Kinetics of the release of poliomyelitis virus from single cells. Virology **1**, 128–139 (1955).
— TOURNIER, P.: The classification of viruses. Ann. Rev. Microbiol. **20**, 45–74 (1966).
MACLEOD, R.: An interpretation of the observed polymorphism of potato yellow dwarf virus. Virology **34**, 771–777 (1968).
— BLACK, L. M., MOYER, F. H.: The fine structure and intracellular localization of potato yellow dwarf virus. Virology **29**, 540–552 (1966).
MACPHERSON, I., STOKER, M.: Polyoma transformation of hamster cell clones–an investigation of genetic factors affecting cell competence. Virology **16**, 147–151 (1962).
MAGNI, G.: Il comportamento di un carattere citoplasmatico negli individui e in populazioni. Ric. Sci. **26** (Suppl.), 1–18 (1956).

MAKINO, S., JENKIN, H. M., YU, H. M., TOWNSEND, D.: Lipid composition of *Chlamydia psittaci* grown in monkey kidney cells in defined medium. J. Bact. **103**, 62–70 (1970).
— YOSHIDA, T.: A critical study of the chromosomes of *Lymantria dispar* L. in relation to the question of racial difference. Cytologia **14**, 145–147 (1949).
MARAMOROSCH, K.: Multiplication of aster yellows virus in *in vitro* preparations of insect tissues. Virology **2**, 369–376 (1956).
— Leafhopper-transmitted plant viruses. Protoplasma (Wien) **52**, 457–466 (1960).
— Present status of insect tissue culture. Proc. Intern. Congr. Entomol., 11th. Vienna, 1960, **2**, 801–807 (1962).
— Arthropod transmission of plant viruses. Ann. Rev. Entomol. **8**, 369–414 (1963).
— Insect infection caused by a plant tumor virus. Wld Rev. Pest Control **9**, 29–41 (1970).
— GRANADOS, R. R., HIRUMI, H.: Etiology of yellows-type diseases. Advanc. Virus Res. **16**, 135–193 (1970).
— MITSUHASHI, J., STREISSLE, G., HIRUMI, H.: Animal and plant viruses in insect tissues in vitro. Bact. Proc., p. 120 (1965).
— SHIKATA, E., GRANADOS, R. R.: Structures resembling mycoplasma in diseased plants and in insect vectors. Trans. N.Y. Acad. Sci., Ser. II **30**, 841–855 (1968).
— — — The fate of plant-pathogenic viruses in insect vectors: electron microscopy observations. In: Viruses, vectors, and vegetation (K. MARAMOROSCH, ed.), p. 417–431. New York: Interscience Publ. 1969.
— — HIRUMI, H., GRANADOS, R. R.: Multiplication and cytopathology of a plant tumor virus in insects. Nat. Cancer Inst. Monogr. **31**, 493–507 (1969).
MARKS, E.P.: Regenerating tissues from the cockroach *Leucophaea maderae:* Effects of humoral stimulation *in vitro*. Gen. comp. Endocr. **11**, 31–42 (1968).
— The action of hormones in insect cell and organ cultures. Gen. comp. Endocr. **15**, 289–302 (1970).
— LEOPOLD, R. A.: Cockroach leg regeneration: Effects of ecdysterone *in vitro*. Science **167**, 61–62 (1970).
— REINECKE, J. P.: Regenerating tissues from the cockroach, *Leucaphaea maderae:* The effect of endocrine glands *in vitro*. Gen. comp. Endocr. **5**, 241–247 (1965).
— — LEOPOLD, R. A.: Regenerating tissues from the cockroach *Leucophaea maderae:* Nerve regeneration *in vitro*. Biol. Bull. **135**, 520–529 (1968).
MARTENSSON, E.: Glycosphingolipids of animal tissue. Progress in the chemistry of fats and other lipids (R. T. HOLMAN, ed.), vol. 10, 365–407. London: Pergamon Press 1970.
MARTIGNONI, M. E.: Problems of insect tissue culture. Experientia (Basel) **16**, 125–128 (1960).
— Insect tissue culture: a tool for the physiologist. Biol. Colloq., Oregon State Chapter Phi Kappa Phi **23**, 89–110 (1962).
— Insect tissue culture: A tool for the physiologist. In: Insect physiology, p. 89–110 (V. J. BROOKES, ed.). Corvallis: Oregon State University Press 1963.
— SCALLION, R. J.: Multiplication *in vitro* of a nuclear polyhedrosis virus in insect amoebacytes. Nature (Lond.) **190**, 1133–1134 (1961).
MARTIN, H. M., VIDLER, B. O.: *In vitro* growth of tick tissues (*Rhipicephalus appendiculatus* Neumann, 1901). Exp. Parasit. **12**, 192-203 (1962).
MAZZONE, H. M.: Utilization of hemolymph in insect tissue cultures. Second Int. Colloq. Invertebrate Tissue Culture, p. 14–21. Milano: Ist. Lombardo Sci. Lettr. (Baselli Foundation) 1968.

McLean, D. M., Chamberlain, R. W., Hammon, W. McD., Johnson, K. M., Russell, P. K., Shope, R. E., Casals, J., Nathanson, N.: Arbovirus names. Amer. J. Trop. Med. Hyg. **18**, 731–734 (1969).

Miggiano, V., Nabholz, N., Bodmer, W.: Hybrids between human leukocytes and a mouse cell line: production and characterization. In: Heterospecific Genome Interaction. Wistar Inst. Symposium. Monogr. No 9, 61–76 (1969).

Millaria, P.: Recherches sur la cytologie des leucocytes d'insectes et leur survie en millieu artificiel. C. R. Acad. Sci. (Paris) **140**, 1006–1008 (1946).

Mitsuhashi, J.: *In vitro* cultivation of the embryonic tissues of the green rice leafhopper, *Nephotettix cincticeps* Uhler (Homoptera: Cicadellidae). Japan. J. appl. entomol. Zool. **9**, 107–114 (1965a).

— Preliminary report on the plant virus multiplication in the leafhopper vector cells grown *in vitro*. Japan. J. appl. entomol. Zool. **9**, 137–141 (1965b).

— Tissue culture of the rice stem borer, *Chilo suppressalis* Walker (Lepidoptera: Pyralidae). I. Cell migration from the explanted tissues of diapausing larvae. Jap. J. appl. entomol. Zool. **9**, 217–224 (1965).

— Tissue culture of the rice stem borer, *Chilo suppressalis* Walker (Lepidoptera: Pyralidae). II. Morphology and *in vitro* cultivation of hemocytes. Appl. entomol. Zool. **1**, 5–20 (1966).

— Chromosome numbers of the green rice leafhopper, *Nephotettix cincticeps* Uhler (Homoptera: Cicadellidae). Appl. entomol. Zool. **1**, 103–104 (1966a).

— Appearance of iridescence in the tissues of the rice stem borer larvae, *Chilo suppressalis* Walker, infected with *Chilo* iridescent virus (Lepidoptera: Pyralidae). Appl. entomol. Zool. **1**, 130–137 (1966b).

— Multiplication of *Chilo* iridescent virus in the *Chilo suppressalis* tissues cultivated *in vitro* (Lepidoptera: Pyralidae). Appl. entomol. Zool. **1**, 199–201 (1966c).

— Establishment of an insect cell strain persistently infected with an insect virus. Nature (Lond.) **215**, 863–864 (1967a).

— Infection of leafhopper and its tissues cultivated *in vitro* with *Chilo* iridescent virus. J. invert. Path. **9**, 432–434 (1967b).

— Plant-pathogenic viruses in insect vector tissue culture. In: Viruses, vectors, and vegetation (K. Maramorosch, ed.), p. 475–503. New York: Interscience Publ. 1969a.

— Preliminary report on the primary culture of smaller brown planthopper cells *in vitro* (Hemiptera: Delphacidae). Appl. entomol. Zool. **4**, 105–113 (1969b).

— Grace, T. D. C.: Adaptation of established insect cell lines to a different culture medium. Appl. entomol. Zool. **4**, 121–125 (1969).

— — Waterhouse, D. F.: Effects of insecticides on cultures of insect cells. Entomol. exp. appl. **13**, 327–341 (1970).

— Maramorosch, K.: Aseptic cultivation of four virus transmitting species of leafhoppers (Cicadellidae). Contr. Boyce Thompson Inst. **22**, 165–173 (1963).

— — Leafhopper tissue culture: Embryonic, nymphal and imaginal tissues from aseptic insects. Contr. Boyce Thompson Inst. **22**, 435–460 (1964).

— Nasu, S.: An evidence for the multiplication of rice dwarf virus in the vector cell cultures inoculated *in vitro*. Appl. entomol. Zool. **2**, 113–114 (1967).

Montes de Oca, F., Macy, M. L., Shannon, J. E.: Isoenzyme characterization of animal cell cultures. Proc. Soc. exp. Biol. (N.Y.) **132**, 462–469 (1969).

Moorhead, P. S., Nowell, P. C., Mellman, W. J., Battips, D. M., Hungerford, D. A.: Chromosome preparations of leukocytes cultures from human peripheral blood. Exp. Cell Res. **20**, 613–616 (1960).

Morgan, J. F., Morton, H. J., Parker, R. C.: Nutrition of animal cells in tissue culture. I. Initial studies on a synthetic medium. Proc. Soc. exp. Biol. (N.Y.) **73**, 1–8 (1950).

MURRAY, M. R.: Nervous tissues *in vitro*. In: Cells and tissues (E. N. WILLMER, ed.), vol. 2, p. 373–455. New York: Academic Press 1965.

NAGARAJ, A. N., SINHA, R. C., BLACK, L. M.: A smear technique for detecting virus antigen in individual vectors by the use of fluorescent antibodies. Virology **15**, 205–208 (1961).

NAGLE, S. C., JR., CROTHERS, W. C., HALL, N. L.: Growth of moth cells in suspension in hemolymph-free medium. Appl. Microbiol. **15**, 1497–1498 (1967).

NAKANE, P. K., PIERCE, G. B. JR.: Enzyme-labelled antibodies for the light and electron microscopic localization of tissue antigens. J. Cell Biol. **33**, 307–318 (1967).

NASU, S.: Electron microscopic studies on transovarial passage of rice dwarf virus. Japan. J. appl. entomol. Zool. **9**, 225–237 (1965).

— Electron microscopy of the transovarial passage of rice dwarf virus. In: Viruses, vectors, and vegetation (K. MARAMOROSCH, ed.), p. 433–448. New York: Interscience Publ. 1969.

— SUGIURA, M., WAKIMOTO, T., IIDA, T.: On the etiologic agent of rice dwarf disease (in Japanese). Ann. phytopath. Soc. Japan **33**, 343–344 (1967).

NAUCK, E. G. VON, WEYER, F.: Versuche zur Züchtung von Rickettsien in explantiertem Läusegewebe. Zbl. Bakt., I. Abt. Orig. **147**, 365–376 (1941).

NICHOLS, W. W., BRADT, C., BOWNE, W.: Cytogenetic studies on cells in culture from the class Insecta. This volume, p. 61–69.

NIELSON, M. W.: The leafhopper vectors of phytopathogenic viruses (Homoptera, Cicadellidae): taxonomy, biology, and virus transmission. U.S. Dept. Agriculture, Tech. Bull. 1382. 386 pp. (1968).

NIU, M. C., TWITTY, V. C.: The differentiation of gastrula ectoderm in medium conditioned by axial mesoderm. Proc. nat. Acad. Sci. (Wash.) **39**, 985–989 (1953).

OBERLANDER, H., BERRY, S., KRISHNAKUMARAN, A., SCHNEIDERMAN, H.: RNA and DNA synthesis during activation and secretion of the prothoracic glands of Saturniid moths. J. exp. Zool. **159**, 15–32 (1965).

O'BRIEN, R. D., WOLFE, L. S.: Radiation, radioactivity and insects, p. 88. New York: Academic Press 1964.

O'CONNOR, T. E., FISCHINGER, P. J.: Titration patterns of a murine sarcoma-leukemia virus complex: Evidence for existence of competent sarcoma virions. Science **159**, 325–329 (1968).

ODELL, T. M., ROLLINSON, W. D.: A technique for rearing the gypsy moth, *Porthetria dispar* L., on an artificial diet. J. econ. Entomol. **59**, 741–742 (1966).

OHANESSIAN, A., ECHALIER, G.: Multiplication of *Drosophila* hereditary virus (σ virus) in *Drosophila* embryonic cells cultivated *in vitro*. Nature (Lond.) **213**, 1049–1050 (1967).

OHANESSIAN-GUILLEMAIN, A.: Étude de facteurs génétiques contrôlant les relations du virus Sigma et de la *Drosophile* son hôte. Ann. Génét. **5**, 1–64 (1963).

OHSHIMA, K.: On the function of the polar filament of *Nosema bombycis*. Parasitology **29**, 220–224 (1937).

OKADA, T. S.: Immunohistological studies on the reconstitution of nephric tubules from dissociated cells. J. Embryol. exp. Morph. **13**, 299–307 (1965).

OTTAVIANO-GOTTARDI, A.: Karyotype selection in cultures with different initial cell concentrations (*D. melanogaster*). Second Int. Colloq. Invertebrate Tissue Culture, p. 189–200. Milano: Ist. Lombardo Sci. Lettr. (Baselli Foundation) 1968.

OZAWA, Y., NAKATA, G.: Experimental transmission of African horse sickness by means of mosquitoes. Amer. J. Vet. Res. **26**, 744–748 (1965).

— — SHAD-DEL, F.: Transmission of African horse sickness by a species of mosquito, *Aedes aegypti*, Linnaeus. Amer. J. Vet. Res. **27**, 695–697 (1966).

PANTHIER, R., HANNOUN, C., OUDAR, J., BEYTOUT, D.: Isolement du virus West Nile chez un cheval de Camargue atteint d'encéphalomyelite. C. R. Acad. Sci. (Paris) **262**, 1308–1310 (1966).

PARODI, A. S., COTO, C. E., BOXACA, M., LAJMANOVICH, S., GONZALEZ, S.: Characteristics of Junin virus. Etiological agent of Argentine hemorrhagic fever. Arch. ges. Virusforsch. **19**, 393–402 (1966).

PAUL, S. D., SINGH, K. R. P.: Comparative sensitivity of mosquito cell lines, Vero cell line and infant mice to infection with arboviruses. Curr. Sci. **38**, 241–242 1969).

— — BHAT, U. K. M.: A study on the cytopathic effect of arboviruses on cultures from *Aedes albopictus* cell line. Indian J. med. Res. **57**, 339–348 (1969).

PAVRI, K. M., GHOSH, S. N.: Complement-fixation tests for simultaneous isolation and identification of dengue viruses, using tissue cultures. Bull. Wld Hlth Org. **40**, 984–986 (1969).

PEARSE, A. G. E.: Histochemistry, theoretical and applied. London: J. A. Churchill Ltd. 1960.

PELEG, J.: Growth of *Aedes aegypti* embryonic cells and tissues *in vitro*. Experientia (Basel) **22**, 555–556 (1966).

— Growth of arboviruses in primary tissue culture of *Aedes aegypti* embryos. Amer. J. trop. Med. Hyg. **17**, 219–223 (1968a).

— Growth of arboviruses in monolayers from subcultured mosquito embryo cells. Virology **35**, 617–619 (1968b).

— Behaviour of infectious RNA from four different viruses in continuously subcultured *Aedes aegypti* mosquito embryo cells. Nature (Lond.) **221**, 193–194 (1969a).

— Inapparent persistent virus infection in continuously grown *Aedes aegypti* mosquito cells. J. gen. Virol. **5**, 463–471 (1969b).

— TRAGER, W.: Growth of West Nile virus in insect tissue cultivated in vitro. Ann. Épiphyties **14**, 211–212 (1963a).

— — Cultivation of insect tissues *in vitro* and their application to the study of arthropod-borne viruses. Amer. J. trop. Med. Hyg. **12**, 820–824 (1963b).

PEROTTI, M. E., BAIRATI, A., JR.: Ultrastructure of the melanotic masses in two tumorous strains of *Drosophila melanogaster* (*tu* B_3 and Freckled). J. invert. Path. **10**, 122–138 (1968).

PETERS, D., BLACK, L. M.: Infection of primary cultures of aphid cells with a plant virus. Virology **40**, 847–853 (1970).

PLUS, N., DUTHOIT, J. L.: Un nouveau virus de *Drosophila melanogaster*, le virus *P*. C. R. Acad. Sci. (Paris) **268**, 2313–2315 (1969).

PORTERFIELD, J. S.: Cross neutralization studies with group A arthropod-borne viruses. Bull. Wld Hlth Org. **24**, 735–741 (1961).

POULSON, D. F.: The effect of certain X-chromosome deficiencies on the embryonic development of *Drosophila melanogaster*. J. exp. Zool. **83**, 271–325 (1940).

— Histogenesis, organogenesis, and differentiation in the embryo of *Drosophila melanogaster* Meigen. In: The biology of *Drosophila* (M. DEMEREC, ed.) New York: John Wiley & Sons 1950.

— Cytoplasmic inheritance and hereditary infections in *Drosophila*. Appendix III in: Methodology in basic genetics (W. J. BURDETTE, ed.), p. 404–424. San Francisco: Holden-Day 1963.

— SAKAGUCHI, B.: Hereditary infections in *Drosophila*. Genetics **46**, 890 (1961a).

— — Nature of the sex-ratio agent in *Drosophila*. Science **133**, 1489–1490 (1961b).

PRICE, W. H.: The epidemic climate. Publ. Hlth Rep. (Wash.) **71**, 125–138 (1956).

PUCK, T. T., KAO, F. T.: Genetics of somatic mammalian cells. V. Treatment with 5-bromodeoxyuridine and visible light for isolation of nutritionally deficient mutants. Proc. nat. Acad. Sci. (Wash.) **58**, 1227–1234 (1967).

Puck, T. T., Marcus, P. I., Cieciura, S. J.: Clonal growth of mammalian cells *in vitro*. J. exp. Med. **103**, 273–284 (1956).

Rahman, S. B., Perlman, D., Ristich, S. S.: Growth of insect cells in tissue culture. Proc. Soc. exp. Biol. (N.Y.) **123**, 711 (1966).

Rao, T. R., Singh, K. R. P., Dhanda, V., Bhatt, P. N.: Experimental transmission of Chandipura virus by mosquitoes. Indian J. med. Res. **55**, 1306–1310 (1967).

Raymond, S.: Acrylamide gel electrophoresis. Ann. N.Y. Acad. Sci. **121**, 350–365 (1964).

Reed, L. J., Muench, H.: A simple method of estimating fifty per cent endpoints. Amer. J. Hyg. **27**, 493–497 (1938).

Rehacek, J.: Preliminary report on tick tissue culture. Acta virol. **2**, 253–254 (1958).

— An improved method of tick tissue cultures. Acta virol. **6**, 188 (1962).

— Propagation of tick-borne encephalitis (T.E.) virus in tick tissue cultures. Ann. Épiphyties **14**, 199–204 (1963).

— Relation between viruses and tick tissues *in vitro*. Proc. XII Int. Congr. Ent. London, p. 774–775 (1965a).

— Cultivation of different viruses in tick tissue cultures. Acta virol. **9**, 332–337 (1965b).

— Preparation of tissue cultures from the tick *Hyalomma dromedarii* Koch. J. med. Entomol. **2**, 161–164 (1965c).

— Arboviruses in mosquito cells "in vitro". Second Int. Colloq. Invertebrate Tissue Culture, p. 241–246. Milano: Ist. Lombardo Sci. Lettr. (Baselli Foundation) 1968a.

— The growth of arboviruses in mosquito cells in vitro. Acta virol. **12**, 241–246 (1968b).

— Persistent infection of mosquito cells grown in vitro with Murray Valley encephalitis and Japanese encephalitis viruses. Acta virol. **12**, 340–346 (1968c).

— Brezina, R.: Propagation of *Coxiella burneti* in tick tissue cultures. Acta virol. **8**, 380 (1964).

— — Majerska, M.: Multiplication of rickettsiae in tick cells *in vitro*. Acta virol. **12**, 41–43 (1968).

— Brzostowski, H. W.: A tick tissue culture medium based on analyses on tick hemolymph. J. Insect Physiol. **15**, 1431–1436 (1969a).

— — The utilization of amino acids and sugars by tick cells cultivated *in vitro*. J. Insect Physiol. **15**, 1683–1686 (1969b).

— Kozuch, O.: Comparison of the susceptibility of primary tick and chick embryo cell cultures to small amounts of tick-borne encephalitis virus. Acta virol. **8**, 470–471 (1964).

— — The use of *Hyalomma dromedarii* tick tissue cultures for the isolation of tick-borne encephalitis virus from its natural foci. Acta virol. **13**, 253 (1969).

— Pesek, J.: Propagation of Eastern equine encephalomyelitis (EEE) virus in surviving tick tissues. Acta virol. **4**, 241–245 (1960).

— Rajcani, J., Gresikova, M.: Relationship of Tribec virus to tick cells and tissues. Acta virol. **13**, 439–442 (1969).

Rinaldini, L. M.: A quantitative method for growing animal cells *in vitro*. Nature (Lond.) **173**, 1134–1135 (1954).

Roth, T. F., Porter, K. R.: Yolk protein uptake in the oocyte of the mosquito *Aedes aegypti* L. J. Cell Biol. **20**, 313–332 (1964).

Rothblat, G. H., Hartzell, R., Mialhe, H., Kritchevsky, D.: Cholesterol metabolism in tissue culture cells. In: Lipid metabolism in tissue culture cells (G. H. Rothblat and D. Kritchevsky, eds.). Philadelphia: The Wistar Inst. Press 1967.

— Kritchevsky, D.: The metabolism of free and esterified cholesterol in tissue culture cells. A review. Exp. molec. Path. **8**, 314–329 (1968).

Sanford, K. K., Earle, W. R., Likely, G. D.: The growth *in vitro* of single isolated tissue cells. J. nat. Cancer Inst. **9**, 229–246 (1948).

SASTRODIHARDJO, S.: Competence of *Cecropia* ovaries *in vitro*. J. Cell Biol. **27**, 93a (1965).
— The competence of the ovary of *Hyalophora cecropia* (Lepidoptera) *in vitro*, with special reference to the "intermediate cell". Doct. Dissert., U. of Illinois (1967).
SCHALLER, F., MEUNIER, J.: Resultats de cultures organotypique du cervaeu et du ganglion sous-oesophagien d'*Aeschna cyanea* Mull. (Insecte-Odonate). Survie des organs et evolution des elements neurosecreteurs. C. R. Acad. Sci. (Paris), **261**, 1441–1444 (1967).
SCHARRER, B.: The prothoracic glands of *Leucophaea maderae*. Biol. Bull. **95**, 186–198 (1948).
— Neurosecretion. XI. The effects of nerve section on the intercerebralis-cardiacum-allatum system of the insect *Leucophaea maderae*. Biol. Bull. **102**, 261–272 (1952).
— The fine structure of Blatterian prothoracic glands. Z. Zellforsch. **64**, 301–326 (1964a).
— Histophysiological studies on the corpus allatum of *Leucophaea maderae*. IV. Ultrastructure during normal activity cycle. Z. Zellforsch. **62**, 125–148 (1964b).
SCHERRER, K., DARNELL, J. E.: Sedimentation characteristics of rapidly labelled RNA from HeLa cells. Biochem. biophys. Res. Commun. **7**, 486–490 (1962).
SCHIN, K. S., CLEVER, U.: Lysosomal and free acid phosphatase in salivary glands of *Chironomus tentans*. Science **150**, 1053–1055 (1965).
SCHNEIDER, I.: Differentiation of larval *Drosophila* eye-antennal discs *in vitro*. J. exp. Zool. **156**, 91–104 (1964).
— Histology of larval eye-antennal discs and cephalic ganglia of *Drosophila* cultured *in vitro*. J. Embryol. exp. Morph. **15**, 271–279 (1966).
— Insect tissue culture. In: Methods in developmental biology (F. H. WILT and N. K. WESSELLS, eds.), p. 543–554. New York: Thomas Y. Crowell Co. 1967.
— Establishment of three diploid cell lines of *Anopheles stephensi* (Diptera: Culicidae). J. Cell Biol. **42**, 603–606 (1969).
SEECOF, R. L., UNANUE, R. L.: Differentiation of embryonic *Drosophila* cells *in vitro*. Exp. Cell Res. **50**, 654–660 (1968).
SENGEL, P., MANDARON, P.: Aspects morphologiques du development *in vitro* des disques imaginaux de la *Drosophile*. C. R. Acad. Sci. (Paris) **268**, 405–407 (1969).
SEN GUPTA, K.: Studies on insect tissue culture. I. Culture of tissues from the wax moth, *Galleria mellonella*, L. *in vitro*. Folia biol. **7**, 400–408 (1961).
— Studies on the specificity of haemolymph from different insects for the culture of tissues from other insects. Proc. Indian Acad. Sci. **59**, 103–109 (1964).
SHADDUCK, J. A.: *Nosema cuniculi: In vitro* isolation. Science **166**, 516–517 (1969).
SHAW, C. R., KOEN, A. L.: Starch gel zone electrophoresis of enzymes. In: Zone electrophoresis, vol. 2 (IVOR SMITH, ed.), 2nd ed. New York: J. Wiley 1968.
SHIKATA, E., MARAMOROSCH, K.: Electron microscopic evidence for the systemic invasion of an insect host by a plant pathogenic virus. Virology **27**, 461–475 (1965).
— — An electron microscope study of plant neoplasia induced by wound tumor virus. J. nat. Cancer Inst. **36**, 97–116 (1966).
— — Electron microscopy of wound tumor virus assembly sites in insect vectors and plants. Virology **32**, 363–377 (1967).
— — Electron microscopy of insect-borne viruses *in situ*. In: Viruses, vectors, and vegetation (K. MARAMOROSCH, ed.), p. 393–415. New York: Interscience Publ. 1969a.
— — Mycoplasmalike bodies in sieve pores of yellows diseased plants and in fat-body cells of two insect vectors. Phytopathology **59**, 1559 (1969b).
— — LING, K. C.: Presumptive mycoplasma etiology of yellows diseases. FAO Plant Protect. Bull. **17**, 121–128 (1969).
— ORENSKI, S. W., HIRUMI, H., MITSUHASHI, J., MARAMOROSCH, K.: Electron micrographs of wound-tumor virus in an animal host and in a plant tumor. Virology **23**, 441–444 (1964).

SHINEDLING, S. T., GREENBERG, B.: Culture of cells of the flesh fly, *Sarcophaga bullata*. This volume, p. 12–19.
SIMIZU, B., RHIM, J. S., WIEBENGA, N.: Characterization of the Tacaribe group of arboviruses. I. Propagation and plaque assay of Tacaribe virus in a line of African green monkey kidney cells (Vero). Proc. Soc. exp. Biol. (N.Y.) **125**, 119–123 (1967).
SINGH, K. R. P.: Cell cultures derived from larvae of *Aedes albopictus* (Skuse) and *Aedes aegypti* (L.). Curr. Sci. **36**, 506–508 (1967).
— PAUL, S. D.: Multiplication of arboviruses in cell lines from *Aedes albopictus* and *Aedes aegypti*. Curr. Sci. **37**, 65–67 (1968a).
— — Susceptibility of *Aedes albopictus* and *Aedes aegypti* cell lines to infection by arbo and other viruses. Indian J. med. Res. **56**, 815–820 (1968b).
— — Isolation of dengue viruses in *Aedes albopictus* cell cultures. Bull. Wld Hlth Org. **40**, 982–983 (1969).
SINHA, R. C.: Sequential infection and distribution of wound-tumor virus in the internal organs of a vector after ingestion of virus. Virology **26**, 673–686 (1965).
— Localization of viruses in vectors: serology and infectivity tests. In: Viruses, vectors, and vegetation (K. MARAMOROSCH, ed.), p. 379–391. New York: Interscience Publ. 1969.
— REDDY, D. V. R.: Improved fluorescent smear technique and its application in detecting virus antigens in an insect vector. Virology **24**, 626–634 (1964).
SMITH, C. N., ed.: Insect colonization and mass production, p. 618. New York: Academic Press 1966.
SOHI, S. S., SMITH, C.: Effect of fetal bovine serum on the growth and survival of insect cell cultures. Canad. J. Zool. **48**, 427–432 (1970).
STANLEY, M. S., VAUGHN, J. L.: Marked sensitivity of insect cell lines to Fungizone (Amphotericin B). J. Insect Physiol. **13**, 1613–1617 (1967).
— — Origin of migrating cells in cultures of moth ovarian tissue based on developmental stages producing optimum outgrowth. Ann. entomol. Soc. Amer. **61**, 1067–1072 (1968).
STAY, B.: Protein uptake in the oocytes of the cecropia moth. J. Cell Biol. **26**, 49–62 (1965).
STERN, C.: Somatic crossing-over and segregation in *Drosophila melanogaster*. Genetics **21**, 625–730 (1936).
STEVENS, T. M.: Arbovirus replication in mosquito cell lines (SINGH) grown in monolayer or suspension culture. Proc. Soc. exp. Biol. (N.Y.) **134**, 356–361 (1970).
STOLLAR, V., STEVENS, T. M., SHENK, T.: RNA of uninfected and Sindbis virus infected *A. albopictus* cells. This volume, p. 164–169.
STRAUS, W.: Lysosomes, phagosomes and related particles, p. 239–319. In: Enzyme cytology (D. B. ROODYN, ed.). New York: Academic Press 1967.
STREISSLE, G., ROSEN, L., TOKUMITSU, T.: Host specificity of wound tumor virus and reoviruses and their serologic relationships. Arch. ges. Virusforsch. **22**, 409–416 (1967).
SUITOR, E. C., JR.: Growth of Japanese encephalitis virus in Grace's continuous line of moth cells. Virology **30**, 143–145 (1966b).
— Plaque formation by an arbovirus in a mosquito cell line. J. gen. Virol. **5**, 545–546 (1969).
— CHANG, L. L., LIU, H. H.: Establishment and characterization of a clone from Grace's *in vitro* cultured mosquito (*Aedes aegypti* L.) cells. Exp. Cell Res. **44**, 572–578 (1966a).
— PAUL, F. J.: Syncytia formation of mosquito cell cultures mediated by Type 2 Dengue virus. Virology **38**, 482–485 (1969).
SUZUKI, N.: Purification of single and double-stranded vector-borne RNA viruses. In: Viruses, vectors, and vegetation (K. MARAMOROSCH, ed.), p. 557–578. New York: Interscience Publ. 1969.

SWEET, B. H., DUPREE, L. T.: Growth of cells derived from *Culiseta inornata* and *Aedes vexans* in tissue culture. A preliminary note. Mosquito News **28**, 368–373 (1968).

— McHALE, J. S.: Characterization of cell lines derived from *Culiseta inornata* and *Aedes vexans* mosquitoes. Exp. Cell Res. **61**, 51–63 (1970).

SYKES, J. A., MOORE, E. B.: A simple tissue culture chamber. Tex. Rep. Biol. Med. **18**, 288–297 (1960).

SZYBALSKA, E. H., SZYBALSKI, W.: Genetics of human cell lines. IV. DNA-mediated heritable transformation of a biochemical trait. Proc. nat. Acad. Sci. (Wash.) **48**, 2026–2034 (1962).

TAYLOR, R. M., comp.: Catalogue of arthropod-borne viruses of the world. Publ. Hlth Serv. Publ. No 1760 (1967).

THOMSON, H. M.: Some aspects of the epidemiology of a microsporidian parasite of the spruce budworm *Choristoneura fumiferana* (Clem.). Canad. J. Zool. **36**, 309–316 (1958).

— A microsporidian parasite of the forest tent caterpillar, *Malacosoma disstria* Hbn. Canad. J. Zool. **37**, 217–221 (1959).

THOMSON, J. A., GRACE, T. D. C.: Cytological observations on cell strains established in culture from insect ovarian tissue. Aust. J. biol. Sci. **16**, 869–876 (1963).

TOKUMITSU, T., MARAMOROSCH, K.: Survival of aphid cells *in vitro*. Exp. Cell Res. **44**, 652–655 (1966).

— — Cytoplasmic protrusions in insect cells during mitosis in vitro. J. Cell Biol. **34**, 677–683 (1967).

— — Bubbling of dividing leafhopper cells "in vitro". Second Int. Colloq. Invertebrate Tissue Culture, p. 10–13. Milano: Ist. Lombardo Sci. Lettr. (Baselli Foundation) 1968.

TRAGER, W.: Cultivation of the virus of grasserie in silkworm tissue cultures. J. exp. Med. **61**, 501–513 (1935).

— The relation of cell size to growth in insect larvae. J. exp. Zool. **71**, 489–508 (1935a).

— The hatching of spores of *Nosema bombycis* Nägeli and the partial development of the organism in tissue cultures. J. Parasit. **23**, 226–227 (1937).

— Multiplication of the virus of equine encephalomyelitis in surviving mosquito tissues. Amer. J. trop. Med. **18**, 387–393 (1938).

— Tsetse fly tissue culture and the development of trypanosomes to the infective stage. Ann. trop. Med. Parasit. **53**, 473–491 (1959).

VAGO, C.: Invertebrate tissue culture. In: Methods in virology (K. MARAMOROSCH and H. KOPROWSKI, eds.), vol. 1, p. 567–602. New York: Academic Press 1967.

— CHASTANG, S.: Culture *in vitro* d'un tissu nymphal de lépidoptère. Experientia (Basel) **14**, 426–427 (1958a).

— — Obtention de lignées cellulaires en culture de tissus d'invertébrés. Experientia (Basel) **14**, 110–111 (1958b).

— — Culture de Borrelinavirus dans les organes d'insectes en survie. C. R. Acad. Sci. (Paris) **251**, 903–905 (1960).

— — Cultures de tissus d'insectes à l'aide de sérum de mammifères. Entomophaga **7**, 175–179 (1962a).

— — Culture de tissus d'insectes à l'aide d'extrait d'embryon de poulet en l'absence d'hémolymph. C. R. Acad. Sci. (Paris) **255**, 3226–3228 (1962b).

— FLANDRE, O.: Culture prolongée de tissues d'insectes et de vecteurs de maladies en coagulum plasmatique. Ann. Epiphyties **14** (no. hors série III), 127–139 (1963).

VANDENHEUVEL, F. A.: Structural studies on biological membranes: The structure of myelin. Ann. N.Y. Acad. Sci. **122**, 57–76 (1965).

VANDEN HEUVEL, W. J. A., HAAHTI, E. D. A., HORNING, E. C.: A new liquid phase for gas chromatographic separations of steroids. J. Amer. chem. Soc. **83**, 1513–1514 (1961).

Van der Kloot, W. G.: The control of neurosecretion and diapause by physiological changes in the brain of the cecropia silkworm. Biol. Bull. **109**, 276–294 (1955).

Vanderzant, E. S., Richardson, C. D., Fort, S. W., Jr.: Rearing of the bollworm on artificial diet. J. econ. Entomol. **55**, 140 (1962).

Varma, M. G. R., Pudney, M.: The growth and serial passage of cell lines from *Aedes aegypti* (L.) larvae in different media. J. med. Entomol. **6**, 432–439 (1969).

— — Tick cell culture: A review of recent work. Proc. 2nd Intern. Congr. Acarol., p. 637. Budapest: Akademiai Kiado 1969.

— Wallers, W.: An improved method for obtaining *in vitro* uniform cell monolayer sheets from tissues of the tick *Hyalomma dromedarii* (Ixodidae). Nature (Lond.) **208**, 602–603 (1965).

Vaughn, J. L.: A review of the use of insect tissue culture for the study of insect-associated viruses. In: Current topics in microbiol. and immunol. Insect viruses, vol. 42, p. 108–128. Berlin-Heidelberg-New York: Springer 1968.

— Faulkner, P.: Susceptibility of an insect tissue culture to infection by virus preparations of the nuclear polyhedrosis of the silkworm (*Bombyx mori* L.). Virology **20**, 484–489 (1963).

Vesell, E. S., Philip, J., Bearne, A. G. J.: Comparative studies of the isoenzymes of lactic dehydrogenase in rabbit and man. J. exp. Med. **116**, 797–806 (1962).

Vickerman, K.: Genetic systems in unicellular animals. Sci. Progr. **54**, 13–26 (1966).

Watkins, J. F., Chen, L.: Immunization of mice against Ehrlich ascites tumor using a hamster/Ehrlich ascites tumor hybrid cell line. Nature (Lond.) **223**, 1018–1022 (1969).

Waymouth, C.: Nutrition and metabolism of animal tissue cultures. J. nat. Cancer Inst. **19**, 495–500 (1957).

Weathersby, A. B.: The role of the stomach wall in the exogenous development of *Plasmodium gallinaceum* as studied by means of humoral injections of susceptible and refractory mosquitoes. J. infect. Dis. **91**, 198–205 (1952).

Weinstein, D.: Comparison of pronase and trypsin for detachment of human cells during serial cultivation. Exp. Cell Res. **43**, 234–236 (1966).

Wenyon, C. M.: Protozoology, vol. 1. New York: Hafner Publishing Co. 1926.

Weyer, F.: Vergleichende Untersuchungen an Rickettsien. Z. Naturforsch. **2** b, 349–358 (1947).

— Explantationsversuche bei Läusen in Verbindung mit der Kultur von Rickettsien. Zbl. Bakt., I. Abt. **159**, 13–22 (1952).

Whitcomb, R. F., Davis, R. E.: *Mycoplasma* and phytarboviruses as plant pathogens persistently transmitted by insects. Ann. Rev. Entomol. **15**, 405–464 (1970).

White, J. F.: A study of the development and competence of the cynthia ovary *in vivo* and *in vitro*. J. Cell Biol. **27**, 113a (1965).

— Histochemical observations on granules in Grace's *Antheraea eucalypti* cell strains. Proc. Tissue Culture Ass. p. 56 (1969).

— Larsen, J. R.: Histochemical and electron microscope studies of silkmoth ovarian cells *in vivo* and *in vitro*. J. Cell Biol. **31**, 122a (1966).

— — Insect muscle differentiation *in vitro*. Proc. Tissue Culture Ass. p. A 16 (1970).

Whitney, E., Jamnback, H., Means, R. G., Roz, A. P., Rayner, G. A.: California virus in New York State. Isolation and characterization of California encephalitis virus complex from *Aedes cinereus*. Amer. J. trop. Med. Hyg. **18**, 123–131 (1969).

Whitten, J. M.: Hemocytes and the metamorphosing tissues in *Sarcophaga bullata. Drosophila melanogaster*, and other Cyclorrhaphous Diptera. J. Insect Physiol. **10**, 447–469 (1964).

Wicker, R., Avrameas, S.: Localization of virus antigens by enzyme-labelled antibodies. J. gen. Virol. **4**, 465–471 (1969).

Wigglesworth, V. B.: The action of moulting hormone and juvenile hormone at the cellular level in *Rhodnius prolixus*. J. exp. Biol. **40**, 231–245 (1963).

WILLIAMS, C. M.: Ecdysone and ecdysone-analogues: their assay and action on diapausing pupae of the cynthia silkworm. Biol. Bull. **134**, 344–355 (1968).

WILLIAMS, M. C., WOODALL, J. P., PORTERFIELD, J. S.: O'nyong-nyong fever: An epidemic virus disease in East Africa. V. Human antibody studies by plaque inhibition and other serological tests. Trans. roy. Soc. trop. Med. Hyg. **56**, 166–172 (1962).

WILLIAMSON, D. L.: Kinetic studies of sex ratio spirochetes in *Drosophila melanogaster* Meigen females. J. invert. Path. **7**, 493–501 (1965).

WOKE, P. A., ROSENBERGER, C. R., JR.: Survival of mouse poliomyelitis virus in living invertebrates. J. infect. Dis. **100**, 12–16 (1957).

WOOD, D. E., SUITOR, E. C., JR.: *In vitro* development of microfilariae of *Macacanema formosana* in mosquito (*Aedes aegypti*) cell cultures. Nature (Lond.) **211**, 868–870 (1966).

WRIGHT, T. R. F.: The genetic control of an esterase in *Drosophila melanogaster*. Genetics **48**, 787–801 (1963).

WYATT, S. S.: Culture *in vitro* of tissue from the silkworm, *Bombyx mori* L. J. gen. Physiol. **39**, 841–852 (1956).

YAMADA, K., TOKUMITSU, T., SHIKATA, E.: Embryonic tissue culture of *Inazuma dorsalis* Motschulsky (Homoptera: Delphacidae). [In Japanese.] Jap. J. appl. entomol. Zool. **13**, 159–161 (1969).

YANG, Y. J., STOLTZ, D. B., PREVEC, L.: Growth of vesicular stomatitis virus in a continuous culture line of *Antheraea eucalypti* moth cells. J. gen. Virol. **5**, 473–483 (1969).

YASAMURA, Y., KAWAKITA, Y.: Study on SV_{40} virus in tissue culture [in Japanese]. Nippon Rinsho **21**, 1201–1219 (1963).

YERGANIAN, G., NELL, M. B.: Hybridization of dwarf hamster cells by UV-inactivated Sendai virus. Proc. nat. Acad. Sci. (Wash.) **55**, 1066–1073 (1966).

YUNKER, C. E., BURGDORFER, W., CORY, J.: Infection of insect cells with typhus and spotted fever group agents. (In preparation.)

— CORY, J.: Growth of Colorado tick fever (CTF) virus in primary tissue cultures of its vector, *Dermacentor andersoni* Stiles (Acarina: Ixodidae), with notes on tick tissue culture. Exp. Parasit. **20**, 267–277 (1967).

— — Infection of Grace's *Antheraea* cells with arboviruses. Amer. J. trop. Med. Hyg. **17**, 889–893 (1968a).

— — Growth of some Group B arboviruses in two established insect cell lines. Proc. 23rd Ann. Int. Northwest Conf. on Diseases in Nature Communicable to Man (19–21 Aug.), p. 79–82 (1968b).

— — Colorado tick fever virus: Growth in a mosquito cell line. J. Virol. **3**, 631–632 (1969).

— ORMSBEE, R. A., CORY, J., PEACOCK, M.: Infection of insect cells with *Coxiella burneti*. Acta virol. **14**, 383–392 (1970).

— VAUGHN, J. L., CORY, J.: Adaptation of insect cell line (Grace's *Antheraea* cells) to medium free of insect hemolymph. Science **155**, 1565–1566 (1967).

ZINDER, N. D.: RNA phages. Ann. Rev. Microbiol. **19**, 455–472 (1965).

Author Index

Page numbers in *italics* refer to the references

Subject Index.

Index to Volumes 40—54

Volume 40

Volume 41

Volume 42

Volume 43

Volume 44

Volume 45

Volume 46

Volume 47

Volume 48

Volume 49

Volume 50

MIX
Papier aus verantwortungsvollen Quellen
Paper from responsible sources
FSC® C105338

If you have any concerns about our products,
you can contact us on
ProductSafety@springernature.com

In case Publisher is established outside the EU,
the EU authorized representative is:
Springer Nature Customer Service Center GmbH
Europaplatz 3, 69115 Heidelberg, Germany

Printed by Libri Plureos GmbH
in Hamburg, Germany